WHY ARE WE HERE?

THE STORY OF THE ORIGIN, EVOLUTION, AND FUTURE OF LIFE ON OUR PLANET

BRUCE BRODIE

iUniverse

WHY ARE WE HERE?
THE STORY OF THE ORIGIN, EVOLUTION, AND FUTURE OF LIFE ON OUR PLANET

iUniverse books may be ordered through booksellers or by contacting:

iUniverse
1663 Liberty Drive
Bloomington, IN 47403
www.iuniverse.com
1-800-Authors (1-800-288-4677)

ISBN: 978-1-5320-4854-8 (sc)
ISBN: 978-1-5320-6797-6 (e)

Library of Congress Control Number: 2019904017

Print information available on the last page.

iUniverse rev. date: 05/23/2019

To my wife, Dora

To my wife, Dot...

CONTENTS

PREFACE

Seven years ago, I retired from a busy cardiology practice and set out on my own journey in an attempt to understand how we got here, how our evolutionary past has shaped human nature, and where we might be headed. *Why Are We Here?* is the culmination of that research and that journey.

I have a general science and biochemistry background. I hold a degree in electrical engineering from Purdue University and a medical degree from Washington University in Saint Louis, and I have had an active and productive career in medicine and interventional cardiology. But I am certainly not an expert in the many fields encompassed by this book, which include cosmology, physics, chemistry, biochemistry, genetics, geology, paleontology, psychology, evolutionary biology, evolutionary psychology, neuroscience, anthropology, computer science, and more. Rather than an expert, I regard myself more as a reporter and a scientific writer. What I have tried to do in more than seven years of study and research is draw from scientific publications and books written by many experts in these diverse fields, develop a synthesis of their knowledge and ideas, and present these ideas with a unified perspective for a general audience.

Eclecticism has limitations in the pursuit of new knowledge and new theories, but it may be a requirement for the synthesis of a field as broad as the origin and evolution of life. My goal in this book is to educate the reader about how life may have started and how the evolutionary process has led us to the diverse life that inhabits our planet today. I hope to give the reader an appreciation of how science has helped us to understand the process of evolution, while leaving the reader with an appreciation of the limitations of our knowledge. I hope to provide a perspective of how our species at this point in time fits into the big picture of a universe that is infinite in time and space, and I hope this story will leave the reader with a lasting sense of wonder at the amazing mystery of life and our universe. An understanding of our evolutionary past and an appreciation of possible scenarios for our future provides a basis for seeking answers to the existential questions of life. Why are we here?

Some of the material in the book is detailed and technical, but I have tried to present it in a way that is understandable, educational, and, I hope, entertaining. The book is divided into six parts. Part 1 provides perspectives on our common heritage and evolutionary theory; part 2 chronicles the origin of the universe and our solar system and the origins of life on Earth; part 3 details the evolution of complex life; part 4 recounts the migration of life from sea to land; part 5 chronicles human evolution; and part 6 deals with the present and future of life on our planet. The text is punctuated with photos, figures, and sidebars to help provide interest and clarity. Each chapter is self-contained and can be read in isolation from the other chapters. So, the book can be taken and digested in pieces, used as a reference, or tackled in continuum from cover to cover. The story of the origin and evolution of life is the greatest, most mysterious, and most miraculous story of all time. Let's get started on the journey.

PART I

PERSPECTIVES

CHAPTER 1

OUR COMMON HERITAGE

> Therefore, I should infer from analogy that probably all the organic beings, which have ever lived on this earth, have descended from one primordial form, into which life was first breathed.
>
> —Charles Darwin, in *On the Origin of Species*

You don't look much like a tree or a worm. But it turns out you have much in common with both. All living things have much in common. We all share common biochemistry: common genes, common proteins, and common biochemical reactions (image 1.1). The common biochemistry shared by us and eight million other current-day species was inherited from our common ancestor who lived three and a half billion years ago. Darwin, with uncanny insight, identified this unifying principle of biology—that all species are related to one another in a vast family tree of life. And he recognized this before anyone knew there was a gene! The implications are mind-boggling. If we could travel back far enough in time, we would find the common ancestor of ourselves and every other living organism, including porcupines, flamingos, and cacti.

All we living creatures are made of cells—tiny structures enclosed by membranes and filled with chemicals dissolved in water. Cells are so small that fifty thousand bacterial cells can fit on a pinhead, but their complexity is enormous. Cells capture energy from the external environment, regulate their own internal milieu, grow, reproduce, and pass on heritable traits to future generations of cells. These qualities define life.

We all know that any system left to itself eventually deteriorates and becomes disordered: cars break down, buildings crumble, and dead organisms decay. This tendency toward decay is expressed in the **second law of thermodynamics**, which states that the universe is always

1

moving from a state of order to a state of disorder. Entropy, or chaos, is always increasing. But life is an island in this ocean of decay. Life is a force that imposes order on our disordered universe by reducing its own internal chaos. Inside your cells are some of the few places in the universe where structures are created rather than destroyed, where order is created from disorder, and where entropy decreases.

Image 1.1: Chimpanzees and *E. coli*. These chimpanzees and *E. coli* bacteria share common biochemistry: common DNA structure and codes, common amino acids and proteins, and common biochemical pathways. Like all living organisms, they have inherited common biochemistry from a universal common ancestor.

But life, like everything else, is subject to the laws of chemistry and physics, and life, too, must obey the second law of thermodynamics. As a cell becomes internally ordered, it releases energy as heat into its surrounding environment, which therefore becomes more disordered and chaotic. So the universe as a whole, the cell and its external environment—life *and* nonlife together—becomes more disordered, even as the cell imposes order on itself.

The creation of order within a cell in the face of a disordered universe is a kind of miracle—not a miracle through some type of divine intervention that disobeys the laws of chemistry and physics but rather a miracle borne of biochemistry. Biochemical processes provide the energy and substance of life. Amazingly, the chemical organization of all life as we know it, from bacteria to chimpanzees, is remarkably and eerily similar. This similarity begins with the common structure and codes contained in our **DNA**.

OUR SHARED DNA

In February 1953, American biologist James Watson and British molecular biologist Francis Crick announced at a pub in Cambridge, England, that they had discovered the "secret of life." Their original paper described the structure of DNA (deoxyribonucleic acid), a three-dimensional double helix capable of reproducing itself (Watson and Crick 1953). We have since learned that DNA contains codes that specify the production of proteins, and that DNA reproduces itself in order to copy genetic information for the next generation.

DNA is a macromolecule located in the nucleus of the cell and is composed of two long intertwining chains that in turn are made up of smaller molecules, known as **nucleotides**, which are linked together end to end (image 1.2). Each nucleotide molecule contains one of four nitrogen bases: adenine (A), thymine (T), guanine (G), or cytosine (C). As nucleotides line up one after the other in strands of DNA, the resulting order of nitrogen bases A, T, G, and C form the genetic code—a code that can be "read" in a single direction to specify the amino acid sequence in proteins that are synthesized in the cytoplasm of the cell. Astonishingly, of hundreds of nitrogen bases on our planet, all living organisms incorporate only four into their DNA: A, T, G, and C. Consequently, the DNA code of all organisms contains only these four "letters."

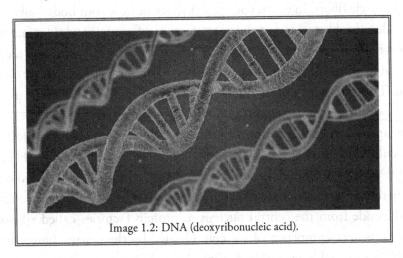

Image 1.2: DNA (deoxyribonucleic acid).

This puzzled Watson, Crick, and their colleagues. Why should all life use only these four nucleotides as codes to direct the activities of the cell? The codes appear to be arbitrary. Why are there no organisms with different coding systems? The logical answer is that all of us, every living thing on Earth, must have inherited this specific DNA coding system from a common ancestor. But that begs one question: Why would our last common ancestor incorporate these four nucleotides into DNA when there are hundreds of nucleotides to choose from? Was this by chance, or did these nucleotides have special properties that caused them to be selected? As it turns out, the four nucleotides chosen have properties that cause them to stick to complementary nucleotides—A to T and G to C—and these properties are what allow these macromolecules to replicate. We'll revisit this in chapter 4.

MAKING PROTEINS

Proteins are often considered to be the single most central compound necessary for life. They are fundamental to our cells' structure and operation. These complex macromolecules, composed of smaller amino acids that link together in an amazing variety of lengths and shapes, function as enzymes that facilitate almost all biochemical reactions—from those that store energy to those that build large molecules. They bind and bring molecules together so they can react, and they jump-start those reactions by drastically reducing the energy required to start them. In this way, protein enzymes can speed up cellular reactions a millionfold. Proteins also form many structural elements of the cell. They are major components of connective tissue, which makes skin elastic; of keratin, which gives hair its threadlike form; and of actin and myosin, which make up muscle fibers. In short, your proteins define how your body looks and functions.

DNA provides the blueprint for the manufacture of proteins. And amazingly, all cells manufacture proteins in the same way. The manufacturing process was described in detail by Francis Crick and given the lofty title **the central dogma of molecular biology** (Voet, Voet, and Pratt 2013, 50). The process can most simply be stated as "DNA makes RNA makes proteins."

The process occurs in two steps. First DNA copies a part of its code to another giant molecule, messenger RNA, in a process called **transcription**. Messenger ribonucleic acid (**RNA**) is a macromolecule similar to DNA composed of four nucleotides linked together in a chain. (It differs from DNA in that it is single-stranded rather than double-stranded, and one of the four nucleotides contains the nitrogen base uracil instead of thymine [U vs T].) Secondly, messenger RNA carries the code from the central nucleus to protein factories called **ribosomes**, which are located in the watery cytoplasm of the cell. Guided by the genetic blueprint read off the messenger RNA, amino acids are assembled into long chains to produce protein molecules in a process called **translation**.

In translation, the ribosome treats each three-letter sequence of nucleotides on RNA as a code (or **codon**), giving instructions to assemble a specific amino acid into the chain. In this way, a long sequence of three-letter codons will determine the sequence of amino acids in the protein chain. Since RNA uses only four nucleotides in its code (A, U, C, and G), a three-letter sequence can provide 4^3, or sixty-four, possible codes, or **codons**, for distinct amino acids. The RNA codon GCA, for example, codes for the amino acid alanine. The synthesis of proteins uses only twenty standard amino acids, so some amino acids have two or three codons that code for them, and there are a few codons that do not code for any amino acid at all (Voet, Voet, and Pratt 2013, 77).

I've gone into some detail here to illustrate the complexity of this system because knowing these intricacies, and knowing that all life and all cells share these intricacies, makes us appreciate our commonality. What is even more amazing is that all organisms use the same three-letter codes on RNA to specify each of the twenty amino acids. GCA codes for alanine in *all* living cells, from bacterial cells to human cells. This coding system is the **universal genetic code** (see sidebar "Universal Genetic Code") (Voet, Voet, and Pratt 2013, 963–68).

Taken together, these three observations of life's universality—the four common nucleotides used in all DNA, the central dogma of molecular biology, and the universal genetic code—provide convincing evidence for the commonality of life on Earth.

But there is more.

LEFT-HANDED LIFE

It is astonishing that of over five hundred amino acids occurring naturally on Earth, only twenty are used by all organisms on our planet to manufacture proteins. (Actually, two additional rare amino acids, known as the twenty-first and twenty-second amino acids in the genetic code, have been recently identified and are used in the synthesis of proteins by some microorganisms [Zhang and Gladyshev 2006].) Perhaps even more astonishing is the fact that all amino acids used to make proteins are left-handed. And we aren't sure why.

Amino acids, like many organic molecules, can exist in two forms: left-handed and right-handed. A left-handed molecule has the same atoms and connections as its right-handed counterpart, but the molecules are geometrically different; they are mirror images of one another. The handedness of a molecule is as important as its atomic composition in determining its chemical properties and the molecules with which it can react. A left-handed glove will fit a left hand but not a right hand. Therefore, we might understand how a biological system might evolve in such a way as to "work" and react with only left-handed or only right-handed amino acids, but not both. Why did life decide to use left-handed amino acids in the first place?

THE UNIVERSAL GENETIC CODE

Each three-letter code (codon) on RNA specifies one of twenty amino acids used in the manufacture of proteins. The universal genetic code is a mapping, usually expressed in table form, of codons to specific amino acids. The universal genetic code determines how the sequence of codes (codons) on messenger RNA are used to determine the sequence of amino acids in the manufacture of proteins.

But why are only twenty standard amino acids used by cells to build proteins when there are hundreds of amino acids to choose from in nature? And how was the universal genetic code determined? Why does the codon GCA on messenger RNA code for alanine, or UGC for cysteine? We can understand why these correlations, once they were chosen, are present in all living cells from bacterial to human cells, since we all descended from a common ancestor. But why were they chosen in the first place? These questions have puzzled and intrigued scientists and philosophers alike. Are these properties a product of chance, or are they a biochemical necessity?

Nobel Prize–winning biochemist Christian de Duve believed these properties evolved as a biochemical necessity. In the earliest cells, small pieces of RNA, with attached amino acids, lined up alongside long pieces of RNA, and facilitated the manufacture of long strands of amino acids, or proteins. De Duve believes those small pieces of RNA, each with a unique three-letter code or codon, had special properties that facilitated their linkage to specific amino acids. Apparently, these small pieces of RNA were able to link with only twenty amino acids out of hundreds in nature, and each small piece of RNA with a unique codon was able to link with only one of these twenty amino acids. This would explain why only these twenty amino acids are used for making proteins, and why specific codons on RNA specify specific amino acids and, as such, define the universal genetic code. This arose out of biological necessity because only selected small pieces of RNA and selected amino acids had special biochemical properties that allowed them join together. This illustrates a more general concept. Much of evolution is driven by and constrained by the biochemical properties of atoms and molecules that populate our planet.

Scientists do not have all the answers, but they have some plausible ideas. Organic molecules, including amino acids, were brought to Earth in its early days by a barrage of meteorites, and examination of these ancient rocks has shown a 5 to 10 percent excess of left-handed organic molecules (Mosher 2008; Choi 2013). Astronomers have recently discovered a nebula (a remnant of a supernova) in our galaxy that emits predominantly right–circular polarized light, which is known to damage right-handed organic molecules (Kwon et al. 2013). So perhaps as amino acids were hitchhiking a ride to planet Earth on a meteorite, the right-handed variety became partially depleted through damage from polarized light. Then, as water evaporated from the meteorites and amino acids paired up—a right-handed one with a left-handed one—to form crystals, the excess left-handed amino acids stood alone. Under this scenario, meteorites brought predominantly left-handed amino acids to Earth, and these were incorporated in first life. Once life began to form proteins with left-handed amino acids, it was committed thenceforth to use only left-handed amino acids.

STORING UP ENERGY: THE CREATION OF ATP

Life needs energy. Without a source of energy, cells cannot do the work needed to grow, move, reproduce, or maintain their internal environment. There are many potential sources of energy in the environment, and organisms differ widely in the way they collect it, but all life stores energy in the same way. We all store energy in the high-energy molecule adenosine triphosphate (**ATP**). ATP is the universal energy currency. It is like a cellular battery; our cells use energy from the environment to produce ATP, and ATP is later broken down throughout the cell to release that energy in measured increments to power chemical reactions that produce movement, growth, and reproduction.

Not only does all life use ATP, but all life also uses the same unique biochemical processes to produce ATP. We will hear more about these complex processes, known as the **electron transport chain** and **chemiosmosis**, in chapter 5, but the fact that all living cells on Earth use ATP as their common energy currency and all use the same processes to produce ATP suggests, once again, that we are all related—that we all inherited these processes from a common ancestor.

THE TREE OF LIFE

Charles Darwin, who observed the curious diversity of life in the Galapagos Islands, was convinced that all life descended from one primordial form. If he could see how much we have discovered since his journey, he would be awestruck. The explosion of research in biochemistry and genetics since Darwin's work has confirmed and expanded our understanding of what he first saw in those Galapagos species. The unity of life is incontrovertible.

New technology allows geneticists to catalog the entire genome of a species—to sequence the millions to billions of nucleotide pairs that make up a complex organism's complete DNA genetic code (Pray 2008). Comparison of genomes between species reveals similarities and differences that provide an estimate of the "genetic distance" between them—the species' level of relatedness. The genetic differences between two species can also help to estimate the time elapsed since the two species diverged from a common ancestor. Combining this information with fossil records, scientists have constructed an entire family tree—a "tree of life." Largely owing to the pioneering work of American microbiologist Carl Woese, all living organisms have been partitioned into three vast domains: **Bacteria**, **Archaea**, and **Eukarya** (Woese, Kandler, and Wheelis 1990). Bacteria and archaea are simple single-celled microorganisms, while eukaryotes are organisms with complex nucleated cells. Eukaryotes include multicellular organisms: fungi, plants, animals, you, and me.

The **last universal common ancestor** of all life (now fondly referred to as **LUCA**) appeared an estimated 3.8 billion years ago and diverged into two separate microorganisms: archaea and bacteria. As we will see later, the first complex eukaryotic cells came into existence when a bacterium merged with an archaeon. Multicellular organisms evolved from that primitive eukaryote and ultimately gave rise to the fungi, plants, and animals living today.

The implications of life's common biochemistry are profound. If all species evolved from one last universal common ancestor, the origin of that life—the origin of that common ancestor—happened once, and *only* once, during the 4.5 billion years of our planet's existence. It was a unique, improbable freak event. It is possible that other forms of life may have arisen independently across the eons, but if they did, they did not survive and leave descendants.

So how did this one-time event happen? And what gave LUCA such a survival advantage over its competitors? I'll take you on a journey from the formation of the first cell, to the merger of an archaeon and bacterium to form the first complex cell, to the evolution of the complex plant and animal life that dominates our planet today.

Let's start at the beginning.

CHAPTER 2

DARWIN'S THEORY OF EVOLUTION

The love for all living creatures is the most noble attribute of man.

—Charles Darwin

In 1831, English naturalist and geologist Charles Darwin, forsaking a career in medicine and theology, embarked on an expedition to South America and the Galapagos Islands as a naturalist on board the HMS *Beagle* (image 2.1). When he and the *Beagle* crew began their exploration of the islands and its diverse wildlife, he embraced traditional creationist views prevalent at the time—that the diversity of species inhabiting our planet were created by God, forever immutable. It was not until he returned home and examined the many specimens he brought with him that he appreciated the significance of his findings and introduced an entirely new and controversial perspective on the origins of species.

The Galapagos Islands were formed by volcanic eruptions and were separated from the mainland and from each other in such a way that the species living on them have been isolated from one another. Darwin recognized, after examining a host of specimens, that species seemed to be isolated to specific islands in the archipelago (Sulloway 2005). He was able to identify from which island a tortoise originated by the shape of its shell. He recognized four separate species of mockingbirds and fourteen separate species of his now famous finches, each confined to a specific island. In an aha moment, according to legend, Darwin postulated that species, when isolated, evolved new traits to adapt to their changing environment, and some transformed into new species. His ideas challenged the fundamental tenet of creationism—the notion that all species are created by the Creator in their present immutable forms (Sulloway 2005).

Darwin's genius was not so much his meticulous collection of data and his keen observations but his courageous willingness to consider new and unconventional ways of thinking. He postulated that new species arise naturally over the course of generations through a process of evolution and natural selection, and he published his theories in 1859 in his now classic book *On the Origin of Species by Means of Natural Selection* (Darwin 1859). His ideas that animals and humans shared common ancestry shocked Victorian society and created great controversy that persists to the present day. But Darwin's theory of evolution has gained broad acceptance by the scientific community and stands as one of the greatest scientific achievements of our time.

DARWIN'S THEORY OF EVOLUTION

Darwin's original theory of evolution had three major parts:

Image 2.1: Charles Darwin (1809–1882).

1. All species exhibit **natural variation**, meaning that each offspring, whether it originated through sexual reproduction or cloning, has slightly different traits from its parents.
2. Species exhibit **superfecundity**. In most species, parents produce more offspring than can survive long enough to reproduce themselves, given the natural hazards of their environment.
3. Individuals survive and reproduce, or they perish, depending on their own particular assortment of traits. Those that have traits that give them survival advantages within their particular environment will be more likely to have young, and they will pass on their winning traits to those young. Over time, more members of the species will have these traits. On the other hand, traits that make an organism less suited for survival are likely to slowly disappear, as those that have these traits often die before they have offspring and don't pass these traits to the next generation. This selection process, in which organisms survive and reproduce or perish based on their particular trait variations, is called **natural selection**.

Darwin's theory provided a comprehensive explanation as to how species evolve to be well adapted to their environments and supports the concept that all living organisms share common ancestry and that we all descended from a common ancestor.

THE THEORY OF EVOLUTION ADAPTS TO CHANGE

The theory of evolution has gained broad acceptance among the scientific community and is regarded as one of the most durable and important theories in the history of science, but as is the case with all theories, there have been modifications made to fit new observations. In one of the most famous of these tests, Darwin wondered why the peacock should have such bright, flamboyant eyespots on its cumbersome tail, which would make the bird quite vulnerable to predators. Such a trait with a negative survival advantage should have long vanished by way of natural selection.

Darwin was puzzled until he observed peacocks gathering together in courtship displays, in which they opened up and spread out their iridescent tail feathers with spectacular eyespots like fans in an attempt to attract mates. The peahens lined up to assess the display and to choose the most formidable peacocks as their mates. These observations led Darwin to modify his theory of evolution and formulate the theory of **sexual selection**, which he outlined in his subsequent book, *The Descent of Man, and Selection in Relation to Sex* (Darwin 1871). According to the updated theory, species are selected not only because they have traits that enhance their ability to survive in their environment but also because they have traits that enhance their ability to attract a mate and reproduce. A peacock with a drab, boring tail may go unnoticed by predators and be able to survive longer than a peacock with a colorful tail, but if his tail is drab, females will never mate with him, and he will not pass on his drab tail to the next generation. In reality, reproduction may be even *more* important than survival for evolutionary success.

A more recent update and modification of Darwin's theory of evolution has been the incorporation of modern genetics. We now know that genes code for traits and that genes and their associated traits are inherited from one generation to the next. Inherited genes either perish with the organism or survive to be transmitted to future generations, depending on natural selection. Richard Dawkins, in his landmark book *The Selfish Gene*, has given genes a personality by describing the selfish gene as "concerned" solely with its own survival. However, this description is metaphoric. Genes don't really have any foresight or conscious purpose; they are blind, unconscious replicators. As DNA replicates, random mutations may occur such that a slightly altered genome passes to offspring. This new genome contains altered genes that code for new traits that may improve or impair survival of the offspring. Mutations that improve survival of the offspring will be passed to future generations, while mutations that impair survival will perish with the offspring. Over time, hereditary mutations in the genome that promote survival and reproduction will accumulate and produce complex organisms that are well adapted to their environments. The incorporation of modern theories of genetic inheritance into Darwin's original theory of evolution has formed an updated theory of evolution known as **neo-Darwinism**.

RENAISSANCE OF LAMARCKIAN INHERITANCE

The human body has an amazing ability to learn and adapt. Imagine you are an elite distance runner. You train for months and years to develop stamina, strength, and speed, culminating in a victory at the world championships. Wouldn't it be terrific if you could pass this skill and training to your children? Imagine how this would accelerate the pace of evolution with each generation building on the last.

French naturalist Jean-Baptiste Lamarck (1744–1829), who preceded Darwin, believed this was possible. He proposed that characteristics acquired during an animal's lifetime could be passed to its offspring. Lamarck explained his theory of **Lamarckian inheritance** through his classic example with giraffes. Giraffes once had short necks, he postulated, but they stretched their necks repeatedly, reaching to feed on leaves of tall trees. Over a lifetime of stretching, their necks gradually lengthened, and when they reproduced, they passed this long-necked trait to their offspring. Darwin embraced the Lamarckian inheritance of learned traits as a part of his theory of variation and natural selection. But Lamarckian inheritance has long been discredited. Neo-Darwinism has taught us that learned behavior does not alter genetic codes and cannot be passed to subsequent generations. Or so we thought.

Recently a number of observations have led to a resurgence of the concept of Lamarckian inheritance. There has been recognition that many traits change very rapidly from one generation to the next—much more rapidly than can be explained by the rate of genetic mutation. In addition, a number of observations have shown that a change in traits in an individual triggered by environmental factors can be transmitted to offspring. One of the first and most famous of such observations came from the experience of the Dutch famine in 1944, known as the *Hongerwinter,* which was a consequence of the Nazi food blockade (Heijmans et al. 2008). The offspring of starving Dutch mothers were small and susceptible to obesity and diabetes. This is not surprising, since environmental factors in utero could affect traits in the offspring. But according to Neo-Darwinian theory, this should not change the genome of the offspring and should not be passed to subsequent generations. Surprisingly, when these offspring had children of their own, the children inherited similar traits. This is a form of Lamarckian inheritance.

In a more recent example, scientists exposed pregnant rats to a fungicide and evaluated their offspring and subsequent male descendants for three generations. The offspring were found to have decreased sperm counts, decreased sperm viability, and an increased incidence of infertility. But surprisingly, in the absence of any continued exposure to the fungicide, subsequent generations of male rats had similar traits of decreased and dysfunctional sperm. And female rats seemed to sense this, because when faced with a choice of descendants of exposed or unexposed males, they overwhelmingly chose unexposed males (Skinner 2016). This illustrates how an environmental exposure that alters traits in one generation can affect the inheritance of traits to subsequent generations.

How does this happen? Biologists and geneticists have learned that environmental factors can turn on or turn off genes without altering the DNA sequences or code (Skinner 2016; Heard and Martienssen 2014; Lim and Song 2012). Activation or deactivation of genes by environmental factors can persist and be passed to subsequent generations and can alter traits in these subsequent generations. The mechanism by which environmental factors can activate or deactivate genes can take several forms, including attaching a methyl group to DNA (methylation) or an acetyl group to histone proteins associated with DNA. The study of how environmental factors can alter gene expression and how changes in gene expression can transmit new traits to future generations without changing DNA sequences is called **epigenetics**.

A recent study from Wayne State University in Detroit documented these environmentally induced molecular changes in the genome of humans. The investigators found that when mothers and their fetuses were exposed to high levels of lead, the fetuses developed DNA methylation, and when the offspring grew up and had children of their own, they also had methylated DNA (Sen et al. 2015). This was one of the first demonstrations that an environmental exposure can have a specific epigenetic effect on the genome that can be passed to two generations in humans. A number of studies have shown that lifestyle factors, such as diet, physical activity, and obesity, as well as toxins and carcinogens, can induce DNA methylation.

Epigenetic inheritance is changing our view of evolution. The process of random mutations of genes and natural selection acts very slowly; it takes many generations for a new beneficial genetic trait to become established in a population. But epigenetic changes initiated by changes in the environment can affect many individuals at once and can be transmitted to the next generation, potentially greatly accelerating evolutionary change. Environmental exposure is especially impactful during embryonic development, and environmentally induced changes in activation and deactivation of genes have been observed in plants, insects, fish, birds, rodents, and humans. In recent generations, scientists have believed that humans are a product of their genes and their environment. What epigenetics has taught us is that environmental factors can act on the genome and alter regulation of the genes and that this altered regulation can persist and be passed to subsequent generations. Epigenetics has now become an integral part of the evolutionary paradigm. Lamarckian inheritance is back in the mainstream.

LAWS AND THEORIES

There has been great controversy and much misunderstanding about Darwin's theory of evolution. How often have you heard a friend say, "Well, that's just a theory; nobody knows if it is really true." But ask a friend about an established scientific law and you almost never hear, "Well, that's just a law." How come? What's the difference between a theory and a law?

In its simplest terms, a law describes *how* nature behaves and *predicts* what will happen in the future under certain conditions. In contrast, a theory tries to provide the best possible explanation for *why* something happened or *why* something is the way it is (LaBracio 2016).

The universal law of gravitation states or describes that two masses will attract one another with a force proportional to the product of their masses and inversely proportional to the square of the distance between them. We can predict the force one object will exert on another if we know the objects' masses and the distance between them. We don't really have any good theories that attempt to explain *why* two masses should attract one another.

Gregor Mendel, a nineteenth-century Moravian monk, performed classic experiments with pea plants and formed theories relating to how inheritance works. He observed that if he crossed purebred white-flowered and purple-flowered pea plants, the offspring was not a blend but rather a purple-flowered plant. He then bred these second-generation plants and found that one in four offspring was a white-flowered plant. Mendel formulated the idea of hereditary units and a theory of inheritance, now known as Mendelian inheritance, to explain these observations. We have since learned that Mendel's hereditary units are genes with dominant and recessive traits.

Likewise, Darwin's theory of evolution, which postulated evolution of all creatures from a common ancestor through random variation and natural selection, provided an explanation as to why and how all creatures have become so well-adapted to their environments and why we all share many common traits.

EVOLUTION AND INTELLIGENT DESIGN

A discussion of the origin of life and biological evolution cannot be inclusive without a consideration of the dichotomy between the theory of evolution and **intelligent design**. The theory of biological evolution has been widely accepted by the scientific community, but there remains a sizable and vocal minority in Western society that embraces the concept of intelligent design. Wherein lies the truth?

Benjamin Kuipers, a professor of engineering at the University of Michigan, has provided an insightful metaphor for this dichotomy in his essay "Why Do We Believe in Electrons, but

Not in Fairies?" (Kuipers 2013). We cannot see electrons and we cannot see fairies, so why would we believe in one and not the other? Although we cannot see electrons, we can measure their effects. Electrons emitted from a cathode ray tube passing between two metal plates and striking a fluorescent screen are deflected in a particular direction when an electric potential is applied across the metal plates. This demonstrates that electrons have a negative charge. Many subsequent experiments have demonstrated that electrons steadfastly follow a set of physical laws that allow us to predict their behavior. The laws have led to the inventions of light bulbs and microprocessors and have helped us further understand our universe.

Fairies are different. As Kuipers writes, "Fairies are much freer. Fairies decide what to do and when to do it. We cannot see fairies, and there are no rules that predict their behavior. And, of course, there are no fairy microprocessors. This does not mean that fairies don't exist, only that there is no evidence to support their existence" (Kuipers 2013).

Evolutionary theory has provided the best scientific explanation for why all living things, from bacteria to human beings, share common DNA codes, common proteins, and common biochemical pathways. And it is the best explanation as to how complex creatures have become so well adapted to their environments. The theory of intelligent design—that life, in all its diversity, was designed and created by an intelligent supernatural entity a few thousand years ago—is not consistent with the accumulated data in the fields of geology, paleontology, genetics, and more.

A wide chasm has developed between religious groups who support intelligent design and scientists who support the principles of evolutionary biology. According to a Pew Research Center analysis performed in December 2013, almost two-thirds of Americans agree with the statement "Humans and other living things have evolved over time," while a third reject the idea of biological evolution, agreeing that "Humans and other living things have existed in their present form since the beginning of time" (Pew Research Center 2013). Beliefs differ by religious group and political party, with a greater proportion of evangelical Protestants and Republicans rejecting evolutionary theory.

The conflict between advocates and opponents of evolutionary theory expanded to the classroom and has been debated and litigated in state legislatures, school boards, and courtrooms for many decades. In 1925, three-time presidential candidate and populist William Jennings Bryan advocated for a Tennessee statute prohibiting the teaching of the theory of biological evolution in public schools in the famous Scopes monkey trial (De Camp 1968). The jury was to decide the fate of John Scopes, a high school biology teacher accused of teaching the theory of evolution in violation of Tennessee statutes. The trial provided great courtroom drama but no resolution. Not until 1968 did the Supreme Court rule in Epperson v. Arkansas in a decision written by Justice Abe Fortas that such bans are unconstitutional and violate the First

Amendment, which prohibits legislation promoting one religious view over others. However, the controversy continues today in local school boards in Kansas, Pennsylvania, and elsewhere, often without resolution.

The views of intelligent design are starkly at odds with what scientific investigations have concluded about the origins of species and the origins of humanity. However, this does not mean that the theory of evolution is incompatible with religious beliefs or with a belief in God. But we'll leave it here. This book is about science, not religion. Darwin's theory of biological evolution has provided the best—the *only*—scientific explanation for complex life that is adapted to its environment. Evolutionary zoologist Richard Dawkins argues that if we eventually discover complex life elsewhere in the universe, it too will most likely be a product of adaptation through natural selection (Bendall 1983). Evolution through variation and natural selection may be universal to all living things in our vast cosmos. Maybe someday we'll be able to test this final hypothesis.

PART II

ORIGINS

CHAPTER 3

IN THE BEGINNING

In the beginning, God created the heaven and the earth.
And the earth was without form, and void; and darkness was upon the face of
the deep. And the Spirit of God moved upon the face of the waters.
And God said, Let there be light: and there was light.

—Genesis 1:1

Imagine all matter in today's universe compressed into a tiny dot infinitely smaller than the dot over the letter *i*—a dot so small that it has no dimensions at all: a singularity. Imagine getting ready to watch the grand spectacle of the universe being born. Unfortunately, there is no place for you to sit because there is no space outside the singularity, and besides, time does not yet exist.

A blinding pulse suddenly shines from this singularity, and there comes an incomprehensible explosion—an instant of glory as the universe expands and fills the void of space. At first there is only energy and heat—ten billion degrees of heat—but within minutes the universe has cooled enough to convert energy into matter: protons, neutrons, and electrons at first, and then the first atoms, grouped in pairs to form the primordial gas hydrogen. As the universe expands, gas and dust congregate into large clouds known as nebulae. Gravity pulls more gas and dust inward, and the nebulae begin to swirl faster and faster, just like a spinning skater does when she pulls her arms to her sides. As the nebulae spin faster, gases are concentrated at the center and heated to extreme temperatures. Hydrogen atoms fuse to become helium. The newborn fusion reactors belch light and heat as their temperatures soar. The first stars are born.

According to the **big bang theory**, this is how the universe began. Most cosmologists and physicists now accept this theory as the best depiction of the origin of the universe 13.7 billion years ago (see sidebar "The Big Bang Theory" and figure 3.1) (NASA 2017).

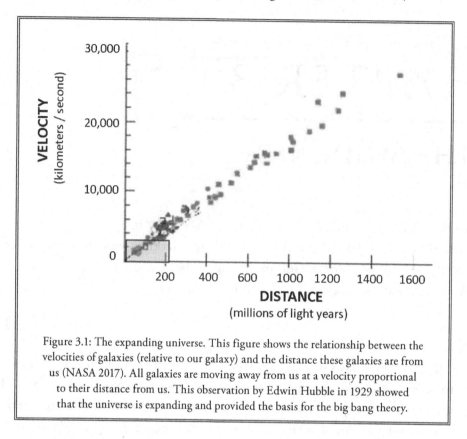

Figure 3.1: The expanding universe. This figure shows the relationship between the velocities of galaxies (relative to our galaxy) and the distance these galaxies are from us (NASA 2017). All galaxies are moving away from us at a velocity proportional to their distance from us. This observation by Edwin Hubble in 1929 showed that the universe is expanding and provided the basis for the big bang theory.

RED GIANTS AND SUPERNOVAS

The universe is forever remaking itself. Stars beget new stars in an endless cycle of death and rebirth. Stars radiate light and heat as hydrogen fuses to form helium. After millions or billions of years, a star's supply of hydrogen gas wanes, its core contracts while its outer shell expands, and the star begins to glow red. The star becomes a **red giant**. The red giant's inner core contracts under its own gravity, producing incredibly high temperatures that cause hydrogen and helium gas to fuse into yet larger atoms: carbon, oxygen, and nitrogen.

As the red giant's life cycle ends, there is a tug-of-war within the star between gravitational forces pulling inward and nuclear forces pushing outward. As its hydrogen fuel is used up, the giant starts to collapse under its own weight, and then it explodes spectacularly as a **supernova**. The immense heat of the supernova explosion facilitates the formation of elements heavier than iron. This cosmic fireworks display outshines all other stars in the sky, and scatters fragments, including the heavy metals, across a wide region of neighboring stars.

Five billion years ago, one such giant supernova exploded and provided material for the birth of hundreds of young stars, including our sun. The explosion scattered the dusty remnants of nearby planets and fragments of the supernova across the abyss. Pieces of celestial scrap—rock, heavy metals, gases, and dust—began to collapse, triggering the formation of new stars and solar systems, including our own.

THE BIRTH OF PLANET EARTH AND ITS UNIQUE LUNAR COMPANION

Earth began as an agglomeration of gases, dust, and solid particles 4.5 billion years ago (see sidebar "How Old Is the Earth?"). As it grew larger, its gravity attracted meteorites, adding the basic ingredients needed for life: vital water, carbon, oxygen, nitrogen, calcium, and phosphorous.

THE BIG BANG THEORY

In 1929, American astronomer Edwin Hubble made the surprising and historic observation that all distant galaxies are moving away from us. He found that light emanating from all galaxies exhibited a redshift, which is a lengthening of the wavelength of light emanating from bodies moving away from an observer (NASA 2017). This is analogous to the lengthening of soundwaves and sudden lowering in pitch of a siren as it passes by and moves away from an observer.

Hubble's observations led him to the astonishing conclusion that all of the universe is expanding—that everything, in all directions, is speeding away from us. Furthermore, Hubble found that the farther distant a galaxy is from us, the more dramatic its redshift—and therefore, the greater its speed. Not only are all galaxies receding from us, but the ones that are farther away are moving away faster than those that are close.

These observations led to the big bang theory. If the universe is expanding at great speed, then the logical conclusion is that at some time in the distant past, the entire universe was compressed into an infinitely small point with an infinitely large mass—a singularity. This was the state of the universe at the moment of the big bang.

Using these data, scientists can also determine how much time has passed since the universe began—the time since the big bang—which turns out to be 13.7 billion years. This was the beginning—or, in the words of Genesis, it was "a day with no yesterday."

As Earth grew by accreting particles and meteorites, one momentous collision changed the fate of our planet. According to the **giant impact theory**, a large protoplanet about the size of Mars collided with Earth shortly after Earth's formation, adding considerable mass to our planet and spitting out remnants that became our moon (Ward and Brownlee 2004, 229–34). The 1969 Apollo mission piqued interest in the origin of our moon and brought home evidence supporting the giant impact theory. The moon, unlike meteorites, contains no volatile elements—such as zinc, cadmium, and tin—probably because they vaporized in the giant impact. Rocks from the lunar surface have the same proportion of heavy versus light oxygen isotopes as rocks found on Earth. These rocks with specific ratios of isotopes are unique to the Earth and the moon and have not been found in specimens from Mars or in meteorites; they

HOW OLD IS THE EARTH?

Before the nineteenth century, only theologians and philosophers, not scientists, were concerned with the age of the Earth. Scientific inquiry into the age of our planet using radiometric dating followed the discovery of radioactivity by French physicist Henri Becquerel in 1896.

Radiometric dating has become an invaluable tool for estimating the ages of minerals and organic remains (US Geological Survey 2019). Here is how it works: In living organisms, radioactive carbon-14 is one trillionth as abundant as its nonradioactive isotope, carbon-12. When an organism dies, its radioactive carbon-14 atoms fade, but nonradioactive carbon-12 does not. Carbon-14 atoms decay with a half-life of 5,730 years, meaning that after 5,730 years there is half as much carbon-14 as originally. So, if the ratio of carbon-14 to carbon-12 in the remains of an organism is one-fourth of its original ratio, we can estimate the organism died 11,460 years ago.

Uranium isotopes can be used to date more ancient samples. The uranium isotope U-235 decays to the lead isotope Pb-207 with a half-life of 704 million years, and U-238 decays to Pb-206 with a half-life of 4.47 billion years. Uranium dating is generally performed on zircon minerals, which are part of volcanic ash. When igneous rocks form as they cool, zircon minerals contain no lead. If an ancient zircon sample has equal parts U-238 and Pb-206, it is estimated to be 4.47 billion years old.

Movement and collision of tectonic plates and weathering have destroyed the oldest rocks on Earth, so they are not available for dating. Cosmologists and physicists have used rocks from meteorites as surrogates, based on the assumption that the Earth and its entire solar system, including meteorites, formed at the same time. Radiometric dating using uranium isotopes from the meteorite that hit Earth near Canyon Diablo, Arizona, 50,000 years ago has calculated the age of these ancient rocks to be 4.5 billion years old.

So we believe Earth is 4.5 billion years old. This is science in action. Know a few things about the way atoms work, and you can measure the age of the Earth.

probably formed during the turbulent mixing of the giant impact. These shared Earth–moon rocks are tokens of our cosmic partnership.

Our moon is something of a freak because of its large size relative to our planet. The moon is one-third the size of Earth and has unique properties that have shaped the natural history of our home planet. We've gotten pretty used to our twenty-four-hour days, but when Earth was young, the days were much shorter. The moon's gravitational pull on the oceans creates a tidal friction that slows Earth's rotation ever so slightly—by just a fraction of a second every year (figure 3.2) (Ward and Brownlee 2004, 227–28). Over millions and billions of years, these small changes add up. Days that lasted only six to eight hours early in Earth's history have now stretched to twenty-four hours as a result of the heavy gravitational embrace of our moon.

The giant impact and the creation of our large moon raises many what-ifs. For example, what if the rotation of the moon around Earth were not in the same direction as the Earth's rotation? In this scenario, tidal forces would slow the moon's velocity and put us on a collision course with our moon rather than causing it to gradually move away from us. What if the giant impact occurred earlier in the course of Earth's formation, when there was not enough mass to "hold on" to the moon? The moon would have been lost, flung into deep space. Or, on the other hand, what if the giant impact occurred later on, when Earth's mass was greater? This would not have allowed the moon to be ejected into orbit. Or what if the giant impact never even occurred?

As Ward and Brownlee asked in their book *Rare Earth*, "If the Earth's formation were replayed 100 times, how many times would it have such a large moon?" We might not have had any moon at all, and where would we be then? Indeed, we are lucky to have our moon, and we are lucky to be here to look up at it.

HADES ON EARTH

The first half billion years of Earth's life are called the Hadean Eon (4.5 to 4.0 billion years ago) after Hades, the Greek god of the Underworld. During this period, Earth underwent a harrowing baptism. The planet blazed with the fire of endless volcanic eruptions. Debris from space—relics from the formation of our solar system—relentlessly bombarded Earth. Earth was initially too small and its gravity too weak to hold on to atmospheric gases, but as it approached its present size and its gravitational pull became stronger, it was able to retain a primordial atmosphere that included methane, water vapor, ammonia, and neon—an atmosphere toxic to most life on our planet today. There was no oxygen. Hydrogen

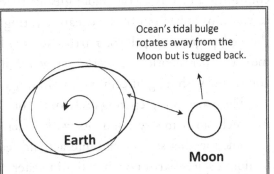

Figure 3.2: Schematic representation of the moon's tidal forces on the Earth. The moon's gravitational pull on the oceans creates a tidal bulge. As the Earth rotates, the tidal bulge moves away from the moon, and the moon tugs at the bulge, slowing Earth's rotation ever so slightly—by just a fraction of a second every year. Over millions and billions of years, these small changes add up; seconds turn to minutes, minutes turn to hours, and hours turn to days. Days that lasted six to eight hours early in Earth's history have now stretched to twenty-four hours (Ward and Brownlee 2004, 227–228).

and helium gases were very light and had escaped from Earth's gravitational pull. Volcanic eruptions added carbon dioxide and sulfur dioxide to the mix.

Our young sun was much weaker and emitted less light than it does today, and dust and debris blocked much of that feeble sunlight. The sky was gray and reddish because nitrogen—the abundant gas responsible for our modern blue sky—was present only in small quantities.

This was no place for life. Surface temperatures were scalding. There was no liquid water, and there were no oceans. The scorching temperatures would have extinguished any newly formed organisms within seconds, and the lack of oxygen would certainly have suffocated today's oxygen-breathing creatures within minutes. Earth had to change—and change it did.

EARTH COOLED

As the Hadean Eon came to an end, the Archaean Eon ushered in a new beginning. The meteorite barrage began to abate, and Earth cooled. The landscape changed dramatically but still had no resemblance to today's terrain. Water vapor condensed, forming oceans that covered most of the Earth. There were no mountain ranges and no deep ocean valleys. Volcanic activity continued and was considerably greater than today. The only land above water was formed as volcanos poked out of the ocean, creating small islands, such as the Hawaiian Islands and the Galapagos archipelago. Earth's surface hardened, creating a thin crust sitting atop the molten upper mantle. This set the stage for plate tectonic activity, which was to create our first continents with magnificent mountain ranges and deep valleys.

The planet was preparing itself for the origin of life. Life as we know it needs an environment in which biochemistry can do its work. There must be a solvent allowing atoms and molecules to come together so they can interact and react. In short, life as we know it requires water in its liquid state. Earth now had liquid water, but there were a number of other conditions our planet needed to support life. Our planet has been called "Rare Earth" by geologist Peter Ward and astrologist Donald Brownlee in their book by the same name because of the conditions it has to support complex life. Let's take a look.

THE GOLDILOCKS ZONE

Scientists have given the name **Goldilocks zone** to the thin zone around a star in which temperatures on the surface of a planet are just right to support liquid water and life as we know it (figure 3.3). The Goldilocks zone analogy comes from the story "Goldilocks and the Three Bears." When Goldilocks came into the house of the bears, she found three cups of porridge: one too hot, one too cold, and one just right. Our planet Earth is just the right temperature.

The temperature on the surface of a planet orbiting a star depends on the energy produced by the star and the planet's distance from that star. We are fortunate that there is a particularly lucky relationship between the sun's energy production and the distance between the Earth and the sun so that we have moderate temperatures on the surface of our planet. We are also lucky that Earth's orbit around the sun is nearly circular. If Earth had an elliptical orbit, we would freeze when very far from the sun and fry when very close to it. Earth's nearly circular orbit allows consistently moderate temperatures.

Moderate temperatures are not the only requirement for a habitable zone. Habitable conditions on a planet require specific properties of the planet's star. Large stars with diameters more than 1.5 times that of our sun have lives too short to support the slow processes necessary for life to evolve. It's taken 3.5 billion years on our planet; life is not a quick day's work. Smaller stars, with diameters less than one-half that of our sun, generate less heat. Their planets must be very close to their stars in order to have moderate temperatures, and planets this close may become tidally locked to their stars, such that the same side of the planet always faces the star. Tidally locked planets may not permit life to evolve, because the dark side of the planet facing away from its sun is as cold as dead space, and the bright side facing the sun perpetually roasts.

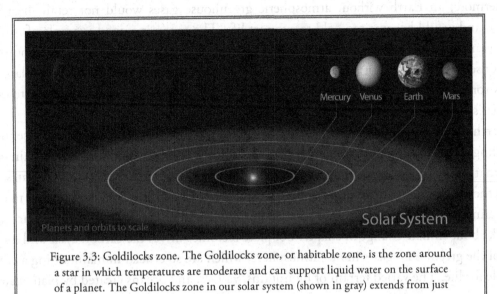

Figure 3.3: Goldilocks zone. The Goldilocks zone, or habitable zone, is the zone around a star in which temperatures are moderate and can support liquid water on the surface of a planet. The Goldilocks zone in our solar system (shown in gray) extends from just outside Venus's orbit through Earth's orbit, reaching to just beyond Mars' orbit.

We had long thought that the existence of these unlikely conditions was rare, but recent discoveries have identified planets in other solar systems that are in Goldilocks zones of their own. In a universe with an estimated over two trillion visible galaxies, each with over a hundred billion stars, there are certainly many such planets (Hille 2017). Given sufficient numbers, even the most improbable circumstances are common.

Our sun is intermediate in size as stars go and is capable of providing a habitable zone with moderate surface temperatures. But this will not last forever. Stars become hotter as they age. It is estimated that our sun has heated up approximately 30 percent since its origin 4.5 billion years ago and will heat up another 10 percent over the next billion years (Ward and Brownlee 2004, 32). As the sun gets hotter, the moderate-temperature habitable zone will move outward beyond Earth's orbit, and Earth will no longer be suitable for life.

RARE EARTH

Earth is special because it is in a habitable zone, but Earth has a number of other unique properties allowing it to support complex life. Earth is indeed a rare planet. Earth is just about the right size. A planet that is too small will not have enough gravity to hold on to an atmosphere. Without an atmosphere, oceans evaporate, water vapor escapes, and there is no protection from cosmic ultraviolet and ionizing radiation—lethal emissions from our sun. Furthermore, an Earth without atmospheric greenhouse gases would not retain heat from the sun and would become too cold to support life. The moon is a good example of this too-small planetary problem. With its weak gravitational forces, it cannot retain an atmosphere, and consequently, it cannot support life. At the other extreme, a planet that is too large has such strong gravity that mobility would be difficult. Plants and animals would need to evolve stronger support structures to support their weight.

Earth has a remarkable process known as **plate tectonics** that appears to be essential in providing conditions that are suitable for complex life. Earth has a solid, relatively thin, outer crust that lies atop a hot, viscous upper mantle. Large regions of Earth's crust, called plates, slide across the viscous upper mantle and collide with neighboring plates, buckling upward to create mountain ranges and buckling inward to create valleys and ocean beds. As we will learn when we talk about climate change in chapter 26, plate tectonics drives the carbon cycle and helps to control the greenhouse gases in our atmosphere, which are so crucial in maintaining a climate suitable for life. The varied terrain of mountains, valleys, and oceans created by plate tectonics also provides a variety of environments that promote biodiversity, a major defense against mass extinctions. Earth's distinctive geological layers with a thin crust atop a molten upper mantle provide conditions that allow plate tectonic activity, which is unique in our solar system.

EARTH'S INVISIBLE PROTECTIVE SHIELD

Outer space is not a friendly place for living creatures, and it poses obstacles to the inception and evolution of life on all planets. Earth and other planets in our solar system are incessantly bombarded by solar wind—a cosmic stream of small charged particles, mostly electrons and protons—emanating from the sun. These particles can slam into atoms in the upper layers of the atmosphere, sending these atoms into space and eroding the atmosphere. As the atmosphere dissipates, water evaporates and eventually the oceans disappear. Without some protection, this bombardment of solar wind would strip Earth's atmosphere, eliminate its oceans, and extinguish life on our planet. Fortunately, the Earth has an extraordinary protective barrier against this attack: its magnetic field.

As Earth formed by accretion of matter from outer space, heavy molten metals sank down to a central core under their own weight. The Earth's core is searing hot, with temperatures reaching 6,000°C (11,000°F) (Anzellini 2013). Frictional heating of core material as it moves with Earth's rotation, and heat generated by the decay of radioactive elements, such as uranium, maintain these blistering extremes.

Earth's inner core is solid, but the outer core is liquid and contains molten iron—a great conductor of electricity. Within this outer core, molten iron moves by convection, rising as it warms and sinking as it cools. Since moving electric charges create magnetic fields, scientists believe that charged particles within the molten iron of the outer core are generating Earth's magnetic field. Our magnetic field extends beyond the outer reaches of our atmosphere and protects us from the barrage of solar wind by deflecting incoming charged particles. As a result, very few of them actually penetrate our atmosphere. The magnetic field deflects the solar wind toward the poles, where the charged particles collide with oxygen and nitrogen atoms at high altitude, often producing spectacular light shows—the aurora borealis in northern regions and the aurora australis in southern regions (image 3.1).

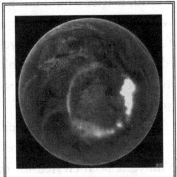

Image 3.1: *Aurora australis.* This image was captured by NASA's IMAGE satellite in 2005 and digitally overlaid onto *The Blue Marble*—a composite image of Earth. This natural light display happens when charged particles from the sun are deflected by Earth's magnetic field and collide with atoms in the upper atmosphere at the southern pole.

Earth's magnetic field is a result of special properties of our planet not shared by its neighbors. Venus has an iron core similar to that of Earth, but Venus rotates on its axis only once every 243 Earth days, and some speculate that this may not be fast enough to generate the churning and convection needed to ceate a strong magnetic field. Mars has an iron core that is solid rather than liquid, so it cannot generate moving electric charges and has little or no magnetic field. Mars has had much of its atmosphere stripped away by solar wind, owing to its lack of a protective magnetic field and because it has reduced gravitational forces due to its small size.

OUR IMPROBABLE UNIVERSE

Are you feeling lucky yet to be alive on this Earth? So many things could have gone wrong, and yet here we are on this miraculous blue planet—a vessel of warmth and protection floating in a sea of hostile, empty space. All around us are barren, inhospitable planets—rocks blasted by cosmic winds, icy cold and infernally hot. We are alone, blessedly and improbably safe.

At least that's how it seems.

Scientists have learned much about the nature of our universe, but perhaps the most remarkable finding is its regularity and predictability. Albert Einstein once eloquently reflected that "the most incomprehensible thing about the universe is that it is comprehensible." Our universe is characterized by laws of physics and by measurable natural constants—the speed of light (c), the gravitational constant (G), the charge of an electron (e⁻), and Planck's constant (h)—which quantify the unique relationships that give our universe its distinct character. Even the slightest variations in these universal constants would drastically change the nature of our universe, making life as we know it impossible. For instance, if gravity were less powerful, matter would not have congealed into galaxies and stars, and our universe would be very cold and devoid of solid objects. If gravity were *more* powerful, our stars would be much larger, and their life cycles too short to allow life to evolve. If the nuclear forces that hold atoms together were a bit stronger, most atoms would be unstable, and biochemistry would not be possible. Thus, we live in an improbable universe that seems particularly suited to support our type of life.

We all, scientist or not, ponder the nature of our existence. It seems so incredibly improbable that life exists, given that there are so many improbable conditions required to support it. Has the universe been created for the express purpose of supporting life and the development of intelligent beings—the design hypothesis? Or is it just a lucky break that we exist?

Some scientists have responded to the questions about the conditions of our existence by citing the **weak anthropic principle**. Philosopher Brandon Carter presented this idea in 1973 at a symposium in Poland honoring Copernicus's five hundredth birthday, and the late cosmologist and physicist Stephen Hawking has promoted the concept more recently. Briefly paraphrased, the weak anthropic principle states that the observed values of all physical and cosmological constants in the universe *must* allow for the origin and evolution of carbon-based intelligent life—because if they were different, we would not be here to observe anything in the first place! Carter believes that our existence and our observations of the constants that determine the nature of our universe represent a type of selection bias. This line of reasoning imagines an almost infinite statistical population of universes, all with different properties and physical constants. An independent transcendent observer would recognize that almost all of these universes would not have properties that support carbon-based life, and intelligent life would not be there to observe those universes and their properties. We on Earth have a statistical bias of observing only one of these universes. If our universe did not have properties to support carbon-based life, we, of course, would not be here and would not observe this universe.

We can use the same reasoning to explain the extraordinary improbability of how we, as individuals, are here today. The number of events leading to your or my existence today, starting with that one-in-a-million sperm that fertilized our mother's egg, is nearly infinite, and if any one of these events were different, we would not be here. But because we are here, we have a selection bias of recognizing each of these unique events responsible for our being. If any of these events were different, we would not be here to recognize the events. We are indeed lucky to be here. This is a theme that will resonate throughout this story.

CHAPTER 4

CHEMISTRY COMES TO LIFE

Life, a chemical process, is to be understood in terms of chemistry. It started through the spontaneous formation and interaction of small organic molecules widely distributed in the universe. Given the physical-chemical conditions that prevailed on prebiotic Earth, these molecules were caught in a reaction spiral of growing intricacy, eventually giving rise to the nucleic acids (RNA and DNA), proteins, and other complex molecules that dominate life today.

—Christian de Duve, in *Vital Dust*

It was a cool Sunday morning in a small rural town in southern Australia. The year was 1969. Some folks were still sleeping, while others sat and enjoyed the cool air with cups of hot coffee on their front porches. Children giggled at silly cartoons, and churches prepared for their morning services. Then a strange object suddenly emerged from the heavens and lit up the morning sky. All eyes were aimed skyward, focused on a bright orange fireball with a silvery rim and a tail of blue smoke that was zipping above the clouds. As the citizens of Murchison gathered to wonder what this new celestial presence might be, the thing exploded into thousands of pieces with an enormous sonic boom. People took cover, terrified of something so obviously out of their control. Fragments hit their homes and crashed into fields—one as heavy as forty-five kilograms punctured a shed roof—but there were no serious damages or casualties.

Fear quickly turned to fascination as observers explored the scattered shards left by the ball of fire. The source of these fragments would come to be known as the Murchison meteorite. No one watching could have known it on that day, but these rocks held clues to the origin of life.

Geologists later told us, through radiometric dating, that this meteorite was estimated to be 4.5 billion years old and that it had formed at the beginning of our solar system, when Earth congealed through the accretion of space rocks and solar dust. Unlike terrestrial Earth, the contents of the Murchison meteorite were not altered by geological and biological influences, and scientists believe the contents represent what Earth and our solar system were like before life began. Later examination of the meteorite, with new technologies, identified a plethora of organic molecules, including simple sugars, nitrogenous bases, and more than ninety amino acids—some of the most important building blocks of life (Schmitt-Kopplin et al. 2010). Meteorites similar to the Murchison meteorite bombarded our planet for the first five hundred million years of its existence, and the Murchison meteorite has provided evidence that these meteorites brought with them a host of organic molecules. It appears clear that organic molecules were plentiful on our planet long before life emerged.

THE VITALISM DOCTRINE

One of the greatest scientific minds of the nineteenth century, French microbiologist Louis Pasteur—who is best known for his works with vaccinations and pasteurization—was a strong supporter of the **vitalism doctrine**—the idea that living organisms are fundamentally different from nonliving matter and are governed by different principles. Fermentation is a biological reaction that converts sugar into carbon dioxide and alcohol. Pasteur was unsuccessful in producing fermentation inside a test tube in the absence of living yeast cells and concluded that a "vital force" was required for this and other biological reactions. This view of life prevailed until the turn of the century, when German chemist Eduard Buchner provided new evidence that transformed these doctrines (Kohler 1971). He demonstrated that the biological reaction of fermentation could occur in the absence of cells—in the absence of life—using a protein extract from yeast. His demonstration of enzyme-facilitated fermentation gained him the Nobel Prize in 1907 and shattered the myth that living organisms did not obey the physical laws of nature.

In the 1950s, a young chemist named Stanley Miller and his mentor, Harold Urey, set out to explore the question of how life began (Miller 1959). They simulated conditions on early lifeless Earth by recreating its primitive atmosphere and sending sparks of electricity through this primordial soup to mimic lightning storms. They sat back and let the cauldron bubble. After several weeks of incubation, they found they had produced five different amino acids—the building blocks of proteins. Their discovery was featured on the cover of *Time* magazine in 1953 and ignited great excitement and public awareness about the possible origins of life. They had demonstrated that organic molecules could be produced from inorganic molecules in the absence of life.

Life is composed primarily of eight elements—carbon, nitrogen, oxygen, hydrogen, calcium, phosphorus, potassium, and sulfur, which make up 98.3 percent of human body weight

(Voet, Voet, and Pratt 2013, 2). These are the elements (or atoms) that form the most stable relationships with each other. Carbon, the friendliest atom of them all, is at the center of these relationships. Carbon is unique in its gregariousness. It is able to share electrons and form bonds with many other elements, and it is better at this than any other atom. Because of these properties, carbon provides the backbone for all organic molecules. This is what we mean when we describe ourselves and all living organisms on Earth as "carbon-based life."

The lesson we can take away from the experiments of Miller and Urey and others is that organic molecules can be created from inorganic molecules simply through the natural laws of physics and chemistry. Organic chemistry is just carbon chemistry; it no more mysterious than that. Prebiotic Earth was a violent place with volcanic eruptions, lightning, and bombardment by meteorites, and this environment was conducive to chemical reactions. Complex organic molecules were created from simpler inorganic molecules in prebiotic oceans. This process of prebiotic **chemical evolution** provided the ingredients needed for the origin of life.

HOW OLD IS LIFE?

In the shallow waters of sparkling Hamelin Pool, at the base of Shark Bay off the southwest coast of Australia, rests wave after wave of mushroom-shaped "rocks" that are teeming with bacteria. Primitive colonies of cyanobacteria trap calcium carbonate and sediment from the ocean water, slowly building laminated cauliflower-like limestone structures called **stromatolites**

(image 4.1). Modern living stromatolites, such as the ones in Hamelin Pool, are very rare, but in the early days of Earth, they flourished, and they have left a number of fossil remains.

In 1993, paleobiologist J. W. Schopf discovered fossils of stromatolites formed by ancient ancestors of cyanobacteria near Shark Bay (Schopf 1993). These perfectly preserved relics were eerily similar to the stromatolites of Hamelin Pool. Dated 3.47 billion years ago, they were the earliest evidence of life on Earth that had been found at that time.

Recently, scientists at University College London unearthed fossils of ancient filamentous microorganisms on the coast of Hudson Bay in Quebec, Canada, dated to 3.8 billion years ago

Image 4.1: Stromatolites, Hamelin Pool, Shark Bay, Australia. Stromatolites are mushroom-shaped rocks that are teeming with bacteria. Today they are very rare, but in the early days of Earth, they flourished, and they have left a number of fossil remains documenting some of the earliest bacterial life on Earth.

(Dodd et al. 2017). The site of these fossils houses some of the oldest sedimentary rock in the world, which accumulated about four billion years ago in hydrothermal vents on the ocean floor and subsequently rose to the surface through geological processes. The fossils contain iron oxide, suggesting the microorganisms were likely bacteria that used iron as an energy source, similar to modern day iron-oxidizing bacteria living in deep-sea hydrothermal vents. These fossils represent the earliest evidence of life on our planet and provide strong evidence that life on Earth first arose in hydrothermal vents on the ocean floor.

The observation that life arose 3.8 billion years ago, very soon after Earth had cooled, raises some interesting questions. If life with complex biochemistry can evolve so easily and quickly, why didn't it evolve multiple times? And if it did, why did only one lineage survive to the present day? And if life evolved so easily here, what about elsewhere in our solar system and the rest of the universe? At the time these microorganisms lived, Mars and Earth both had liquid water on their surfaces. Unless we are a very special exception, we would also expect to find evidence for life on Mars four billion years ago. We'll revisit this later.

THE RNA WORLD

Of all the mysteries in the universe, after the big bang, the origin of life is perhaps the most intriguing and the most elusive. How did randomly generated lifeless molecules suddenly come alive? Darwin gave us a powerful explanation of how life on Earth evolved over billions of years, but even he would not touch the question of how it all got started. "One might as well speculate about the origin of matter," he quipped (Darwin 1911).

In the 1960s, Carl Woese, Leslie Orgel, and Francis Crick—all distinguished scientists working independently of one another—proposed what has since been called the **RNA world hypothesis**. These investigators proposed that primitive self-replicating molecules similar to present-day RNA spontaneously formed from existing organic molecules and became the fundamental ingredient in the origin of life.

How did these complex replicating polymers come to exist? We know that the building blocks of RNA (nucleosides) have been found in meteorites and were present in prebiotic Earth. But one of the unanswered questions about the origin of life has been how prebiotic organic molecules could come together to form the macromolecules so essential to life. Connecting nucleosides to phosphate groups to form nucleotides and joining nucleotides into long chains to form RNA is very problematic and today requires the help of protein enzymes. Enzymes did not exist in prebiotic Earth so how could this happen? We now may have some answers. Chemists at Scripps Research Institute in Los Angeles think they may have found the "missing link" that facilitated this process (Gibard, Bhowmik, Karki, Kim, and Krishnamurthy 2017). They identified a nonprotein compound (diamidophosphate or DAP) that can phosphorylate (add a phosphate group to) nucleosides, forming nucleotides, which are then capable of joining together in long chains to form RNA macromolecules. Scientists believe DAP or a similar molecule may have been present on prebiotic Earth and facilitated the synthesis of early RNA, as well as other macromolecules that helped to jump-start life.

RIBOZYMES

Current-day RNA is composed of long chains of nucleotide building blocks. Only four nucleotides are used in RNA (abbreviated A, U, G, C), and these have special properties allowing them to stick to complementary nucleotides—A to U and G to C. This has given RNA the ability to form new complementary strands of RNA. If RNA is placed in a solution with free-floating nucleotides (A, U, G, and C), the free-floating nucleotides will stick to complementary nucleotides on the RNA strand, connect together, and then split off to form a complementary strand of RNA. If the process is repeated, a new complementary strand will form that is identical to the original strand—replication completed. The chemical properties of nucleotides—their complementary nature and their steadfast stickiness—are the defining characteristics that promote replication of RNA.

Current-day RNA replication requires protein enzymes to facilitate the reaction, but RNA is needed to produce protein enzymes, and protein enzymes were not available. It's a chicken-and-egg problem. In 1982, Thomas Cech, at the University of Colorado, and Sidney Altman, at Yale University, working independently, discovered properties of particular RNA that provided a potential resolution to this conundrum (Cech 2018). They found that special RNA could fold into complex shapes and was capable of catalyzing biochemical reactions—a function previously thought to be restricted to protein enzymes. These catalytic RNAs—known as **ribozymes**—are capable of catalyzing their own replication and provide an explanation how the earliest replicators could replicate without the help of protein enzymes. This discovery boosted credibility for the RNA world hypothesis, and Cech and Altman were awarded the Nobel Prize in Chemistry in 1989 for their work.

But there was still another problem. Replication by ribozymes appeared to be self-limiting. If ribozyme-like replicators were to be able to jump-start life, they had to be capable of sustained replication. In 2009, graduate student Tracey Lincoln and her mentor, Professor Gerald Joyce, out of Scripps Research Institute in La Jolla, California, set out to investigate the problem (Lincoln and Joyce 2009). They placed naturally occurring ribozymes in test tubes with free-floating nucleotides. As in previous experiments, ribozymes reproduced sporadically, attracting complementary nucleotides and forming new chains; but after a cycle or two, the replications would cease. Most of the time, the ribozymes would copy accurately, but very occasionally mistakes were made—mutations in the code. Some of the copying errors made the ribozymes capable of more sustained replication. Initially the molecules would copy themselves only once and then stop, but now the mutants were able to copy again and again, multiplying wildly. The most "fit" replicators, of course, quickly dominated the mixture. Eventually some of the ribozymes evolved to the point that their replication simply never came to a stop; they had become "immortal" molecules, copying ceaselessly, with no end in sight. These molecules displayed a rudimentary type of Darwinian molecular evolution. Commenting on the quest to synthesize living organisms, Joyce mused in the Scripps Research Institute's *News and Views*, "We're knocking on that door, but of course we haven't achieved that" (Schrope 2009).

The evolution of these early replicators provides an instructive example as to how evolution is constrained by physical and chemical properties of atoms and molecules. The four nucleotides used in RNA were selected from hundreds of nucleotides present in nature because of their special properties that allow them to stick to complementary nucleotides. The late Belgian evolutionary biologist Christian de Duve believed that the sticky, complementary nature of these four nucleotides explains why they, and they alone, are used as building blocks in the synthesis of RNA in all cells in all forms of life on Earth (de Duve 1995, 85). De Duve believed that it is very likely that life evolving in other solar systems and in other galaxies may have replicators similar to RNA, with the same four nucleotides used as building blocks.

FIRST PROTEINS

The Lincoln–Joyce experiments provided evidence that was needed to solidify the plausibility of the RNA world hypothesis. But the evolution of replicators was destined for a dead end without something more. Replicators are great at replicating, but to what purpose? A mutating replicator can only get better or worse at copying itself. The nucleotide code, lacking an interpreter, goes unread. There is no communication with the broader world. For these replicators to continue their journey toward life requires that the replicators begin to direct the construction and operation of a cell. This begins with the manufacture of proteins.

Proteins control nearly every biological process. Protein enzymes facilitate the majority of metabolic reactions, and proteins provide essential structural components for all cells, such as the stretchy myofibrils of muscle cells and the collagen in connective tissue. Protein molecules are long chains of amino acids that, given their attractive and repulsive properties, naturally fold into numerous shapes and sizes.

Francis Crick of DNA fame first described how cells make proteins (figure 4.1). As we discussed in chapter 1, DNA codes are copied (transcribed) onto another large molecule, messenger RNA, in a process known as **transcription**. Messenger RNA then travels from the nucleus to ribosome factories in the cytoplasm of the cell, where proteins are made. The code on mRNA is used to line up specific amino acids to form peptide chains, the basic units of proteins, in a process known as **translation**. It's a stunningly neat cellular assembly line, and the process is happening right now in every one of your cells and the cells of all living organisms.

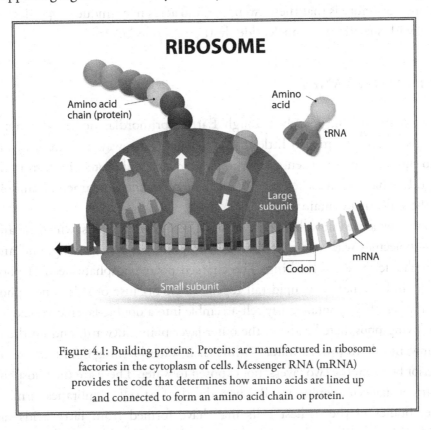

Figure 4.1: Building proteins. Proteins are manufactured in ribosome factories in the cytoplasm of cells. Messenger RNA (mRNA) provides the code that determines how amino acids are lined up and connected to form an amino acid chain or protein.

How could this have started in the earliest cells? Christian de Duve provided a potential scenario in *Vital Dust* (de Duve 1995, 69–71). Originally there was no DNA, only primitive RNA, so we can skip that first step of transcription. Small pieces of RNA must have bound to specific amino acids based on their chemical properties. These short RNA–amino acid complexes would, by their nature, tend to stick to long strands of RNA at complementary nucleotides, and the attached amino acids would line up and join to form strands of amino acids—just as they do in modern-day translation. The code on long strands of RNA would determine the sequence of amino acids and the properties of the protein. Mutations in the long strands of RNA would result in new proteins, and if these new proteins provided adaptive advantages that improved survival of the organism, such as functioning as catalysts for important biochemical reactions, the code would be passed to the next generation and be preserved. There are, of course, many unknowns in this scenario, but biochemists are putting the pieces of the puzzle together and gradually gaining a better understanding of how protein synthesis may have started.

One question that has puzzled biochemists for decades is why, with over 150 amino acids available, only twenty were chosen to be used in protein synthesis. We don't know the answer to this, but one possibility is that these twenty amino acids had unique properties that allowed them to bind with specific three-nucleotide short pieces of RNA (tRNA).

FIRST CELL MEMBRANES

RNA and proteins drifting loosely through Earth's primordial oceans did not constitute life. Biochemistry can accomplish little if its products float away and are lost. Life needs a boundary to concentrate its nutrients, contain its products, and protect itself from the external environment. Life had to have a cell membrane. The cell is the fundamental unit of life and is defined by the walls that contain it.

Cell membranes as we know them now are built from phospholipids, which are **amphiphilic molecules**—molecules with one part that is hydrophilic, or water-loving, and another part that is hydrophobic, or water-loathing. The charged polar phosphate head of phospholipids attracts water, and the nonpolar lipid tail repels water. Because of this, when phospholipids are placed in water, they spontaneously self-assemble into a double-layered vesicle (figure 4.2). Their water-loving phosphate heads on the outer layer point outward, and on the inner layer they point inward—both toward water—while their water-loathing lipid tails are tucked into the oily interior between the two layers, not exposed to water. These are the phospholipids that comprise current-day cell membranes. But how did the first cell membranes form?

The membranes of the earliest cells may have formed from fatty acids rather than phospholipids. Chemists have shown that fatty acids can be produced from hydrogen and carbon monoxide gases emanating from hydrothermal vents, suggesting they may have been abundant in prebiotic Earth (Exploring Life's Origins 2018). Fatty acids are amphiphilic molecules, just like phospholipids, and form bilayers that spontaneously assemble into spherical vesicles when placed in water. The amphiphilic properties and likely ready availability of fatty acids make them prime candidates to form membranes for the earliest cells. It is likely that cell membranes composed of phospholipids evolved later.

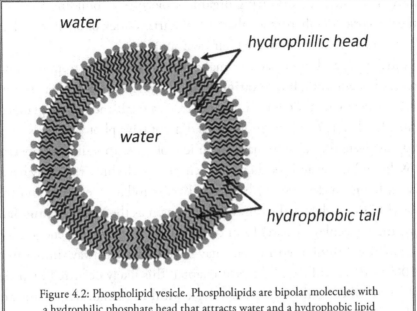

Figure 4.2: Phospholipid vesicle. Phospholipids are bipolar molecules with a hydrophilic phosphate head that attracts water and a hydrophobic lipid tail that repels water. Because of these properties, they spontaneously form two-layered vesicles when placed in water. Molecules similar to phospholipids are thought to have formed the first cell membranes.

Proteins embedded in current-day lipid cell membranes provide vital functions as channels to selectively permit ions and larger molecules to cross the membrane. In addition, protein enzymes catalyze pumps that actively transport ions across the cell membrane, forming a voltage gradient, much like a battery, which enables a number of critical processes. There is a lot we don't know about the evolution of these processes, but it seems likely, based on the universality of the proteins and pumps in current-day organisms, that they evolved early and concomitantly with the phospholipid membranes.

MANUFACTURING THE BUILDING BLOCKS OF LIFE

Macromolecules—carbohydrates, lipids, proteins, and nucleic acids—are the stuff cells are made of. These giant molecules are formed from the building blocks of life—smaller organic molecules that include simple sugars, fatty acids, amino acids, and nucleotides. These smaller organic molecules were present on primordial Earth before life emerged. They had been brought to Earth by meteorites and were also produced in prebiotic Earth from inorganic molecules through chemical reactions, as shown by Miller and Urey and others. The earliest forms of life almost certainly used naturally occurring organic molecules as building blocks to construct cells, but as the meteorite bombardment abated and Earth's stock of organic molecules became depleted, cells soon had to learn to make their own supplies.

Today, plants, algae, and some bacteria make organic molecules from carbon dioxide and water in a process known as **carbon fixation**, using photosynthesis and energy from the sun. We animals are unable to make organic molecules from scratch and rely on plants to provide us with our organic food. The first microorganisms did not have photosynthesis, so investigators have pondered how were they able to make organic molecules to build and operate their cells.

Botanist William Martin and geochemist Michael Russell think they may have the answer. Microorganisms, living in deep-sea hydrothermal vents today (acetogens and methanogens) utilize the Acetyl-CoA biochemical pathway (also known as the **Wood–Ljungdahl pathway**) to produce organic molecules (acetate) from carbon dioxide, capturing energy from hydrogen gas. Martin and Russell think the first cells may have used a pathway similar to this (Martin and Russell 2007). As we will see in the next chapter, this likely occurred in ancient deep-sea hydrothermal vents, where carbon dioxide was plentiful in primordial oceans and hydrogen gas spewed abundantly from the vents. Current Acetyl-CoA pathways require iron–sulfur protein enzymes, which were not available to the earliest cells, but the iron–sulfur linings of microcavitations within limestone structures in the vents may have served as primitive catalysts. This pathway is exemplary of the transition from geochemistry to biochemistry.

While the Wood–Ljungdahl pathway is used by some microorganisms today, nature has found much better ways to produce organic material. Today plants and algae produce organic products through photosynthesis and supply the entire animal kingdom and most of life with essential supplies of organic material.

CAN SCIENCE EXPLAIN THE ORIGIN OF LIFE?

Science cannot yet explain how life originated, but scientists are systematically putting the pieces together. The process of chemical evolution has shown that organic molecules can be created from inorganic molecules and that complex macromolecules—replicators, proteins, and phospholipid membranes—can be created from simple organic molecules. There are many gaps in our knowledge of how these macromolecules came together in an extraordinary event to create first life, and we may never know for sure how life emerged on our planet 3.8 billion years ago. But it is clear that life is a chemical process, and with the evidence so far accumulated, scientists have little doubt that chemistry gave rise to vibrant life.

PANSPERMIA: SEEDING LIFE ON EARTH

The one-time chance occurrence of the formation of the first replicator and the origin of life is hard for many to accept. What may be even more remarkable is that this chance event may have happened over a relatively short time span. Earth formed about 4.5 billion years ago but was subjected to heavy bombardment by asteroids until the end of the Hadean Eon about 3.8 billion years ago. During this time, Earth was molten, and scientists believe it was too hot to support life. Geological evidence suggests that life began about 3.8 billion years ago. If this is true, the highly improbable events leading to the origin of life happened once over a relatively short time span. And if it happened once, why didn't it happen more than once? Why didn't replicators originate on another occasion, creating a second lineage for life on our planet? That leads us to consider another possibility.

Maybe our ancestors were aliens. It sounds like an absurd proposition, but many respected scientists—including astronomer Fred Hoyle, Frances Crick of DNA fame, and even the late brilliant physicist Stephen Hawking—have voiced support for the concept of **panspermia**—the idea that Earth could have been seeded with cells from extraterrestrial sources (Line 2007; Crick and Orgel 1973; Hawking 2008).

Given that a plethora of organic compounds were carried to Earth on the backs of meteorites, like the Murchison meteorite, it is possible that our planet was "infected" with cells in the same way. Many bacteria, when exposed to harsh environmental conditions, begin a process of spore formation, in which they shut down metabolic processes and surround themselves with multiple protective coats, providing protection against radiation, toxins, and intense heat and cold. Bacterial spores can survive under such punitive conditions in a dormant state for prolonged time periods without any nutrients. In one famous example, bacterial spores isolated from the Permian Salado sedimentary rock formation in New Mexico, which were over 250 million years old, were found to be alive and viable (Vreeland, Rosenzweig, and Powers 2000). We can imagine that bacterial spores buried within the crusty layers of meteorites might have survived the long, tumultuous journey from distant solar systems until they finally impacted planet Earth and found a new home.

In 1984, while most Americans were enthralled with the fourth launch of *Challenger*, Ronald Reagan's presidency, or the newest episode of *Night Court*, US government scientists were busy examining newly found remnants of a meteorite that had crashed into Antarctica thirteen thousand years before. The meteorite was called, rather unceremoniously, rock 84001. The remnants of the meteorite were estimated to be four billion years old, and evidence indicated that the meteorite had broken off from Mars and traveled on a sixteen-million-year journey to Earth.

Years after its discovery, and long into a new presidency, further studies suggested that the meteorite might contain fossils of nanobacteria. The findings, while controversial, were published in the prestigious journal *Science* and received great public attention, including a televised press conference in 1996 by President Bill Clinton. At his press conference, Clinton waxed philosophical:

> Today, Rock 84001 speaks to us across all those billions of years and millions of miles. It speaks of the possibility of life. If this discovery is confirmed, it will surely be one of the most stunning insights into our universe that science has ever uncovered. Its implications are as far-reaching and awe-inspiring as can be imagined. Even as it promises answers to some of our oldest questions, it poses still other questions even more fundamental. (Clinton 1996)

While there remains controversy over whether the microscopic etchings found on Rock 84001 are true fossils of bacterial life or mere artifacts, the discovery popularized panspermia as a possible source for the origin of life on Earth. If the findings *are* true, they raise many more questions than answers, as President Clinton suggested. The search for where and how life began then becomes even more complex.

CHAPTER 5

THE UNIVERSAL ENERGY SYSTEM

Anyone who is not thoroughly confused just doesn't understand the problem.

—Efraim Racker, Polish biochemist, speaking about respiration in the
mitochondrial membrane, 1960

Life creates order in our disordered universe. It builds, grows, moves, reproduces, puts molecules together, and breaks them apart again; and life needs energy to do all this work. Cellular energetics is just as important to living organisms as cellular informatics—the use of DNA to direct action. Until the last century, we did not understand how cells converted energy from sunlight or from inorganic and organic molecules into a form they could use to power the cell. However, several decades ago the mechanisms were uncovered, and the findings were totally unexpected. As with our genetic code, they were found to be universal to all life.

RESPIRATION: HOW WE MAKE ATP

For most of us, respiration means breathing. Biochemists, however, speak of **cellular respiration** as the process by which organisms convert energy from the environment into usable ATP (adenosine triphosphate), the high-powered molecule used to run cells. ATP stores energy within the bonds that hold the molecule together. Whenever your cells need a shot of energy, they can get that energy by breaking down those bonds. Plants convert energy from sunlight into ATP, some microorganisms convert energy from inorganic molecules into ATP, and we animals convert energy from organic food into ATP. All organisms convert energy from the environment into ATP. All organisms respire.

Here is how it works in animals: Respiration occurs in the **mitochondria**—organelles within cells that serve as little power factories. The cells ingest and metabolize glucose or other organic molecules, from which they capture energized electrons. The energized electrons are transported like an electric current passing down a wire along an **electron transport chain**—a series of proteins embedded in the mitochondrial membrane. As the electrons are passed between the protein complexes, they incrementally release their energy, and the cell uses this energy to pump protons (hydrogen ions) across the mitochondrial membranes. This building up of protons on one side of the mitochondrial membrane creates a proton concentration gradient—a gradient with energy that can drive chemical reactions.

The final stage of respiration, **chemiosmosis**, is the process of converting the energy of the proton gradient into chemical energy in the form of ATP. To understand this, we need to understand **ATP synthase**, the complex protein enzyme that makes the process work. ATP synthase is a nanoscale molecular machine that is a true wonder of the molecular world. This complex protein, embedded in the mitochondrial membrane, has a rotor-like structure and shaft that rotates as protons pass through the mitochondrial membrane, driven by the electrochemical gradient. The mechanical energy of the rotating shaft is converted to chemical energy by joining ADP and phosphorous to form energy-rich ATP.

This process is analogous to the conversion of the potential energy of water behind a dam into electrical energy. When water behind a hydroelectric dam is released and allowed to flow through a floodgate, it hits the blades of a turbine, causing the turbine to spin. A shaft connects the turbine to a generator, which converts the mechanical energy of the rushing water and the spinning turbine into electrical energy.

The electron transport chain and chemiosmosis sound like science fiction, don't they? The electrons dancing across your mitochondrial membranes, the pumping of protons from one place to another, the reservoir, the dam, and tiny molecular turbines working to give your cells the energy they need—it's a wild tale. The concept of converting the mechanical energy of rotating molecular turbines into chemical energy is totally bizarre.

When British biochemist Peter Mitchell published his peculiar theory of cellular respiration in the prestigious journal *Nature* in 1961, the reaction from the scientific community was skeptical, to say the least (Mitchell 1961). Mitchell's new theory of chemiosmosis introduced probably the most dramatic paradigm shift in the annals of biochemistry. He professed that he had solved the mystery of how cells produce ATP, but his outlandish idea was a totally new model—unlike anything biochemical societies had ever considered before. How could protons flow across a membrane, turn a molecular turbine in a large protein complex, and convert this mechanical energy into chemical energy in the form of high-energy molecular bonds? It was too much!

Mitchell had conducted his research outside of accepted academic circles and had never been in the mainstream of the scientific community. He had inherited a considerable family fortune, drove around in a Rolls Royce, and wore his hair long in emulation of Beethoven (Roskoski 2004). He had an eccentric lifestyle and personality and stood apart from the academic elites.

Despite early reservations, after ten years of acrimonious debate, most biochemists did, in fact, accept Mitchell's hypotheses as being true. He didn't endear himself to his colleagues by maintaining a chart showing the dates when each of his rivals converted to his point of view. But Mitchell had the last word when he received the Nobel Prize for Chemistry in 1978 for his work, and his findings have stood as one of the landmark achievements in the annals of biochemistry.

WHAT WE "EAT" AND WHAT WE "BREATHE"

The human body has about thirty-seven trillion cells. Each cell houses up to a thousand mitochondria, and each mitochondrion has thousands of complexes producing ATP through the electron transport chain and chemiosmosis. Your cells are *very* busy. In fact, the cells of all organisms, from bacteria to plants to animals, are doing this same hard work; we all use this system to create ATP. This bizarre process—the electrons jumping, the damming of protons, the tiny turbines turning—is as much a part of our evolutionary heritage as the universal genetic code. The electron transport chain and chemiosmosis, like DNA, are universal to life.

However, the "food" consumed to run this universal energy system varies across the spectrum of life. Plants, animals and bacteria use different sources of energy—different kinds of "food," to power the electron transport chain. Plants use energy directly from the sun; animals, a few steps up the food chain, and many microorganisms obtain energy from organic materials; and some primitive microorganisms actually obtain energy from inorganic chemicals. In each case, the energy source is used to release activated electrons that can initiate and power the electron transport chain.

The molecules that organisms "breathe" are the molecules they use as the final electron acceptor in the electron transport chain, and these vary from organism to organism. We animals breathe oxygen. Oxygen, because of its high affinity for electrons, can extract the last bit of energy from electrons. This gives animals the most energy-efficient metabolism of all living things—**aerobic metabolism**. We can generate a whopping thirty-eight molecules of ATP from just one molecule of glucose that we get from our food. Compare this with the measly two molecules of ATP produced from that same glucose molecule using **anaerobic metabolism** without oxygen.

Some environments, such as deep vents on the ocean floor and dark cavities in the intestines of animals, simply do not contain oxygen, but life must go on. Microorganisms lived without oxygen for billions of years and know how to "breathe" alternatives, such as nitrates, carbon dioxide, and sulfate, which serve as the final electron acceptors in their electron transport chains.

The membranes in which the electron transport chain operates and across which the process of chemiosmosis occurs differ between animals, plants, and bacteria. As we have seen, in animals these processes occur in the membranes of mitochondria. In plants they occur in membranes of organelles called **chloroplasts**, and in bacteria, they occur across the cell membrane. Remember these details for later, when we discuss where mitochondria and chloroplasts might have come from in the first place. (A hint: Millions of years ago, mitochondria and chloroplasts probably had free-swimming lives of their own.)

HOW DID THE FIRST CELLS RESPIRE?

Until recently, many investigators had thought that **fermentation**, a common metabolic process that converts the energy in organic foods into ATP in the absence of oxygen, was the primary form of energy metabolism in early cells. But now most investigators believe this is highly unlikely (Lane 2015). First of all, organic molecules, such as sugary glucose, were in short supply on primordial Earth. Many organic molecules came to Earth via asteroids and meteorites, and they may have been depleted within a short period of time after the extraterrestrial bombardment ceased. Fermentation also requires multiple complex protein enzymes, and it's hard to imagine how the most primitive life could have produced those enzymes without a preexisting energy supply needed to produce RNA and proteins. It is a chicken-or-egg problem. In addition, genetic studies suggest that different types of fermentation evolved multiple times independently across the tree of life. Archaea, bacteria, and eukaryotes all use fermentation, but the reactions and enzymes are different and do not appear to be related. Since our fermentation processes are so different, it's very unlikely that all current life inherited fermentation from an early common ancestor.

Given all of these issues, we can conclude that our most ancient ancestors hadn't learned to ferment anything yet. So how did those early cells produce the energy they needed to multiply and inherit the Earth? The answer seems inescapable. The electron transport chain and chemiosmosis are universal to all life and must have been inherited from a common ancestor. They must have been present in the earliest cells. But this begs the larger question: How did such a weird process as the electron transport chain and chemiosmosis ever get started?

ORIGINS IN THE VENTS

The electron transport chain and chemiosmosis operate across cell membranes and rely on membranes to function. If these processes were present in early cells, there must have been cell membranes present to allow the processes to operate. However, as British biochemist Nick Lane explains in *Power, Sex, Suicide*, there is a conundrum (Lane 2015). Eugene Koonin, a Russian-born geneticist working at the National Institute of Health, analyzed the genes of bacteria and archaea with some astonishing results suggesting that the evolution of cell membranes may have followed an unusual pathway.

As you may remember, archaea are microorganisms similar to bacteria but different enough to be classified as one of three separate domains (bacteria and eukaryotes are the other two domains). Koonin found that bacteria and archaea share many traits and much biochemistry, and there is no doubt that they descended from a common ancestor. But the cell membranes of bacteria and archaea are wholly different. Cell membranes of all organisms are made of two layers of phospholipids, but the structure of the phospholipids, the enzymes that facilitate the synthesis of these phospholipids, and the bonds that join the phospholipid layers together are entirely different in archaea and bacteria (Sojo, Pomiankowski, and Lane 2014). So here is the conundrum: the electron transport chain and chemiosmosis, which operate in the cell membrane, were present in the common ancestor of archaea and bacteria, but a common cell membrane was not. This presents the astonishing possibility that the electron transport chain and chemiosmosis evolved before there was a cell membrane, and that our last universal common ancestor was housed in something other than a cell membrane (Lane 2015, 83–85).

So what in the world did this nonmembraned protocell look like? How was it contained? How did it survive? No one knew the right answers to these questions until one day geologist Michael Russell at the University of Glasgow found a model for the protocells in his laboratory lying quietly and unobtrusively in the viewfinder of his microscope.

The specimens he was examining are the now famous 350-million-year-old fossils from the hydrothermal vents near Tynagh, Ireland. Looking at slices of the bubbly rocks from these vents with an electron microscope, he discovered tiny compartments the size of organic cells, each lined with iron–sulfur minerals (Lane 2009, 20–21). He was able to reproduce these structures in the laboratory by bubbling an alkaline solution similar to the alkaline solution bubbling from ocean floor vents into an acidic solution similar in composition to primordial oceans. The rocky structures he produced contained microscopic compartments lined with iron–sulfur minerals similar to the tiny compartments in the vents at Tynagh.

Could these tiny pockets in the rocky structures of the Tynagh fossils be a model for the incubators of the first life on Earth? Russell thought so (Russell et al. 1994). And Nick Lane thinks so too, as he states in his recent book *The Vital Question* (Lane 2015, 102–9). The ingredients for life were right there at the hydrothermal vents, deep on the ocean floor. The hydrogen gas that bubbled from the vents in abundance could provide energy, and the plentiful carbon dioxide that was dissolved in the ocean waters could provide a source of carbon from which to build organic molecules. Those microscopic pockets embedded in bubbly rocks could provide a shelter or compartment in which organic molecules could be concentrated and used in primitive biochemical reactions. The warm temperatures of 60 to 90°C in the vents were ideally suited for biochemical reactions, and the iron–sulfur minerals lining the pockets could serve as primitive catalysts for those reactions. These metal sulfides have been preserved over eons of evolution and are embedded deep within many protein enzymes today.

In addition to all these resources, the microcompartments had a natural hydrogen ion gradient (or proton gradient)—acidic on the outside from the oceanic carbon dioxide, and alkaline on the inside from the bubbly issuance of the vents. We already know, of course, that such proton gradients, as would have existed across those tiny iron–sulfur membranes, can be used as a source of energy, given the presence of the right turbine, such as ATP synthase, to capture energy from the proton flow.

So there we have it. In this scenario, all the protocells had to do was produce a primitive protein complex—something like ATP synthase—to convert the energy from the natural proton gradient into ATP or some other high-energy compound. This would be the primordial beginning of modern chemiosmosis. The natural electrochemical gradient would have provided the protocells a limitless supply of energy—but it also would have made them dependent on the vents.

When Russell published his findings in 1994, his ideas were met with hearty skepticism from his colleagues—partly because at that time, alkaline hydrothermal vents had not yet been discovered (Russell et al. 1994). (Alkaline hydrothermal vents are to be distinguished from the previously discovered and more common "black smokers," which are much hotter and spew forth black-colored water high in sulfur content.)

Image 5.1: Lost City. This image of a calcium carbonate chimney in a deep-sea hydrothermal vent in the Atlantic Ocean Ridge was taken during the 2005 Lost City Expedition. Deep-sea alkaline hydrothermal vents such as this may have been the ideal hatcheries for the origin of life. Courtesy National Oceanic and Atmospheric Administration (NOAA).

But a few years later, in 2000, a National Science Foundation expedition exploring the depths in the middle of the Atlantic Ocean discovered a spectacular previously unknown field of alkaline hydrothermal vents on the ocean floor (image 5.1) (Boetius 2005). The vents were surrounded by a host of multipinnacled white towers, accreted from mineral deposits and reaching heights of up to two hundred feet. The magnificent alkaline vents were given the name "the Lost City" in poetic reference to the myth of Atlantis.

A new expedition using the three-person submarine Alvin was launched to explore the vents. Alvin dived to crushing depths of three miles to document the chemistry and life in the vents. Scientists found that the upwelling mineral-rich hydrothermal fluids from the vents reacted with cooler seawater above to produce those carbonate towers—chimneylike structures riddled with tiny microcavitations. The pockets were lined with iron–sulfur membranes, and the microcells were able to maintain an alkaline internal environment separate from the acidic external ocean. Sound familiar? The Lost City microcavitations bear an uncanny similarity to the fossils at Tynagh and the structures created in the laboratory by Mike Russell.

The hydrothermal vents—without light, without oxygen, and crushed by three miles' worth of hydrostatic pressure from the ocean waters above—seem immensely inhospitable. It is hard for us to imagine that life could exist there. But Alvin's expedition quickly changed our perceptions. Life does not merely cling to the vents; it flourishes there. The vents teem with microorganisms in profuse quantity, and they grow and multiply there quickly.

The vents release chemically energetic hydrogen gas. Primitive microorganisms like methanogens and sulfur bacteria thrive in the vents today. They feed off hydrogen gas to generate their energy, and they capture carbon dioxide from the cold ocean to produce organic compounds. In Earth's cold, deep, primordial sea, these magma-hot fissures may have been the only nurseries available to the earliest cells. It may seem strange, but perhaps this was the ideal hatchery for the origin of life.

ESCAPE FROM THE VENTS

Let's assume that Mike Russell is right and that life began in ancient hydrothermal vents, with primordial biochemistry and replicator molecules sheltered in microcavities of rock lined with iron and sulfur. We still have one problem with our story. The protocells would have been dependent on the vents. They would have required energy from the natural proton gradient, created by the alkaline vents and acidic ocean, to help power their chemical reactions. How did they become independent and venture out into the world?

Evolutionary biochemists Nick Lane and Bill Martin believe they have an answer (Lane and

Martin 2012). Under their scenario, the protocells first generated an enzyme system to convert the energy from the natural proton gradient into ATP. This enzyme system is the molecular turbine, ATP synthase. Amazingly, ATP synthase has persisted and is present in membranes of all cells today, with remarkably little variation between species. This all serves as evidence that those protocells did in fact evolve an ATP synthase molecule very early on.

Next the protocells developed their own cell membranes—fatty acid bilayers—*within* the sanctuary of the iron–sulfur microcavitations. As you may remember, archaea and bacteria evolved cell membranes separately after they had already evolved ATP synthase and chemiosmosis. For the cells to break free from their energy dependence on the hydrothermal vents, they had to develop their own electron transport chain embedded in their new membranes so they could create their own proton gradient. Using hydrogen gas as their electron donor and source of energy, and carbon dioxide as their final electron acceptor, and evolving enzymes to facilitate electron transport, they were able to generate energy for transporting protons across the cell membrane, much in the same way methanogens living in hydrothermal vents do today.

The protocells, having evolved their own cell membranes and having achieved energy independence, were now ready to escape from the vents and venture out into the world. As they left the vents, they carried with them the remarkable processes of cellular respiration—electron transport and chemiosmosis. These are the footprints of our origins. The traces of our humble beginnings at the deep-ocean vents are apparent in all life as we know it today.

PERSPECTIVES

In 1871, Darwin speculated in a letter to his friend Joseph Hooker that life on Earth may have started in a "warm little pond." This now appears very unlikely. A "warm little pond" would not likely have had the prerequisites for the birthplace of life, beginning with a source of energy and compartments for concentrating organic molecules. Early cells on the surface of Earth would have been vulnerable to damage to their RNA and other cellular contents from ultraviolet radiation emitted by the sun before our atmosphere had an ozone layer to protect them. Most investigators agree that ancient hydrothermal vents had all the ingredients necessary for life and stand as the most likely candidate for the place where life began.

Chemiosmosis was born from the natural proton gradient that existed across microcavitations in carbonate spires of ancient vents, and this strange process has persisted for over three billion years. The electron transport chain and chemiosmosis have provided an efficient system for capturing energy from the environment, especially when oxygen is used to soak up the last bit of vigor from activated electrons. The electron transport chain can accommodate many types of electron donors, including inorganic molecules, such as hydrogen gas; organic molecules; and solar-activated electrons embedded in the chlorophyll pigments of plants. The electron transport chain can utilize many types of final electron acceptors, including the strongest electron acceptor of all, oxygen. And perhaps most importantly, the electron transport chain can capture energy in increments with little waste. Energy from activated electrons is used to pump protons one or two at a time across a membrane until enough energy is produced in the form of an electrochemical gradient to produce a molecule of ATP. No longer does energy have to be produced by reacting whole molecules with whole molecules, resulting in inherent waste. The electron transport chain and chemiosmosis have transcended chemistry. Nick Lane has speculated in *The Vital Question* that just as electron transport chains and chemiosmosis have evolved out of necessity here on Earth, they may have evolved out of necessity elsewhere in the cosmos and may be universal to life wherever it is found (Lane 2015, 121).

CHAPTER 6

SOLAR-POWERED BACTERIA

Oxygen is without a doubt the most precious waste imaginable. Yet not only is it a waste product, it is also an unlikely one. It is quite feasible that photosynthesis could have evolved here on Earth, or Mars, or anywhere else in the universe, without ever producing any oxygen at all. That would almost certainly consign any life to a bacterial level of complexity …

—Nick Lane, in *Life Ascending*

Bacteria ruled the world for two billion years. They might still rule the world today in unicellular hegemony were it not for the genesis of **photosynthesis**—arguably the single most important invention of biological evolution.

Imagine what a dull place Earth would be without photosynthesis. Algae and plants use photosynthesis to make the food we eat and provide the oxygen we breathe. Without reactive oxygen to cleanse the dust and haze from our atmosphere, the sky would be polluted and gray. There would be no oxygen-based ozone layer to protect life from damaging ultraviolet radiation, and microorganisms would live in hiding, away from the light of the angry sun. Earth's landscape would be colorless, rocky, and sparse, void of the lush green plants that blanket our planet today. Those bacteria that are able to live without oxygen would have drifted through a limited existence on the ocean floor. Put simply, without oxygen and organic food supplied by photosynthesis, there would have been no evolution of the complex life we cherish today. Needless to say, in a world without photosynthesis, we would not be here.

A NEW KIND OF AIR

Englishman Joseph Priestley was a man of many talents. He was a renowned chemist, a Unitarian minister, a prolific theologian, a political philosopher, and the inventor of soda water, but he is best remembered for his discovery of oxygen. Like many scientists throughout history, Priestley was something of a dark horse in his day. He alienated many of his British compatriots with his outspoken support of the American and French revolutions and his criticism of the Anglican Church. He was held in contempt by many and was forced to flee his home in 1791 after an angry mob torched his house and church. The homeless Priestley fled, along with his wife, to London and then to America, where he spent the rest of his life.

Priestley's prolific work as a scientist, however, was never interrupted by the turmoil of his philosophical activism. In 1774, while still in Britain, he isolated an "air" that appeared to be completely new (American Chemical Society International Historic Chemical Landmarks 2017). He conducted a series of spectacular experiments in discovering this new "air." In his initial experiments, he observed that the flame of a burning candle, enclosed in a sealed glass container, went out long before its wax was gone. He later showed that a mouse died after a short time within the sealed container. There was evidently something in the container that both the burning candle and the mouse were consuming—something both needed to survive.

In his most famous experiment, Priestley used a lens to focus sunlight onto a pile of red-orange powder (mercuric oxide) within an airtight container. The light heated the powder and caused it to release some kind of gas. This gas was the mysterious new "air" that we now know as oxygen. It caused the candle flame to burn more intensely and allowed the poor mouse to survive much longer in the jar. Putting a green plant in the container had much the same effect. When the plant was exposed to sunlight, it would refresh the trapped air, permitting the flame to burn continuously and the mouse to breathe and survive. Whatever the candle and the mouse were consuming, the plant was replenishing. These surprisingly simple experiments were groundbreaking. It was only much later that scientists began to tease apart the process by which plants release life-giving oxygen into the atmosphere. The process, of course, is photosynthesis.

PHOTOSYNTHESIS

When we speak of photosynthesis today, we generally speak of **oxygenic photosynthesis**—the process by which plants, algae, and cyanobacteria use energy from the sun to produce organic molecules from carbon dioxide and water, releasing oxygen as a by-product.

Photosynthesis, like cellular respiration in animals and many microorganisms, uses an electron transport chain and chemiosmosis to produce ATP, but unlike cellular respiration, photosynthesis uses energy from sunlight rather than from organic or inorganic molecules. Chlorophyll pigments, embedded within membranes in **chloroplast organelles** of plants and algae, absorb sunlight and release activated high-energy electrons that enter the electron transport chain.

In a separate process known as the **Calvin cycle**, plants, algae, and cyanobacteria produce organic molecules from carbon dioxide and water, using energy from ATP produced through photosynthesis (Voet, Voet, and Pratt 2013, 644–48) (figure 6.1). The enzyme that facilitates this reaction, **RuBisCO**, is perhaps the most abundant and possibly the most important protein on Earth. RuBisCO is responsible for facilitating the manufacturing process that supplies food for our entire planet. Look around; wherever you see green plants, they are at work producing our food supplies.

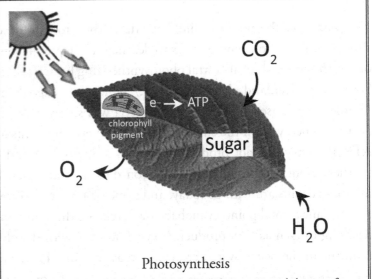

Photosynthesis

Figure 6.1: Photosynthesis. Solar radiation releases activated electrons from chlorophyll pigments within cells in the leaves of plants, and the energy is used to produce the high-energy compound ATP (adenosine triphosphate). Oxygen is released as a by-product. Plants absorb CO_2 from the atmosphere and H_2O from the soil and use energy from ATP to produce sugar and other organic products in what is known as the Calvin cycle. All this takes place in chloroplast organelles within cells in plant leaves.

The process of photosynthesis in plants, algae, and cyanobacteria is nearly identical except for where it occurs within the cells. In plants and algae, photosynthesis takes place within membranes of chloroplast organelles located in the watery cytoplasm of the cells, while in cyanobacteria, it occurs in the cell membranes.

Photosynthesis in plants and cellular respiration in animals are complementary processes. They achieve the same goals but go about it in different ways—and, curiously, they support each other. Plants convert carbon dioxide and water into organic molecules using energy from the sun and releasing oxygen as a waste product. Animals use organic molecules produced by plants as a source of energy and, with the help of oxygen, create ATP, releasing carbon dioxide and water as waste products. It is this balance that forms the basis of a major part of Earth's ecology. Photosynthesis provides us with the organic food we eat and the oxygen we breathe. We animals provide plants with the carbon dioxide needed to make organic food. We need each other.

ORIGINS OF PHOTOSYNTHESIS

Not long after the origin of life, bacteria learned to capture energy from the sun through a simple form of photosynthesis. As we learned in chapter 4, some of the earliest evidence of life on Earth was fossils of ancestral photosynthesizing bacteria discovered in Western Australia and Southwest Greenland, dated 3.5 and 3.7 billion years ago (Schopf 1993, Nutman et al. 2016). Photosynthesis used by these early bacteria was surely very primitive. These microorganisms captured energy from the sun with photosynthetic pigments and used this energy to produce ATP. But they did not produce oxygen as a by-product and did not have a Calvin cycle to synthesize organic products from carbon dioxide and water. Descendants of these primitive microorganisms are with us today, and some still use this simple type of photosynthesis.

At some pivotal point, cyanobacteria developed the ability to split water with photosynthesis, releasing oxygen as a by-product. Oxygen reacted with dissolved molecules of iron and other minerals in the ocean water, creating a mass rusting. Heavy oxides precipitated to the ocean floor, where they were deposited in sedimentary layers, resulting in **banded iron formations** (image 6.1). Some of these layers of iron oxide, which have been found in sedimentary rock beds around the world, were deposited as long as 3.7 billion years ago. Because of this and other evidence, most experts believe that cyanobacteria began producing oxygen not long after the origin of life itself (Buick 2008).

The mass rusting continued for roughly a billion years, until there was no more iron and no other minerals to "soak up" the oxygen released by cyanobacteria, at which point, about 2.3 billion years ago, there was a sharp rise in atmospheric

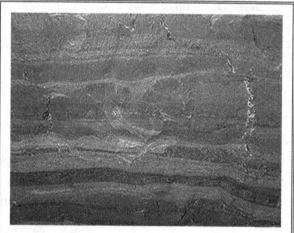

Image 6.1: Banded iron formation. This 2.7-billion-year-old banded iron formation was found near Temagami, Ontario. Banded iron formations are distinctive sedimentary rocks with bands of iron-rich sedimentary layers. The iron-rich dark gray layers formed when oxygen, released by cyanobacteria, reacted with iron in the oceans, forming iron oxides that precipitated as sediment to the ocean floor.

oxygen concentration (Holland 2006). To us this sounds like the beginning of a new, hospitable Earth. Finally, a planet that we might actually inhabit! Finally, air that we might begin to breathe someday! But for microbes living on Earth at the time, the arrival of this new gas was nothing short of a disaster. Scientists refer to this moment in history as the **great oxygen catastrophe**.

Think about it. Until this moment, nascent life on our planet had already been evolving and adapting to survive in Earth's primitive environment. That environment consisted of oceans, barren landscapes, and an atmosphere with no oxygen. And then, in a seeming geological overnight, their planet was gassed. Oxygen is highly reactive, and it was toxic to most of those primitive bacteria; it reacted wildly with their cell membranes, ripping up the structure. Even today, oxygen is toxic to many anaerobic bacteria that don't use oxygen and live in oxygen-deprived environments, such as animal intestines, deep sea vents, and soil. The great oxygen catastrophe caused the near extinction of all microbial life—all but for a few tenacious bacteria that somehow managed to survive. Perhaps they clustered around the deep-sea vents, where they were insulated from the toxic effects of oxygen. Sheltered there, they would have had time to adapt. Fortunately for us, our ancestors were among the chosen few who found a place to hide.

EVOLUTION OF PHOTOSYNTHESIS

The processes of oxygenic photosynthesis in plants, algae and cyanobacteria are strikingly similar. The DNA found in chloroplasts of plants and algae (yes, DNA resides outside the nucleus in some organelles) have many codes in common with the DNA in cyanobacteria. There is little doubt that all three descended from a common photosynthetic ancestor, and that ancestor was certainly an ancient relative of cyanobacteria. Remember that in plants and algae, photosynthesis takes place in chloroplast membranes, while in cyanobacteria it takes place in the cell membrane itself. It turns out that the ancestors of chloroplasts almost certainly were ancient free-swimming cyanobacteria. Modern-day chloroplasts still actually look like bacteria! It is generally accepted that chloroplasts formed when ancient eukaryotic cells—the complex nucleated cells that make up plants and animals and that we will hear about in the next chapter—ingested and "captured" cyanobacteria, co-opting their photosynthetic capability for use in powering their cells. These ancient eukaryotic cells with their newly acquired chloroplasts were the ancestors of all current-day algae and plants.

And what about the pigments used in photosynthesis? Why do cyanobacteria, algae, and plants all use **chlorophyll pigments**, coloring our landscapes green? Chlorophyll belongs to a class of pigments called porphyrins, which are a group of deeply colored molecules of which heme, a component of hemoglobin in red blood cells, is perhaps the best known. Looking at the heme molecule and the chlorophyll molecule leaves no doubt that they are related. They both have four-complex carbon–nitrogen rings linked together surrounding a central metallic atom—iron for heme and magnesium for chlorophyll. Heme is part of cytochrome c, an essential component of the electron transport chain, which is used in cellular respiration and is ubiquitous in the cells of plants, animals, and microorganisms. It almost certainly evolved first. But it wasn't too big a step for evolution to copy and modify the process of producing heme to produce chlorophyll and initiate the process of photosynthesis that changed the course of life on our planet (Blankenship 2010).

Chlorophyll absorbs light mainly in the blue and red wavelengths and reflects green light, which is why most plants are green. This is puzzling, because much of the energy from the sun is transmitted in the green spectrum. So why didn't cyanobacteria evolve pigments that absorbed a broader spectrum of light from the sun, including the green spectrum, and then pass these on to algae and plants? We live in a monochromatic world of green trees and shrubs, when we could be surrounded by red plants or blue ones. In fact, the most efficient system would be a pigment that absorbs light of all wavelengths, in which case plants would be black!

We don't know the answer. Some have speculated that green chlorophyll evolved because purple bacteria may have dominated early Earth (Sparks, DasSarma, and Reid 2006). Current-day photosynthesizing purple halobacteria utilize a pigment that absorbs green light and reflects blue and red light, giving them a purple color. Ancestors of these purple bacteria may have dominated the planet and covered the ocean's surface with a purple hue. Cyanobacteria, which may have emerged later beneath a blanket of purple microorganisms, may have selected chlorophyll pigments because they absorbed what light was available—red and blue light that was not absorbed by ancient purple bacteria.

We must remember that evolution does not design like an engineer. Evolution makes random changes and waits to see if they work. The new designs may not be perfect, but they may be the best available—and chlorophyll apparently was the best available pigment. It has survived and has had a lasting legacy. When you step into a forest's shade, you are sheltered beneath a canopy of comforting green leaves—not red, purple, blue, or black. Chlorophyll is responsible for the greening of Earth.

WHY OXYGEN?

Why cyanobacteria produce oxygen in the process of photosynthesis is one of the greatest mysteries in the natural history of life. Why should it happen? It seems so improbable. Oxygen provided no survival benefit to cyanobacteria and in fact was highly toxic. Cyanobacteria certainly had no altruistic motives to provide oxygen for future complex cells and creatures; cyanobacteria were not thinking of us! Furthermore, making oxygen from water is *not* easy.

When ancient bacteria developed the capacity for photosynthesis, the process was pretty simple. Sunlight activated electrons from pigments located in the bacterial cell membrane, and the high-energy electrons were used to produce ATP. The pigments had to replenish their lost electrons to repeat the cycle. Back then these electrons were obtained from hydrogen sulfide or ferrous ions that were readily available, floating freely in primordial oceans. Stealing electrons from hydrogen sulfide or ferrous ions is not difficult.

The transition to oxygenic photosynthesis may have occurred out of necessity (Lane 2009, 81–83). Supplies of hydrogen sulfide and ferrous ions in the oceans were diminishing, and cyanobacteria needed a substitute. Water was readily available, but water is a stable, chemically inactive molecule, to which the persistence of our oceans attests. Splitting and stealing electrons from water is very problematic. The chlorophyll pigments that had lost their electrons after being bombarded by the sun wanted those electrons back. They had a very strong attraction for electrons and were good candidates to steal electrons from water, but something more was needed for this to happen.

An unusual hybrid compound that was part mineral and part biological facilitated the process. The complex today consists of a cluster of four manganese atoms and a calcium atom bound in an oxygen lattice and entwined with various proteins. It is stuck to the edge of chlorophyll pigments, where it serves as a vise to hold and position water molecules in such a way that they can react with the chlorophyll ion and be split, donating an electron and releasing oxygen as a by-product. It is appropriately called the **oxygen-evolving complex**.

The origin of this unusual metal cluster–protein complex is a mystery. Geologist Mike Russell has suggested that the metal cluster portion of the complex may have been stolen from inorganic mineral clusters in primitive deep ocean vents. Its complex mineral component couldn't have been a product of biology. Ancient cyanobacteria must have evolved protein complexes to join with the metal cluster to produce this bizarre hybrid enzyme complex. Whatever its origin, the incorporation of the oxygen-evolving complex into the process of photosynthesis to facilitate the splitting of water has all the appearances of a lucky accident (Lane 2009, 86). This was a highly improbable event that happened only once in history. We know this because all the descendants of cyanobacteria that use oxygenic photosynthesis utilize an oxygen-evolving complex that is remarkably similar. There is little variation. One ancient cyanobacterium somehow captured and evolved this peculiar hybrid enzyme complex, and billions of years later here you are, living on the oxygen it produced.

It is quite possible that photosynthesis could have evolved without ever producing oxygen as a by-product. And if there were no oxygen, life on our planet would consist of only microorganisms; complex life would not be possible. The unlikely evolution of oxygenic photosynthesis is one more example of how chance plays such a dominant role in the natural history of life on our planet.

OXYGEN BEGETS AEROBIC RESPIRATION

Oxygen has transformed our planet. The flooding of our oceans and atmosphere with oxygen by cyanobacteria has led to the evolution of aerobic metabolism. Because oxygen has such a strong attraction for electrons, it can extract the maximal amount of energy from organic food. As a result, aerobic respiration is the most efficient system for producing energy in the form of ATP to power cells. Aerobic respiration has provided a clear survival advantage over anaerobic respiration and has evolved as a universal trait in the animal kingdom.

Before oxygen was available, microorganisms produced ATP through anaerobic respiration. These anaerobic microorganisms had all the machinery to produce ATP but were not very efficient, because the final electron acceptor of the electron transport chain was not a strong attractor of electrons, as was oxygen, and could not extract the maximum energy from activated electrons. Once oxygen was available, and after enzymes were produced to facilitate the chemical reactions, many ancient archaea and bacteria transitioned from anaerobic to aerobic respiration. We will see in the next chapter how complex cells (eukaryotes) and animals inherited the capacity for aerobic respiration, and how this led to the diverse animal life that dominates our planet today.

CHANCE VS. BIOLOGICAL NECESSITY

This is a good opportunity to discuss the role of chance versus biological necessity in the evolution of life. Did things evolve because they *had* to be that way for life to survive? Or did things just happen by accident? In *Vital Dust*, Christian de Duve championed the position that biochemical necessity determined much of evolution (de Duve 1995, 85). We saw in chapter 4 how he argued that the nucleotides adenine (A), thymine (T), cytosine (C), and guanine (G) were selected from among the many nucleotides in our environment for incorporation into DNA because they alone have special properties. The four nucleotides are complementary and stick specifically one to another—A with T and C with G—and this special property facilitates DNA replication. DNA evolved the way it did out of biochemical necessity. If life evolved again under similar conditions, he argued, the same nucleotides would be chosen, and likely that alternative life would eventually be very similar to what it is today.

On the other hand, the incorporation of the oxygen-evolving complex into photosynthesis appears to have been a chance occurrence—but what an effect it has had on our planet! Without those little metal atoms, there would be no oxygen in our atmosphere; the mass rusting would never have happened, and the oceans would be soupy with iron. Those countless microorganisms that died in the great oxygen catastrophe would have lived on. If this chance occurrence had not happened, life would be very different indeed. The renowned late paleontologist and evolutionary biologist Stephen Jay Gould was a strong proponent of the role of chance in the evolution of life. He reiterated his famous metaphor in his book *Wonderful Life*: "Rewind the tape" of life and "each replay of the tape will yield a different set of survivors and a radically different history" (Gould 1989, 50). As we will see again and again as we continue our journey forward through time, the natural history of life is in many ways a story of the balance between chance and biological necessity.

PART III

EVOLUTION OF COMPLEX LIFE

CHAPTER 7

THE GREAT HISTORIC RENDEZVOUS

There is one cataclysmic event, arguably the most decisive event in the history of life, which really was a rendezvous, literally a historic rendezvous.... This was the origin of the eukaryotic (nucleated) cell....

—Richard Dawkins, in *The Ancestor's Tale*

Two billion years ago, Earth's landscape was barren. The oceans teemed with microorganisms, but there was nothing that walked, swam, or crawled on our blue planet; no life was yet visible to the naked human eye. It was not until the birth of the complex nucleated cell, the eukaryote, that life had a chance to grow bigger. Eukaryotic cells allowed for the development of multicellular organisms and promoted the evolution of the complex, diverse life that enriches our planet today. There is little argument that the emergence of the first eukaryotic cell was a seminal event in the evolution of life—but there is a great deal of controversy about how it happened.

THE EUKARYOTIC CELL

If asked to separate all life into types, most of us would start with "animals" and "plants"—the obvious, visible choices. A particularly thoughtful person might include "bacteria," giving a nod to creatures we can't see. But there is more to life than meets the eye. Genetic studies pioneered by American microbiologist Carl Woese and others have reframed our thinking and classified life into three major domains: bacteria, archaea (microorganisms that differ from bacteria in a few key ways), and eukaryotes (Woese, Kandler, and Wheelis 1990). Bacteria and archaea are microscopic organisms without a nucleus; they are often grouped together and called prokaryotes. Eukaryotes, on the other hand, dwarf the much smaller bacteria and archaea, reaching sizes of up to one hundred times that of the prokaryotes. They contain nuclei within the cells and are orders of magnitude more complex. Eukaryotes have evolved into multicellular organisms, giving rise to the kingdoms of plants, animals, and fungi.

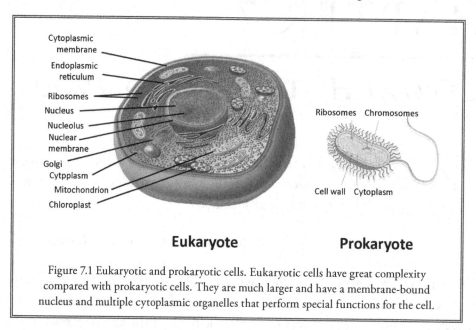

Figure 7.1 Eukaryotic and prokaryotic cells. Eukaryotic cells have great complexity compared with prokaryotic cells. They are much larger and have a membrane-bound nucleus and multiple cytoplasmic organelles that perform special functions for the cell.

If you were to miniaturize yourself to the size of a small molecule and step into a eukaryotic cell, you would find yourself in a wild, futuristic city (figure 7.1). The membrane-bound nucleus is the control center here; it holds endless coils of DNA wrapped in complex proteins called histones, all of which are neatly packaged into chromosomes. Outside the nucleus in the watery expanse of the cytoplasm, there are numerous organelles—structural centers that perform special functions for the cell. **Mitochondria** are the cell's power plants; they harness organic compounds and use oxygen to churn out energy-packed ATP molecules. Ribosomes are minute factories that produce proteins, which then serve as reaction-enabling enzymes and basic building blocks for the cell. The endoplasmic reticula are also factories; they produce more proteins, lipids (or fats), and steroids. Lysosomes are garbage disposal plants that digest worn-out cell parts. Golgi apparati are distribution centers that package and export proteins to places where they are needed within the cell—or even to far-off places outside of the cell altogether. And the cytoskeleton, while functioning as a scaffold to support the structure of the cell, is a transportation network for our cell-city, like a system of roads and rails that shuffle molecules along microtubules and microfilaments.

BACTERIAL SLAVES

Mitochondria and chloroplasts, the energy-producing organelles in eukaryotic cells of animals and plants, have puzzled biologists for a long time. Mitochondria have their own DNA, can produce their own proteins, and divide by binary fission, as if they themselves were autonomous organisms. They have properties of bacteria, and they *look* like bacteria. And yet they function as power plants within our own cells.

The best explanation for this mitochondrial and chloroplastic oddity is a wild tale of cellular capture and exploitation. It is now generally accepted that mitochondria and chloroplasts are descendants of free-living bacteria—bacteria that were ingested by eukaryotic cells early in the course of evolution, somehow were not digested, survived, and became slaves to their eukaryote hosts. Over time, the captured bacteria transferred most of their DNA to the nucleus of the host cell, relinquished control, and entered into a symbiotic relationship, becoming dependent organelles performing specialized functions for the host cell. The entire process is known as **endosymbiosis**. Eukaryotes have the unique capacity for Pac-Man-like ingestion of smaller cells by a process known as **phagocytosis** that has made endosymbiosis possible (figure 7.2). Neither smaller bacteria nor archaea possess this capacity.

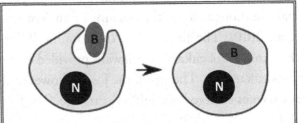

Figure 7.2: Phagocytosis and endosymbiosis. Eukaryotes are capable of Pac Man–like ingestion of smaller cells by an action known as phagocytosis. Surprisingly, ingested bacteria were sometimes not digested and were transformed into organelles that performed specialized functions for the host cell—a process known as endosymbiosis.

The theory of endosymbiosis is nearly incontrovertible. Even the littlest details are in place. For example, mitochondria have a double membrane. You can probably picture how the outer membrane is derived from the vesicle that originally carried the bacteria into the eukaryotic cell, while the inner membrane is derived from the bacteria's own cell membrane. Here's the clincher: sequencing of mitochondrial DNA suggests that mitochondria are descendants from ancient ancestors of the modern-day pathogen *Rickettsia*, a genus of proteobacteria responsible for typhus and Rocky Mountain spotted fever. It is perhaps ironic that our life-giving mitochondria's closest relative is a microorganism that can so easily harm us. The story of the **chloroplast** is very similar. Sequencing of chloroplast DNA has shown that chloroplasts are descended from ancient ancestors of photosynthesizing cyanobacteria.

While there is general agreement that eukaryotic cells acquired organelles through phagocytosis and endosymbiosis, how the smaller prokaryotic cells (bacteria and archaea) were able to grow large enough to morph into eukaryotic cells capable of phagocytosis remains a mystery. Let's explore some possibilities as to how this might have happened and when the first eukaryotes may have emerged.

FIRST EUKARYOTES

Remember: based on anatomical and chemical evidence in the fossil record, scientists estimate that life on Earth began about 3.8 billion years ago. For the next two billion years, life remained microscopic. Until recently, it was thought that complex eukaryotic cells did not evolve until about one billion years ago. But new fossil findings have challenged that hypothesis. In 2010, an international team of paleontologists discovered thumb-sized fossils in Gabon, West Africa, that are thought to be the remains of ancient multicellular eukaryotic organisms (El Albani et al. 2010). The fossils' size, their multicellularity, and the presence of sterol compounds characteristic of eukaryote cell walls provided evidence suggesting that these fossils were formed from eukaryotes. The age of the fossils, however, did not mesh with the prevailing theory that eukaryotes emerged one billion years ago. The fossils of these organisms were 2.1 billion years old!

These findings remain controversial, but they suggest that eukaryotes may have emerged much earlier than previously thought. Exactly *how* they originated from the much simpler prokaryotes has remained a great mystery, and there are conflicting theories about how this may have happened. In his insightful book *The Origins of Life: From the Birth of Life to the Origins of Language*, British evolutionary biologist John Maynard Smith has described a scenario in which a prokaryotic cell gradually evolved the size and capacity for phagocytosis (Smith and Szathmáry 1999, 59–78). He proposes that prokaryotes shed their rigid cell walls and evolved cytoskeletons, which are internal scaffoldings that can maintain the form of the cell and enable the cell to alter its shape and create invaginations in the cell membrane. As the cells grew larger, they were able to engage in phagocytosis and could gobble up bacteria. Some of these bacteria were transformed into mitochondria and chloroplasts that provided energy for the rapidly growing cell, transforming the giant protoeukaryotes into full-blown eukaryotes. While this theory sounds pretty straightforward, some details of this scenario are very problematic.

THE FATEFUL ENCOUNTER

If eukaryotic cells arose by Darwinian evolution due to variation and natural selection in a scenario as described by John Maynard Smith, we would expect to find somewhere in nature transitional cells with a variety of internal structures at various stages between prokaryotes and eukaryotes. But as Nick Lane describes in *The Vital Question*, this does not appear to be the case (Lane 2015, 157–91). There is no equivalent of *Tiktaalik*, the genus of fish with legs that provides the missing link between fish and tetrapods. There is an evolutionary gap between prokaryotes and eukaryotes with no transitional figures. Furthermore, if complex cells evolved incrementally by natural selection, we would expect them to evolve multiple times and have many lineages. This is not the case. Genetic analyses indicate that all eukaryotes evolved from one common ancestor. There is only one lineage.

Genetic comparisons between eukaryotes and archaea and bacteria suggest that eukaryotes are related to *both*. In fact, modern-day eukaryotes share the most genes and are most closely related to two current-day microorganisms: methanogens, the methane-producing archaea that we met in the deep-sea vents in chapter 4, and *Rickettsia*, the genus of proteobacteria responsible for typhus and Rocky Mountain spotted fever. Furthermore, DNA found in mitochondria of eukaryotes is similar to DNA in current-day *Rickettsia*, suggesting that they derived from a common ancestor (Anderson et al. 2003). This suggests that the first eukaryote formed from ancient ancestors of methanogens and *Rickettsia*, and that *Rickettsia* transformed into mitochondria (Emelyanov 2001; Martin and Mentel 2010).

John Maynard Smith's scenario would have us believe that ancient methanogens morphed into eukaryotic cells and then engulfed ancient proteobacteria, which turned into mitochondria. There are two problems with this scenario. First, since all current-day eukaryotes contain or at one time contained mitochondria, this would mean that all eukaryotes that did not engulf proteobacteria became extinct, and this seems highly unlikely. Secondly, prokaryotes would require a great deal of energy to grow large enough to morph into eukaryotes capable of phagocytosis. A prokaryote's energy supply is dependent upon the amount of surface area its cell membrane has available for chemiosmosis. As a cell increases in size, its volume increases faster than its surface area, so its energy supply may not be able to keep up with demand. This is why bacteria do not participate in phagocytosis; they simply cannot produce enough energy to grow large enough to do it.

So how did a prokaryote (an ancient methanogen) incapable of phagocytosis ever manage to snag itself a mitochondrion? How did the eukaryote line first begin, if that early archaeon had no way to capture an early proteobacterium—no way to begin endosymbiosis?

The answer is mystical and speculative. Nick Lane believes the origin of the first eukaryote was a freak accident that occurred one time on our planet billions of years ago (Lane 2009, 88–117). Genetic evidence sitting in the nuclei of all eukaryotes points to the conclusion that the fusion of an ancient archaeon and an ancient bacterium happened only once. All eukaryotes descended from a chimera created by the merger of a methanogen and a proteobacterium. But prokaryotes do not have the capacity for phagocytosis and rarely fuse. Furthermore, creation of the first eukaryote not only required a rare fusion event but also required that the bacterium trapped within the archaeon survive digestion and engage in a symbiotic relationship with its host. The probability of transgressing both of these bottlenecks and creating the first eukaryote is infinitesimally low. Nick Lane calls it a "fateful encounter." The uncertainty about how this fusion may have occurred adds to the mystery and intrigue of a one-time event that changed the course of life on our planet.

The creation of this first eukaryote through the synergistic union of a bacterium and an archaeon produced a complex cell endowed with capabilities vastly beyond those of its precursor prokaryotes. This is yet another example of the march of evolution toward greater complexity through synergistic union of simpler parts.

MITOCHONDRIAL JOURNEY

Following the merger of the ancient archaeon and proteobacterium, giving birth to the first eukaryote, the ancestral proteobacterium was transformed into a mitochondrion, which provided energy for the new chimeric cell to grow large enough to develop the capacity for phagocytosis. Mitochondria solved the problem created by the volume–surface area mismatch that accompanied cell growth. By forming multiple convoluted membranes, mitochondria could provide enough surface area for chemiosmosis to produce large energy supplies for the growing cell.

The Earth had little oxygen in the oceans and atmosphere at the time of the formation of the first eukaryote about 2.1 billion years ago. Because of this, the energy metabolism of ancestral proteobacteria as they morphed into early mitochondria was probably metabolically versatile (able to function with or without oxygen). When oxygen levels increased in the environment, eukaryotes and their mitochondria adapted and transitioned to the use of aerobic energy metabolism. As we have learned, aerobic energy metabolism, which involves the use of oxygen, is much more efficient than anaerobic energy metabolism. As eukaryotes evolved into multicellular organisms and grew in size, they needed more energy. Aerobic metabolism provided that energy, produced a great survival advantage, and became the dominant form of energy metabolism for animal life on our planet.

EUKARYOTES POPULATE THE EARTH

Armed with mitochondria equipped with high-powered aerobic respiration, the new eukaryote cell grew in size, dwarfing the smaller archaea and bacteria. We might imagine, if bacteria were sentient beings, the horror created by this new chimeric monster as it rapidly grew, multiplied, and populated the Earth, terrifying and engulfing colony after colony of smaller defenseless bacteria. One such encounter resulted in phagocytosis of a cyanobacterium, producing a single-celled photosynthesizing eukaryote, the ancestor of green algae and of all land plants. Another primitive eukaryote, an ancient relative of the current-day single-celled microorganism choanoflagellate, became the precursor of all animal life today. Eukaryotes, the new rulers of the planet, were equipped and ready to begin the long journey of evolution into complex multicellular life that now dominates the Earth. Survival of the fittest was off and running. All of this started with the highly improbable event that Richard Dawkins, in his insightful and witty treatise *The Ancestor's Tale,* called "The Great Historic Rendezvous" (Dawkins 2004, 536). We are indeed lucky to be here.

CHAPTER 8

THE CAMBRIAN EXPLOSION

Wind the tape of life back to Burgess times, and let it play again. If *Pikaia* does not survive in the replay, we are wiped out of future history—all of us, from shark to robin to orangutan.

—Stephen Jay Gould, in *Wonderful Life:*
The Burgess Shale and the Nature of History

Life began on Earth about 3.8 billion years ago. Bacterial life evolved and came to dominate the seas, and life remained microscopic and invisible for over three billion years. Then all that changed, in the blink of a geological eye. About 540 million years ago, suddenly, from out of nowhere, there appeared in the fossil record of the **Burgess Shale** a diversity of spectacular macroscopic life teeming in the seas. Where there had only been simple microorganisms for billions of years, the oceans were suddenly filled with a diversity of marine animals and algae the likes of which the world had never seen before. This was the **Cambrian explosion**.

JOURNEY TO THE CAMBRIAN

Let's travel back in time five hundred million years to the Cambrian period at the time of the Cambrian explosion. Feel the younger air on your face. It is mild—warmer than in our modern day. There are no polar ice caps and no glaciers. Landmasses are scattered across the face of the planet because of fragmentation of the supercontinent Rodinia. The oceans have risen as a result of the warm climate and retreat of the glaciers, and lowland areas are flooded, creating shallow ocean waters and inland seas.

The landscapes are bare and lifeless, and the shores are mostly barren, with very little soil and only some lichens and algae clinging to the shoreline. But the seas are teeming with abundant marine animal life and algae. Our ancestors are here, but they all live in the sea. Because we are on this journey to discover our heritage, we brought scuba-diving equipment along with us. We step into the cool briny water, into a wild circus of life.

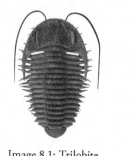

Image 8.1: Trilobite. These extinct arthropods are iconic symbols of the Cambrian period.

The seas are swarming with **trilobites**—ancient **arthropods** about an inch long with a hard shell, multiple segments, and jointed legs (image 8.1). Many crawl along the ocean floor scavenging for food, while others swim or float near the surface. Some are migrating in lines, similar to the migratory queues of modern crustaceans, while others are huddled together in large numbers, shedding their hard exoskeletons, naked after molting, in some sort of mating ritual (Speyer and Brett 1985). Their spiky horns on their heads and curlicue shoulder spines may be some of the first examples of sexually selected traits in animals (Knell and Fortney 2005). These iconic creatures have remained a symbol of the Cambrian Period and the dawn of macroscopic life on our planet.

The sea floor is covered by mats of microbes, some forming coral-like stromatolites. The only plant life we find is colonies of algae resembling seaweed. Forests of sponges rooted to the ocean floor branch upward, appearing more like plants than the primitive animals that they rightly are. A host of bizarre-looking creatures swim or float in the ocean waters, and crawl or burrow in the sediment. Earthworms dominate the soft ocean floor. A fleshy white "penis worm" traversing the seafloor below us, three inches long with a retractable proboscis-like mouthpart, digs into the sediment and vanishes from sight (Gannon 2015). *Wiwaxia*, a strange-looking type of animal about two inches long with a flat bottom and a back ornamented with scales and spines projecting upward, hugs the ocean floor and crawls slowly across the sand. As it has no obvious head or tail, we can tell its front only from its direction of movement.

Image 8.2: *Anomalocaris*. This giant predator dominated the Cambrian seas.

Off to our right we see a fearsome-looking creature about six feet long resembling a giant shrimp with two large compound eyes bulging out on both sides of its head on stalks. It is chasing its prey by undulating large fin-flaps on either side of its body and will grab its victim with two enormous spiked prehensile appendages. This is **Anomalocaris**, the giant predator that dominated the Cambrian period for millions of years but is now extinct with no living descendants (image 8.2).

Soon we come upon **Opabinia**, one of the strangest inhabitants of the Cambrian (image 8.3). This bizarre creature is about five centimeters long and has five eyes pointed in all directions. Its long, flexible proboscis projects forward like a vacuum hose with a menacing claw at the end. Its mouth is positioned under the head and pointed backward, ready to receive food delivered from underneath by the proboscis. *Opabinia,* now long extinct, remains one of the curious oddities of the Cambrian fauna.

Image 8.3: *Opabinia*. With five protruding eyes pointing in all directions and a long, flexible proboscis, *Opabinia* was the strangest creature of the Cambrian period.

Off to our left, an inconspicuous wormlike animal about two inches in length wiggles its flattened body toward us. Its body is divided into segments, faint V-shaped lines outlining muscles on either side of a long central notochord. This is **Pikaia**, the first known **chordate** and the only known chordate in the Cambrian period (image 8.4). This is our ancestor—the mother of all vertebrates, from fish to human beings. We cannot be sure whether *Pikaia* or another chordate is our direct ancestor, but we do know that chordates were sparse in the Cambrian and faced a tenuous future. We are lucky our chordate ancestors managed to survive and continue their lineage. We are lucky to be here.

THE BURGESS SHALE

Shaped by forces of plate tectonics and sculpted by glaciers from the most recent ice age, Yoho National Park sits on the western side of the continental divide in British Columbia, Canada, near the tourist centers of Banff and Lake Louise. High in the Canadian Rocky Mountains, overlooking the limestone-green Emerald Lake and the snow-capped President's Range beyond, there lies a treasure that holds the secrets of our past. Within a single bed of sedimentary rock eight feet thick and a city block long, there are fossils that reveal secrets of the dawn of large-bodied life on our planet

Image 8.4: *Pikaia*. This extinct small fishlike animal was the only known chordate of the Cambrian period and may be the ancestor of all vertebrates.

FOSSILIZATION

When an organism dies, its remains are usually destroyed before they have time to fossilize. Organisms with hard parts, such as exoskeletons or endoskeletons, which are more durable, and organisms buried in sediment are more likely to fossilize.

When an organism dies and is buried in sediment, the soft tissues usually decay, leaving the hard parts, the bones, behind. Mineral-rich water seeps into spaces within the remains where it may crystallize and form a fossil of the skeletal structures. This most common method of fossil formation is called petrification. If soft tissues are slow to decay, fossilization can sometimes preserve detailed anatomy of the organism. Compression fossils are formed in sedimentary rock that has undergone physical compression. The best compression fossils are leaves, since leaves are flat and compression distortion is minimal. In a third type of fossilization, a body may completely decay and leave a hole or mold in the sediment, which then fills with minerals that form a cast fossil of the organism.

The Burgess Shale fossils have provided an invaluable record of the Cambrian explosion, owing in large part to the remarkable preservation of soft body parts, such as guts, gills, and antennae. The soft tissues are so well preserved in this area that most scientists believe that the creatures were covered quickly in sudden mudslides, which trapped the organisms and prevented exposure to oxygen. This froze their bodies in time.

Geologist Robert Gaines and his colleagues have put forth another hypothesis as to how the exquisite preservation of the Burgess Shale fossils may have happened. They observed that layers of calcium carbonate overlaid a number of exceptionally preserved fossil fields—a feature that less-preserved fossil fields did not have. They postulate that calcium carbonate acted as a barrier to flesh-eating microorganisms, preventing the decay of soft tissues and allowing an unusually long time for fossilization and preservation. Massive tectonic plate shifts at the beginning of the Cambrian period are thought to have altered seawater chemistry by dissolving calcium and other minerals in the ocean waters, which then overlaid these particular fossil deposits. Such well-preserved fossil fields are often called *Lagerstätten* (storage places). We can be thankful for this fortuitous preservation of our ancestors, which has given us so many insights into our origins.

540 million years ago during the Cambrian period. This is the **Burgess Shale**.

In 1909, paleontologist Charles Walcott was nearing the end of the season's fieldwork in British Columbia when he discovered a unique fossil bed. These fossils were special; the soft body parts of the strange creatures were preserved in intricate detail (see sidebar "Fossilization") (Gaines et al. 2012). Walcott returned the next year with his wife, sons, and daughter in tow, and established the Walcott quarry. He worked with hammers, chisels, and small explosives from 1910 until 1913, and returned in 1917 at age sixty-seven to finish the job. All in all, Walcott amassed over sixty-seven thousand specimens from the Burgess Shale, which he brought back to the Museum of Natural History at the Smithsonian Institution in Washington, DC—where Walcott was secretary and where they still reside.

Walcott was painstakingly meticulous in his collection of the Burgess Shale specimens, but as the late paleontologist Stephen Jay Gould describes in *Wonderful Life*, he did not understand the special significance of his findings (Gould 1989, 24–25). More than fifty years went by before anyone truly recognized what Walcott had found. In 1971, Professor Harry Blackmore Whittington of Harvard University published a comprehensive reexamination of Walcott's work. Whittington recognized that the Burgess Shale fossils exhibited many different types of body plans, or **phyla**, some of which were quite bizarre and no longer exist on our planet (Whittington 1971). Professor Whittington's findings showed that many of the phyla identified in the Burgess Shale have long gone extinct, and astonishingly, for whatever reason, no new phyla have evolved to take their place. All we living creatures have evolved from phyla that existed in the Cambrian period. This has profound implications for the natural history of life, which we will discuss at the end of the chapter.

WHY THE EXPLOSION?

The cause for the sudden emergence of complex life on our planet 540 million years ago has been a mystery. The lack of fossils in older sediments adjacent to the Burgess Shale suggested there was no complex life before the Cambrian explosion. Creationists argued that this was evidence for divine intervention, and scientists were puzzled and looked for explanations. It is now clear that the diverse creatures of the Cambrian period did not arise from nothing, as some originally thought. The fossil record prior to the Cambrian period is very sparse (probably because the denizens of those ancient seas had few hard skeletal parts to fossilize), but macroscopic complex life has now been documented in the fossil record well before the Cambrian explosion. In the early 1990s, investigators from Cliffs Mining Services Research Laboratory in Michigan discovered ten-centimeter-long ribbonlike fossils in Michigan's Upper Peninsula thought to be imprints of the multicellular algae *Grypania spiralis* dated to 2.1 billion years ago (Han and Runnegar 1992). A few years later, an international team of researchers discovered thumb-sized fossils in Gabon, West Africa, with characteristics of multicellular organisms also dated to 2.1 billion years ago (El Albani et al. 2010). The morphology of these macroscopic fossils was curiously different from previous fossils and from modern-day organisms. Both the *Grypania* fossils and the odd African remains provide evidence that multicellular life may have evolved from unicellular eukaryotes a whopping 1.6 billion years before the Cambrian explosion.

Despite evidence of complex macroscopic life existing well before the Cambrian period, scientists are still left with unanswered questions about the reasons for the sudden emergence of diverse marine life. The most commonly held view is that a rise in oxygen levels at the beginning of the Cambrian period triggered the explosion. Scientists have been able to estimate atmospheric oxygen levels from changes in the isotopic composition of carbon and sulfur in sedimentary rocks and have found that oxygen levels reached modern-day levels for the first time at the beginning of the Cambrian period about 540 million years ago (Berner 1999). The widespread proliferation of photosynthesizing algae, which breathe in carbon dioxide and breathe out oxygen as waste, was probably responsible for this. The rising oxygen levels facilitated the emergence of highly efficient aerobic energy metabolism, and this likely led to the evolution of multicellular organisms with increased body size and complexity. Aerobic metabolism may have provided the energy needed for predatory behavior, and this may have triggered an evolutionary arms race between predator and prey, accelerating the course of evolution and contributing to the diversity of new species that culminated in the Cambrian Explosion (Sperling et al. 2013).

Sexual reproduction may have also been a major contributing factor to the sudden emergence of complex life. During the time period leading up to the Cambrian explosion, the evolution of unicellular eukaryotes and multicellular organisms was accompanied by the evolution of sexual reproduction. Offspring conceived through sexual reproduction had a great diversity of traits, and this allowed the offspring to adapt rapidly to changing environments and accelerated the pace of evolution. This may have been an important driving force for the Cambrian explosion (Va 2007). We'll hear more about the evolution of sexual reproduction in chapter 10.

Another of several hypotheses explaining the causes of the Cambrian explosion is the light switch theory of British zoologist Andrew Parker, which holds that the evolution of eyes started an arms race that accelerated evolution and led to the Cambrian explosion (Parker 2003). Complex eyes evolved not long before the Cambrian period and provided huge adaptive advantages to both predators and prey.

There is yet another possible explanation for the Cambrian explosion. In 1869, geologist John Wesley Powell, during an expedition down the Colorado River along the Grand Canyon, discovered a "Great Unconformity"—a juxtaposition of two different types of rock from remote geological periods. Similar "Great Unconformities" have been found in other distant geographies. This mashing together of sedimentary rocks from remote geological periods can occur only as a consequence of mammoth shifts in Earth's tectonic plates, and it is clear that this happened during the Cambrian period. The collisions of tectonic plates crumpled large terrestrial regions, shifting these regions below sea level and allowing the oceans to come roaring into regions that were previously dry land. The seas dissolved calcium, magnesium, phosphate, and bicarbonate from these newly acquired landscapes, and ocean waters became mineralized.

Geologists have proposed that mineralization of the oceans may have facilitated the formation of mineral-based skeletal structures in marine animal life, allowing a sudden rapid evolution of skeletonized, complex animals (Peters and Gaines 2012). Mollusks and arthropods evolved hard external shells known as **exoskeletons**, and vertebrates evolved internal skeletons composed of bones, known as **endoskeletons**. These hard, rigid structures offered a number of important advantages, including protection, locomotion, and support for increased animal size. These skeletal structures also facilitated fossilization, which has left a lasting record of the explosion of life in the Cambrian period.

Evolution was not able to predict the great advantage of mineralized body parts—teeth, claws, shells, and skeletons, for example—and these would not likely have appeared if not for the mineralization of the oceans. A chance momentous geological event that led to the fortuitous mineralization of the seas may have been the trigger that jump-started the evolution of complex animals. This may be another example of a fortuitous event that shaped the course of evolution—yet another chaotic landslide into the world we know today.

LESSONS FROM THE CAMBRIAN

The Burgess Shale remains are perhaps the world's most significant fossils, not only because of their elegance of form, preservation, and beauty, but also because they so completely revised our view of the natural history of life. The fossils contain many surprises. According to Stephen Jay Gould's interpretation in *Wonderful Life*, the number of phyla, or body plans, found in the fossils far exceeds the number of phyla today. Somehow, the roughly one hundred animal phyla identified in the Burgess Shale have collapsed over the years, leaving us with fewer than forty animal phyla today. Examples of modern phyla include chordates (mostly vertebrates), mollusks (snails, octopuses, mussels, and others), arthropods (spiders, centipedes, crustaceans, and insects) and annelids (earthworms and leeches). All the phyla present today were already present in the Burgess Shale—meaning that many have been lost but no new phyla have been created. But what we have lost in the number of phyla (disparity) we have gained in the number of species (diversity). Within each phylum, there are far more species today than were present in the Burgess Shale. As Gould succinctly put it, "The history of life is a story of massive removal followed by differentiation within a few surviving stocks, not the conventional tale of steadily increasing excellence ..." (Gould 1989, 25).

This finding makes some sense in light of the evolution of multicellular organisms. Once a successful body form has developed, it is very difficult to change it. Huge changes in an animal's body structure are so drastic that they usually are not compatible with survival, and therefore animals normally do not evolve in such ways. This is why the Hox genes, the regulatory genes that control body form, are very similar across a wide variety of organisms. While it is difficult to develop new body plans, it is not difficult to build new parts for those bodies, such as new types of eyes, limbs, and teeth. It is this subtle changing of parts (not the full-scale rearrangement or changing of body plans with the development of new phyla) that has created so many new species since the Cambrian. While there has been a great increase in diversity of species, it appears that the *only* time there was a profusion of new phyla along the tree of multicellular life was at its root, during the Cambrian, when life first figured out how to create a body. It's almost as if the Cambrian creatures, in their scramble toward embodied-ness, raced to produce every kind of body imaginable, to exploit an opportunity that living things would never have again.

The late Stephen Jay Gould argued that the findings from Burgess Shale have truly profound implications for the natural history of life (Gould 1989). The fact that there have been no new body plans since the Cambrian probably means that new phyla—new designs—do not inevitably evolve. This is in sharp contrast to traditional thinking, which holds there is an inevitable progression toward better and better body designs, progressing toward animals that are supremely adapted to their environments. No matter what disruptions might derail the train of evolution, the train will get back on track and produce animals with perfect fits to their environments. Gould agreed that phyla with some of the best body designs may have survived by natural selection, because superior anatomy has made them better adapted; but many others have survived by sheer dumb luck, avoiding destruction by the random catastrophic events that ruined others. Gould championed the concept of **contingent evolution**. Survival of species in this view is indeed a consequence of their adaptations but is also contingent upon the freak accidents that may or may not happen around them. If a catastrophe is big enough, no adaptation will ensure survival. And once a phylum is lost, there is no guarantee it will ever be reinvented.

Pikaia, our Cambrian eellike ancestor, birthed a lineage of chordates from which we ultimately arose. We shouldn't really kid ourselves into feeling superior for surviving this long or discovering calculus or inventing the internet or skyscrapers. We're subject to the same dumb luck as any other lowly creature that has crawled this Earth. "Replay the tape," as Gould put it, and maybe just one more of the Cambrian phyla would have survived to eat all our ancestors. Maybe our evolutionary great-grandparents would die in a meteor strike. Evolution would likely travel down a radically different pathway contingent upon various random catastrophes. We may very well not be here.

But Gould's theories are not dogma. Simon Conway Morris, in his controversial book *The Crucible of Creation,* challenges Gould's position on evolution and provides a different perspective (Conway Morris 1998). Rather than an inescapable randomness of evolution being contingent upon freak accidents, Conway Morris argues for the predictability of evolutionary outcomes. He bases this on many examples of **convergent evolution**—the independent evolution of similar body structures and functions by distantly related species as they adapt to similar habitats. There is no doubt that convergent evolution is powerful; the development of aerodynamic flight by such disparate creatures as moths and hummingbirds and bats is one example of convergence, as is the independent evolution of sophisticated camera-like eyes in vertebrates (like us) and cephalopods (like octopuses and squids). These convergences in bodily form are no accident; these are structures that were so advantageous that they've evolved more than once, and in similar ways. Given the power of convergent evolution, Conway Morris argues that evolution will march toward more complexity and more perfection regardless of bad luck and extinctions, and that we would see relatively similar species evolve no matter how many times we "rewind the tape." The debate between these two giants of paleontology is especially interesting because Conway Morris is a Christian popularly known for his theistic views of evolution, and Stephen Gould had no role for creationism. Renowned paleontologist Richard Fortney, in his review of Conway Morris's *The Crucible of Creation*, opined, "He does not say so in so many words, but one senses the pull of the divine in the Conway Morris version" (Fortney 1998). There may be a hint of a Creator in the title *The Crucible of Creation*.

Today most evolutionary biologists support the concept of contingent evolution as advocated by Gould (York and Clark 2011). Evolution, while subject to a number of biological constraints and steered by natural selection, is contingent on innumerable chance events and has no predestined outcome. We, and other life on our planet, are a product of chance, and we are lucky to be here. Consider the mass extinctions. The dinosaurs were wiped out through no fault of their own; there is no way to adapt to the hammer-strike of a meteorite colliding with the Earth. It hit, and the age of the dinosaurs ended, snuffed out like a candle flame, but for the few lonely theropods that lived on to become birds. Of course, this massive death opened the door for the eventual dominion of mammals. Out with the dinosaurs, in with us. The entire history of life on Earth hinged upon the trajectory of that one meteorite—and others before it. And as we learned from the Burgess Shale, we have seen no new body plans (phyla) since the Cambrian explosion. Once a phylum is lost, there is little opportunity for a new phylum to take its place—the natural history of life must take a different course. "Replay the tape," as Gould used to so often say, and we would almost certainly see a different story (Gould 1989, 311–23).

CHAPTER 9

MULTICELLULAR LIFE

Bacteria made the rules for how multi-cellular organisms work.

—Bonnie Bassler of Princeton University

The transition from unicellular microscopic life to the diverse, complex multicellular life that culminated in the Cambrian explosion was a huge evolutionary leap. We learned in chapters 7 and 8 that multicellular organisms first showed up in the fossil record 2.1 billion years ago, but it took more than a billion years before they burst forth in great numbers in the Cambrian explosion (Han and Runnegar 1992; El Albani et al. 2010). Once conditions were right, multicellular organisms evolved not once but multiple times, producing algae, fungi, and the great animal and plant kingdoms that dominate our planet today. How did this happen? How did cells make the transition from a solitary life to one of intensive cooperation, working with other cells as a single cohesive unit?

BACTERIAL "TALK"

An individual bacterium is the ultimate loner. It floats solo in an enormous world, trying to find nourishment and to avoid the "jaws" of microbial predators. Faced with these pressures, some early bacteria made a "choice"—instead of going it alone, they began to cooperate with their neighbors. This was a prelude to the formation of bacterial colonies, and eventually multicellular organisms. Myxobacteria are a modern-day example of cooperative colonial bacteria. Myxobacteria, also known as "slime bacteria" because of the sugary slime they produce that allows them to glide en masse through their native soil, divide and reproduce, but they remain together in swarms known as "wolf packs" and cooperate for the benefit of the group. Each bacterium secretes degradative enzymes that break down insoluble organic molecules so the products can be absorbed as food. Alone, there is only so much digestive juice a myxobacterium can ooze—not enough to be effective. But in a wolf pack, the colony can drool out a sea of concentrated enzyme. With the help of the swarm, the bacteria have immense digestive power and become unstoppable; they can eat whatever they want. Cooperation definitely has its advantages.

For microorganisms to cooperate with one another, they must first communicate. Molecular biologist Bonnie Bassler at Princeton University is one of the principal investigators who discovered how bacteria communicate through **quorum sensing** (Bassler 2009). Bacteria talk with molecular "words," producing and releasing hormonal molecules that lock onto receptors of similar bacteria. Each bacterial species has its own molecular "language"—its own specific system of hormones and receptors that are not sensed by other bacteria. This allows the bacteria to communicate in a species-specific way. For example, when a bacterial colony invades a host, the bacteria freely multiply and begin to release hormones. With their specialized receptors, each bacterium can "taste" how dense the area is with the signaling hormone. When the bacteria reach a critical number so that sufficient hormone is produced, quorum sensing occurs. The bacteria sense that they have sufficient numbers (a "quorum") to coordinate their activities, and they begin, through a molecular trigger, to act in unison. This provides clear advantages. The colony acting together can produce a toxin strong enough to overwhelm its host, while a single bacterium acting alone would fail. Such synergistic cooperation provides a strong selective survival advantage and promotes the evolution of better and better communication systems.

GREEN ALGAE: MODELS FOR THE EVOLUTION OF MULTICELLULAR LIFE

Green algae are a funny in-between life form. One particular family of photosynthesizing green algae includes *Chlamydomonas* (unicellular green algae), *Gonium* (algae that group in colonies), and *Volvox* (multicellular green algae). These organisms may provide a unique model of how the evolution from single to multicellular organisms may have happened (Miller 2010).

Let's start with unicellular *Chlamydomonas* (image 9.1). *Chlamydomonas* is found all over the world in soil, fresh water, oceans, and even in snow on the tops of mountains. They have an "eye"—a simple eyespot that senses light—and two flagella that allow them to swim toward energizing light with a breaststroke-like motion. These creatures have become a favorite model for researchers in cellular and molecular biology, and have helped to answer many questions: How do cells move? How do cells respond to light? How can unicellular organisms reproduce sexually? We'll hear more about these charismatic algae later.

Gonium are unicellular green algae closely related to *Chlamydomonas* that, unlike their solitary cousins, form colonies. The colonies provide advantages of larger size, including better protection against predators. The colonies consist of four to sixteen cells arranged in a flat plate and held together by molecular glue. Each cell is similar to *Chlamydomonas*, with an eyespot and two flagella. The *Gonium* cells in a colony orient their flagella in the same direction, enabling them to propel their group like a raft through water toward oncoming light. The cells that compose a *Gonium* colony are all of the same type, with no differentiation for specialized functions.

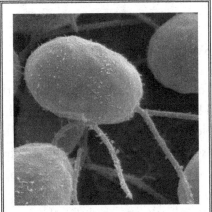

Image 9.1: *Chlamydomonas reinhardtii* (scanning electron microscopic image). These unicellular green algae and their close relative, the multicellular algae Volvox, provide a model of how multicellular organisms may have evolved. Image courtesy of Louisa Howard, Dartmouth College.

Volvox are multicellular green algae, closely related to *Chlamydomonas* and *Gonium,* found in freshwater ponds and puddles. These algae contain fifty thousand or more cells linked together by bridges of connective tissue to form a hollow sphere. Unlike *Gonium*, they have two cell types—somatic cells and germ cells. Somatic cells, which reside in a single layer on the surface of the sphere, have two flagella and an eyespot—nearly identical to *Chlamydomonas*. Their job is to help propel *Volvox* to sunlight so it can photosynthesize. Germ cells are undifferentiated cells without flagella that have the ability to break off and divide to form new multicellular organisms—reproducing *Volvox*. *Volvox* displays two key features that define multicellular organisms: cell "types" that are differentiated for specialized functions, and cooperation between cells to form a functioning organism. This division of labor with the development of distinct cell types delegated to perform specific functions (such as swimming and reproduction) was an essential step in the evolution of multicellularity.

Scientists have uncovered clues regarding the evolution of multicellularity by comparing genomes of unicellular *Chlamydomonas* with multicellular *Volvox* (Miller 2010). Surprisingly, many of the genes that are important in the development of multicellular *Volvox* are also present in *Chlamydomonas*, indicating they were also present in their unicellular common ancestor. Two such genes are the genes that control asymmetrical cell division and embryo inversion. Why should unicellular *Chlamydomonas* contain such genes that are relevant only for multicellularity? The answer lies in an important paradigm of how genes evolve: genetic duplication and divergence. Duplication of genes creates redundancy and provides fuel for innovation. With duplicate genes, one copy can continue to provide its existing function (and preserve the viability of the organism), while the duplicate copy can diverge through mutations and acquire new and distinct functions. The genetic codes of the two genes remain very similar to one another, and there is no doubt about their common heritage.

SWITCHES AND APPS

Your body is composed of about thirty-seven trillion individual cells (Bianconi et al. 2013). There are approximately two hundred different types of them—nerve cells, muscle cells, white blood cells, red blood cells, liver cells, cells of all kinds—and they are all working collaboratively as a single, moving, breathing being: you. These diverse cells perform specialized functions, cooperating in an attempt to support you throughout your life as you weather changes in your environment. They stick together, give up their independence, and curtail their own reproduction to work for the greater good of the organism as a whole. If you're sitting here reading this book, then they've done a pretty good job so far.

Each of us originates from one solitary cell. Anyone who has studied the development of the human embryo from a single-celled **zygote** into a full-term baby appreciates that embryonic development is one of the most complex and awesome phenomena known to man. How does this happen? How can tiny DNA molecules in a *single cell* contain enough information to build a creature consisting of trillions of cells, with hundreds of cell types performing specialized functions, and every part in the right place?

We know that every cell of a given plant or animal of any species contains the same genetic information as every other cell in that organism. Any gardener knows that if he cuts off the tip of a plant shoot and sticks it the earth, it can sprout roots and grow into a clone of the original plant. It can do this only because the tip of the plant shoot contains all the genes of the original plant, including those necessary for root making. Of course, we have now confirmed this fact with genetic analyses; every cell of every organism contains the entire genome of that organism.

As embryotic cells divide, they differentiate into cells with specialized functions that will become part of specialized organs. The specialized structure and function of each cell is under genetic control. Since all cells have the same genetic information, cells can differentiate into specific cell types–for example, muscle cells, nerve cells, and liver cells—only when genes that control those cell types are activated while other genes are suppressed. The cells have learned how to do this. Within the nucleus of all cells, there are specialized genes called **regulator genes** that produce regulator proteins called **transcription factors.** Transcription factors activate or suppress **structural genes** that determine the function of the cells. You can think of regulatory genes as switches and structural genes as complex apps—specialized programs coded in the

DNA directing the formation of specific proteins that determine cell type. The switches turn on or turn off the apps that determine the structure and function of the cell. There are apps to build muscle cells, nerve cells, breast cells, skin cells, and all the rest.

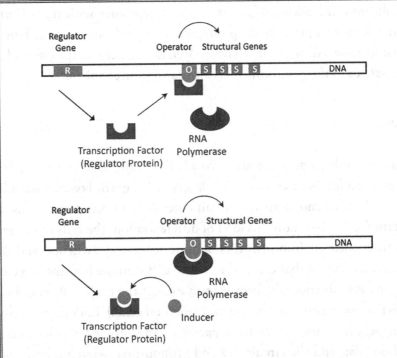

Figure 9.1: How *E. coli* digests food. Nobel laureates Francois Jacob and Jacques Monod studied how the common intestinal bacterium *Escherichia coli* digests its food (Ullmann 2009). Structural genes in *E. coli* produce the protein enzyme lactase, which catalyzes the digestion of the sugar lactose. The production of lactase and all proteins are catalyzed by the enzyme RNA polymerase. *E. coli* is able to turn on and turn off genes that control lactase production depending on when it is needed.

E. coli's regulator genes inhibit the production of lactase by producing a transcription factor (a regulator protein) that blocks the action of RNA polymerase (top panel). The transcription factor binds to the operator (a section of DNA next to the structural genes), preventing RNA polymerase from binding to the operator and blocking lactase production. When lactose is present in the environment, it binds to the transcription factor, preventing it from binding to the operator and freeing the operator to bind to RNA polymerase (lower panel). In effect, lactose turns on the switch to activate lactase production. There's lactose nearby? Now *E. coli* can eat it. This is a classic example of how environmental factors can turn on or turn off genes.

External stimuli can also interact with transcription factors to turn on or turn off structural genes. For example, when the sugar lactose is present in the environment of *E. coli* bacteria, it interacts with a transcription factor to activate structural genes to produce the enzyme lactase so the bacteria can digest lactose (figure 9.1) (Ullman 2009). In another example, a state of starvation will cause the release of cortisol, which combines with transcription factors to activate structural genes that promote the production of sugar from proteins. This is an adaptive mechanism for starvation. All organisms have evolved the capacity to turn on and turn off genes in response to changes in the environment as an adaptive mechanism.

BODY PLANS

In most animals, the embryo differentiates into a hollow cup-shaped structure of cells called a **gastrula**. The gastrula forms three onion-like layers called **germ layers**, each with a different cell type: an inner layer of endodermal cells, an outer layer of ectodermal cells, and a middle layer of mesodermal cells. To attain this level of differentiation, the cells of the gastrula receive various signals from their environment that activate regulatory switches, and those switches trigger apps (structural genes) that direct the cells to differentiate into the three cell types.

Specific organs are subsequently formed from each of the three cell types in response to a variety of signals that switch on specialized apps (structural genes). Endodermal cells differentiate to form the stomach, intestine, colon, liver, pancreas, and urinary bladder; mesodermal cells form muscles, bone, fat, and the circulatory and genitourinary systems; and ectodermal cells form skin, the brain, the nervous system, and most of the eye.

The regulatory genes that control the general body plan are remarkably similar in all symmetrical animals, despite enormous diversity among the creatures themselves. These regulatory genes, called **Hox genes**, are the on/off switches that decide during the course of development when and where on the animal's body to place the head, thorax, abdomen, legs, antennae, wings, and tail. They determine the body plan for all symmetrical animals, from flies to mice to humans. You can think of these genes as tiny molecular contractors shouting instructions as the embryo develops: "Put the legs here! A little to the left … there! And put the tail over there!" These genes don't specify what *type* of legs or tail the new creature will grow; that is to be determined by the structural genes. The Hox genes simply determine when and where the legs and tail will be attached. And they do this by selectively activating specific structural genes at the appropriate time and place in response to environmental conditions.

There are vast evolutionary distances between most animal species alive today—meaning that it has been a long time since those animals diverged from common ancestors. Regulatory

Hox genes are very similar in all symmetrical animals and are believed to have been present in the shared common ancestor of all symmetrical animals (Akam 1995; McGinnis and Krumlauft 1992). This common ancestor is believed to have lived in the Precambrian period about six hundred million years ago. One of the reasons the Hox genes have persisted with so little change over hundreds of millions of years is that variations in these genes would result in great changes in body morphology, which would likely be lethal. As Stearns and Hoekstra describe in their textbook *Evolution: An Introduction*, evolutionary changes in organisms are like bits of restoration occurring to a car while it is moving. There is no opportunity to take the car into the shop for a full overhaul. Organisms must remain viable *while* they are being remodeled or they will simply drop dead. Think about it: An animal born with a mutation that gives it two legs rather than four is not likely to survive to bear young. Evolutionary changes must usually be small and manageable, as large changes will only interrupt function and threaten viability. Because of this, new body types have not evolved and survived for the past five hundred million years—not since the Cambrian explosion, when all current **phyla**, or major body plans, were already present. And until a catastrophic mass extinction wipes out most everything and makes room for something completely new, we do not expect to see the evolution of any novel phyla on our planet as we move forward.

THE TRANSITION TO MULTICELLULARITY MAY NOT HAVE BEEN SO DIFFICULT

The transition from single-celled organisms to complex multicellular organisms was momentous and dramatically changed the natural history of life, but it may not have been as difficult as we imagine (Penissi 2018). The transition happened multiple times—just once in plants, and just once in animals, but multiple times in fungi and algae.

As unicellular organisms joined together to form multicellular organisms, they had to learn how to turn on and turn off genes so they could differentiate to perform special functions. They had to learn sophisticated gene regulation. It turns out, surprisingly, that single-celled organisms had already learned how to do this. Single-celled *Chlamydomonas*, for example, is able to activate genes that produce and energize flagella needed to propel the organism toward light for most of its lifetime. But when the cell needs to reproduce, it deactivates the flagella genes and loses its flagella, while activating the genes needed for cell division and reproduction. *Chlamydomonas* can both swim and reproduce, but not at the same time. *Chlamydomonas* has the capacity to regulate genes, and the ancient common unicellular ancestor of *Chlamydomonas* and *Volvox* surely had the same capacity. When ancient unicellular eukaryotes evolved to create multicellular organisms, they already had the machinery needed to activate and deactivate genes in the process of differentiation. The evolution of multicellularity was not such a big jump as originally thought (Penissi 2018).

Scientists have now been able to demonstrate the evolutionary transition from unicellular to multicellular life in the laboratory (Herron et al. 2018). Investigators have grown populations of *Chlamydomonas* in the presence of a protozoan predator that feeds on small algae. After a year and 750 generations, multicellular organisms began to appear. The evolution of multicellularity seemed to provide protection against these predators, and the adaptive advantage that multicellularity and size provide against predation may have played a major role in the evolution of multicellularity. Indeed, the transition to multicellularity may not have been so difficult.

THE MARCH TOWARD MULTICELLULARITY

As unicellular organisms joined together, and as multicellular organisms evolved, the advantages of division of labor and synergistic cooperation immediately became clear. As we have seen, large colonies of bacteria worked together to produce digestive enzymes and toxins in sufficient quantities to be effective, while single bacteria acting alone were ineffective. Animals evolved muscles, limbs, and wings that gave them mobility, allowing them to find mates and new habitats, and enhancing their ability to evade predators and pursue prey. Creatures evolved complex eyes that enabled them to form detailed images of their environment and to respond rapidly to changes. Plants evolved roots that extended deep into the soil to reach water and nutrients, and they produced leaves to harness energy from the sun. With such obvious, enormous adaptive advantages, it is no surprise that the evolution from unicellular to multicellular organisms occurred not once but multiple times, leading to the kingdoms of fungi, plants, and animals that dominate our planet today.

CHAPTER 10

THE GREAT LOTTERY

> A world without sex is a world without the songs of men and women or birds or frogs, without the flamboyant colors of flowers, without gladiatorial contests, poetry, love or rapture. A world without much interest.
>
> —Nick Lane, in *Life Ascending, The Ten Great Inventions of Evolution*

Sex consumes us all. Our movies, books, songs, TV shows, and conversations are dominated by themes of romantic love and sex. It is the source of our greatest pleasures and our greatest sorrows. Like every other facet of life, the process of sexual reproduction emerged through natural selection, and it stayed with us because it confers important survival advantages. Sexual reproduction evolved alongside the evolution of multicellular life and played a major role in accelerating the pace of evolution, culminating in the Cambrian explosion. But how did it happen? And how does it work?

ASEXUAL REPRODUCTION—CLONING

Mitosis is the process by which most cells reproduce. Your body needs new cells all the time, whether to replace damaged cells or to facilitate growth; and to produce new cells, the old ones simply copy themselves. In preparation for this copying, a cell's DNA furls itself into neat packages called **chromosomes**, which are located in the nucleus of the cell. The chromosomes divide, creating identical copies of themselves, and the cell itself then divides, distributing identical sets of chromosomes to two daughter cells. These daughters are clones of one another, as they each contain exactly the same set of DNA—the complete set of instructions that makes you who you are.

Mitosis is how the body creates new cells, and it is also the root process of **asexual reproduction**, or **cloning**. Organisms that reproduce by cloning produce offspring that are genetically identical to themselves. They contain undifferentiated cells that bud off and then multiply into a mass of new cells. The baby cell mass develops into an adult organism—an organism that is a genetic copy of its parent. Microorganisms (bacteria and archaea), many plants, and some invertebrates (worms and hydras) reproduce by cloning.

SEXUAL REPRODUCTION

Sexual reproduction is a whole 'nother way to make a baby, and it is so complex that living things had to invent an entirely new system of cell division to make it work. Rather than just copying the genome of one parent to make a genetically identical baby, sex allows two organisms—two parents—to both make a genetic contribution. The random mixing of their genes produces an entirely new genome—one that will be used to grow a new, one-of-a-kind progeny.

To reproduce sexually, the female parent produces an undifferentiated female germ cell, an **egg**, and the male parent produces an undifferentiated male germ cell, a **sperm**. These **gametes** merge in a process called **fertilization**, combining genetic material from the mother and father into one cell. This new zygote then develops into an embryo that is not a clone but a new organism, with genes donated from both its mother and father.

MEIOSIS: THE GREAT GENE SHUFFLER

The method by which organisms create gametes is through a complex and extraordinary process of cell division called **meiosis**. In humans, meiosis begins with an undifferentiated precursor cell that has twenty-three pairs of chromosomes, which are strands of DNA wrapped in proteins. Because the cell has paired chromosomes, it is called a **diploid cell**. One of each chromosome pair comes from the father, and one from the mother. The chromosome pairs are **homologous**, meaning they have corresponding genes that code for the same function, although the genes are not identical. You might have a gene from your mother that codes for brown eyes and a corresponding gene from your father that codes for blue eyes—and the interaction between the two genes will determine what your eyes look like.

Meiotic cell division occurs when a diploid precursor cell with twenty-three paired chromosomes undergoes cell division twice, producing four **haploid cells** (gametes) with twenty-three unpaired chromosomes. The cells are converted from diploid to haploid because chromosomes duplicate only once, while the cells divide twice. During the process, there is swapping of some homologous genes between paternal and maternal chromosomes and random distribution of either maternal or paternal chromosomes to the daughter haploid cells (gametes). This shuffling of genes and chromosomes creates literally trillions of possible genetic combinations for each individual gamete.

When a female gamete (an egg) is fertilized by a male gamete (a sperm), a new diploid zygote is formed with a complete genome of twenty-three paired chromosomes, one of each pair from the father and one from the mother, ready to grow into a person. The number of genetic possibilities of the new zygote becomes almost infinite as the product of the trillions of possible genetic combinations of the sperm and egg. Sexual reproduction is a great lottery. Any one offspring will be a completely new organism unto itself. It might be born with genes that make it strong, beautiful, cunning, and virile. Or it might be born with genes that do not allow it to survive.

WHEN DID SEXUAL REPRODUCTION FIRST EVOLVE?

Corals are polypoid marine invertebrates best known to us as reef builders. They reproduce asexually but can also reproduce sexually by a specialized process known as **broadcast spawning**. The corals release large quantities of sperm and eggs into the ocean waters around the time of the full moon, and these gametes undergo multiple simultaneous fertilizations, creating large numbers of larvae of similar age.

This type of sexual reproduction may be the oldest on record. Fossils from ancient wormlike creatures called *Funisia* were discovered in southern Australia and showed closely packed organisms of the same age, suggesting the organisms were reproduced by the same process of broadcast spawning (Droser and Gehling 2008). This fossil record was dated to just before the beginning of the Cambrian period, 565 million years ago, and represents the oldest fossil evidence of sexual reproduction.

However, sexual reproduction may have started much earlier. *Giardia lamblia*, a unicellular intestinal parasite responsible for the common infectious diarrhea giardiasis, diverged early after the formation of the first eukaryote and is among a small minority of eukaryotes that reproduce asexually. However, recent genetic studies have identified genes in *Giardia* that are specific for meiosis, suggesting that its ancient ancestors must have had sex (Ramesh, Malik, and Logsdon 2005). (Yes, unicellular eukaryotes can have sex, and we will see some examples shortly.) Researchers believe that very early eukaryotes—ancient ancestors of *Giardia*—were already using meiosis in sexual reproduction. The earliest eukaryotes appeared in the fossil record 2.1 billion years ago, so it appears sexual reproduction began not long after that.

HOW DID SEXUAL REPRODUCTION EVOLVE?

We don't typically think of bacteria as capable of having sex, but they actually do have something a lot like it, and this may give us clues as to how sexual reproduction may have started. Bacteria reproduce asexually by cloning themselves, but bacteria can also share DNA with each other, giving rise to new genetic combinations. Many bacteria have hairlike appendages several times longer than their length, called pili, attached to their surfaces. Pili from the "male" bacterium ensnare the "female" bacterium, drawing "her" close and forming a bridge, which allows transfer of pieces of DNA from the "male" to the "female" in a process called *conjugation*. The quotation marks imply that these bacteria are not male and female, and that this is not real sex, but the similarity is enough that the pili are often called sex pili. When genetic material from one bacterium is introduced into another bacterium, the homologous genes (genes that code for the same function) are aligned, and the donor gene replaces the recipient's gene in a process called genetic **recombination**. Probably the best-known current-day advantage of gene transfer, from the perspective of the bacteria, is that information about antibiotic resistance can be spread among bacterial populations. That's scary for us but great for the bacteria that are trying to use humans as hosts.

The transfer of DNA from one bacterium to another is an intricate process that requires facilitation by protein enzymes. How did such a complex process ever evolve? As is so often is the case, the answer is that the enzymes had already been produced for another purpose. These are the same enzymes used to repair fractured DNA. When bacteria began the process of transferring DNA, they did not have to make the great leap of forming entirely new enzymes to make it happen, because enzymes for breaking apart and gluing together DNA already existed to make repairs. There was no need to reinvent the wheel; the bacteria could modify what they already had.

The process of DNA transfer in bacteria provides insights into the obscure origins of meiosis. Most of the enzymes used by bacteria during DNA repair and DNA transfer are very similar to enzymes used in meiosis. This is the missing link. If our earliest eukaryotic ancestor was indeed created by the fusion of a bacterium and an archaeon, then this ancestral eukaryote inherited the enzymes for meiosis from the bacterium. This is the bridge from asexuality to sexuality. These bacterial enzymes—first used for DNA repair and then for recombination as bacteria transfer genes—have been modified, through the ages, to be used by eukaryotes in meiosis, in which genes are swapped between maternal and paternal chromosomes. Meiosis has since become the cornerstone of sexual reproduction.

Chlamydomonas, the unicellular green alga with an eyespot and two flagella that we met in the last chapter, provides further clues as to how sexual reproduction may have originated. Here is how *Chlamydomonas* has sex (figure 10.1). Each organism—each cell-- contains a haploid genome with seventeen unpaired chromosomes. When two organisms are ready to mate, they fuse to form a diploid zygote with seventeen paired chromosomes. On exposure to light, the diploid cell undergoes meiotic cell division—duplication of the seventeen paired chromosomes—followed by two cell divisions that form four haploid daughter cells, each with seventeen unpaired chromosomes containing a mixture of genes from both donor cells. After these haploid progeny cells develop into mature algae, they are ready to mate and repeat the cycle. (Unlike us, *Chlamydomonas* exists as a haploid cell for most of its life, and so it misses out on some of the advantages of duplicate chromosomes.) This is an example of **isogamous sex**, meaning that the fusing cells—the fusing gametes—are the same size and are identical in appearance, and that the organism has no distinct sexes.

Chlamydomonas may provide a model for how sex got started. John Maynard Smith, in *The Origins of Life: From the Birth of Life to the Origin of Language*, describes how it may have happened (Smith and Szathmary 1999, 87–91). The first eukaryotic cells were likely haploid, containing just one set of chromosomes, like *Chlamydomonas*. Two of these haploid cells likely fused to form the first diploid cell. This would not be an unusual occurrence, since cell fusion in unicellular eukaryotes is common even today. The next and crucial step was likely the introduction of meiosis, which requires the duplication of chromosomes within

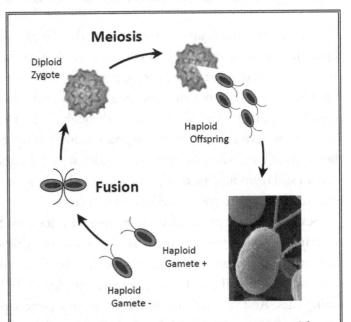

Figure 10.1: *Chlamydomonas* life cycle. The primitive sex life of *Chlamydomonas*, a single-celled green alga, is probably very similar to sexual reproduction in early single-celled eukaryotes. Two haploid *Chlamydomonas* cells, each with seventeen unpaired chromosomes, fuse to form a diploid zygote, which develops a thick cell wall. The zygote undergoes meiotic cell division, producing four haploid offspring, which grow into mature *Chlamydomonas* cells ready to repeat the cycle. The image of *Chlamydomonas* is courtesy of Dartmouth Electron Microscope Facility.

the new diploid cell, the mixing of genes, and two cell divisions to form four haploid gametes. Duplication of chromosomes and cell divisions would not be difficult, since these are normal processes used in mitotic cell division. The mixing of genes entails lining up parallel strands of DNA in paired chromosomes, cutting out and swapping segments, and splicing them in place—all facilitated by protein enzymes. We now know that the enzymes responsible for this process were already available, having been inherited from bacterial enzymes that were used for DNA repair and DNA transfer.

With the evolution of multicellular organisms, gametes were created from undifferentiated cells through meiotic cell division. These gametes were released to fuse with gametes from other multicellular organisms to form zygotes, which can divide and differentiate into a mature multicellular organism.

WHY DID SEXUAL REPRODUCTION EVOLVE?

All eukaryotic organisms—plants, animals, fungi, and protists—reproduce sexually or reproduced sexually in the past. It is almost certain that the common ancestor of all modern-day eukaryotes had sex. Surely the early eukaryotes that evolved from asexual bacteria and archaea reproduced asexually, but they are long extinct. Only eukaryotes that developed the ability to reproduce sexually were able to squeeze through the bottleneck of evolutionary selection. There are some modern eukaryotes that reproduce by cloning, but they also reproduce sexually, and they typically revert to cloning only under stressful circumstances. Of course, bacteria and archaea still reproduce asexually, but without the ability to have sex, these microorganisms have been stuck in a primitive unicellular existence.

Sex is a complex process with some major limitations. Sex requires two partners to produce offspring—it takes two to tango, as they say—and this slows down the whole reproduction process. You could produce offspring much faster if you could just bud a few clones of yourself whenever you felt like it. Finding a mate can definitely be problematic—just look at online dating sites. And why should we contaminate a perfect product, anyway? When we've got it right, why don't we clone ourselves rather than pollute our offspring with potentially imperfect genes from imperfect mates? Why would we want to sully Mozart's genes with input from a lesser source? Why not just clone him? What's so great about sex, anyway?

Scientists have been asking these questions for some time, and they have come up with some convincing answers. In 1973, evolutionary biologist Leigh Van Valen at the University of Chicago proposed the Red Queen hypothesis to explain the advantages of sexual reproduction (Scoville 2017). The term "Red Queen" comes from an analogy to the character of the same name in Lewis Carroll's 1871 novel *Through the Looking-Glass*, in which the Red Queen tells Alice, "It takes all the running you can do, to keep in the same place." This analogy describes the constant evolutionary struggle for organisms "just to keep up" and survive among coevolving enemies.

Consider the example of host–parasite relationships. Parasites are continually evolving new, innovative ways to overwhelm their hosts. For the hosts to survive, they must evolve in response, developing new defenses against their rapidly changing parasites. The parasites, in their turn, will respond to their hosts' new defenses by evolving ever new methods of attack. It's a bit like an evolutionary game of chess between the species, with each opponent attempting to outstrategize the other.

Species that reproduce sexually have a major competitive advantage in this struggle. They create offspring with great genetic diversity, producing a great diversity of traits, and this allows the offspring to adapt quickly to rapidly changing environments. Species that reproduce by cloning have minimal diversity in their offspring (only the diversity supplied by rare random mutations) and cannot adapt quickly to changing environments. Sexual reproduction has provided a major survival advantage in this Red Queen predator–prey arms race, and this explains why it has become the dominant form of reproduction among multicellular organisms.

An example from the human world might make the concept of genetic diversity easier to understand. Imagine that you are unusually financially savvy and have invested in high-tech stocks, and because you've lived during a technology boom, this choice has made you very successful. If you are a clonal organism, all of your offspring will be just like you and invest in high-tech stocks. But if the tech boom collapses, they could suffer devastating losses. However, if you reproduce sexually, genes from your partner will make your children interested in different ventures. Your offspring will have diverse investments—some in tech stocks, some in utilities, some in industry, some in energy, some in financial services, and some in health care. If the tech boom collapses, the offspring who invested in tech stocks may take a hit, but your other offspring will still flourish. Diversification of an investment portfolio is essential in a changing market, and diversification of genes and traits is essential in a changing environment.

Sexual reproduction has other advantages besides diversity in offspring. In clonal reproduction, all genes—including those with deleterious mutations—are copied from one generation to the next. Over many generations, the load of harmful mutations, which are much more common than beneficial mutations, will increase. Eventually the number of dangerous mutations will become so great that these descendants will die out altogether. Using the analogy of a ratchet, a mechanical device that only allows forward movement, this hypothesis was later named **Muller's ratchet** (Crow 2005; Haigh 1978). Sexual reproduction can overcome Muller's ratchet because only half of the genes of each parent are passed on to the next generation, so harmful mutations are not necessarily transmitted to offspring and are often simply eliminated from the line of inheritance. Offspring that do inherit bad genes may not survive, but many other offspring without the deleterious mutations (and some with beneficial mutations) will survive.

Despite all of these safeguards, many offspring are born with destructive genetic mutations. Even in these cases, all is not lost. All sexual organisms—humans included—are diploid (or, more technically, have a diploid phase of their life cycle), with two sets of homologous chromosomes. The paired maternal and paternal chromosomes in your cells give you two copies (two **alleles**) of each of your genes—two different copies of instructions for each function your body needs to perform. This provides you with genetic protection. If your paternal gene is defective, then your maternal gene can still provide the appropriate function. For example, cystic fibrosis is an inherited genetic disease in children characterized by thick mucus secretions and recurrent infections in the lungs, leading inevitably to premature death. The disease results from the production of a defective protein that is important for mucus formation. In the case of cystic fibrosis, both the maternal and paternal genes responsible for making the protein must be defective for the protein to be defective. The genes are **recessive genes**; one normal gene is sufficient to make a normal protein and prevent the disease.

But a normal copy of a gene does not always protect the organism from a defective copy. If a defective gene is dominant, the **dominant gene** will produce a harmful "broken" protein even if the paired gene is normal. This is the case in Huntington's disease, a hereditary neurodegenerative disorder with symptoms of involuntary writhing movements and dementia beginning in the third to fifth decade of life and leading to premature death. Huntington's is caused by a dominant defective gene that codes for an abnormal protein that damages brain cells. Folk singer Woodie Guthrie was a victim of Huntington's disease.

Sexual reproduction helps organisms to combine beneficial mutations and to shed harmful mutations, but remember that none of this is purposeful. The genes transmitted to offspring are selected by lottery. Any baby could have the healthiest genes of both of its parents, or the weakest, most diseased ones. Sexual reproduction simply allows for more genetic diversity than asexual reproduction. With asexual reproduction, all offspring will have the same genes as the parent (barring rare genetic mutations); none of the offspring will be weaker—but none will be stronger either. Sex provides the possibility that while some of your babies will be weaker than you, others will be stronger, more fertile, and more able to weather whatever changes happen in their environment. In this way, sexual reproduction speeds up the process of evolution itself; offspring with sicker genes die out more quickly, and offspring with healthy genes survive and pass their genes to the next generation.

Despite these multiple advantages, sex is not always best. In some situations, cloning can provide an edge. For example, bacteria invading a host organism rely on strength in numbers to overwhelm the host. Cloning allows them to reproduce and colonize more rapidly than sexual reproduction would. For bacteria, then, asexual reproduction is the norm; cloning is their typical method of reproduction. Animals occasionally utilize asexual reproduction too. Female blacktip reef sharks, for example, when deprived of mates in captivity, have "done it on their own," producing eggs that develop into viable offspring without help from a male.

So this gives us some idea of how sexual reproduction may have started. But we have a long way to go to reach the kind of sexual reproduction we see in complex animals today. We'll continue the journey in the next chapter.

CHAPTER 11

SPERM, EGGS, FERTILIZATION, AND SEXUAL SELECTION

> You might think that female birds would go for males with perfect wing length, but in many species, they focus, instead, on some extravagant feature- like the size and beautiful intricacy of a peacock's tail. And, if a peacock with an even bigger tail shows up, he gets all the attention and gets to have the kids.
>
> —Andrew Shaner, MD, UCLA Psychiatrist

Sexual reproduction has provided such great survival advantages that it has become almost universal in the plant and animal kingdoms. But the earliest forms of sexual reproduction were not very efficient. Gametes were the same size and looked identical and were not very specialized to do their work. Life began in water, and marine animals used water as a medium for fertilization, but as animals moved from sea to land, they had to learn new ways to reproduce without readily available water. And through all these changes, animals had to evolve traits enabling them to attract mates in order to enhance their reproductive success. Let's see how the story unfolded.

EVOLUTION OF SPERM AND EGGS

Not long after sexual reproduction burst on the scene, an arms race emerged between species to develop gametes, sex organs, and external genitalia that were able to optimize the union of the sperm and egg and enhance survival of the embryo. These evolving body structures enabling sexual reproduction are what we call **primary sexual characteristics**.

In the earliest forms of sexual reproduction, there were no male and female sexes, and gametes from two mating organisms were similar in size and appearance. We saw an example of this in the last chapter with *Chlamydomonas*. This is termed **isogamous sex**. But specialized differences between male and female gametes soon emerged and provided potential survival advantages. Males evolved sperm that were small, motile, and able to find and penetrate the egg, while females evolved eggs that were large and contained food for the zygote and the developing embryo. The gametes became very different indeed. Today, human eggs are our largest cells, just visible to the naked eye, while sperm are our smallest cells, consisting principally of a nucleus and a long tail. With one ejaculation, millions of sperm compete for the favors of one egg. The evolution of **anisogamous sex**, sex with distinct male and female gametes, has enhanced the ability of species to reproduce.

Animals evolved testes to produce sperm, and ovaries to produce ova, or eggs. These reproductive organs also became the source of sex hormones. Testes produce testosterone, which in human males is responsible for increased height, increased muscle mass, male hair distribution, male sex organ development, and male sexual drive; and ovaries produce estrogens, which are responsible for female body characteristics, such as breasts, the uterus, and female hair distribution. These characteristics that distinguish the sexes of a species but are not directly part of the reproductive system are termed **secondary sexual characteristics.**

EXTERNAL FERTILIZATION

Remember broadcast spawning, one of the earliest forms of sex that was used by corals. During broadcast spawning, the male and female corals simultaneously release sperm and eggs, which form slicks on the surface of water where fertilization occurs and where microscopic larvae soon emerge. The timing of coral spawning is critical because corals are not mobile, and the male and female species cannot swim to one another. Corals in the Florida Keys mysteriously and predictably release their gametes six to eight days after the full moon in August. This creates quite a spectacle,

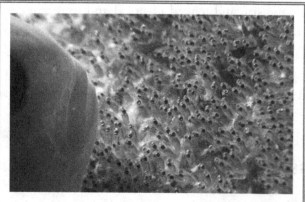

Image 11.1: Mother fish and her spawn. Most fish reproduce by broadcast spawning, a process in which females and males simultaneously release large quantities of eggs (spawn) and sperm into the water, where fertilization can occur.

and scuba divers often gather to watch predators line up for a meal of gametes and larvae. Broadcast spawning is a type of **external fertilization**, meaning that fertilization occurs outside the female's body. During external fertilization, the gametes and larvae are exposed and are at risk from predators. To ensure the survival of their offspring against such odds, the corals release massive numbers of gametes in synchrony. There is strength in numbers. Broadcast spawning is also the type of sexual reproduction used by many fish (image 11.1).

Life began in the water, and water is used as a medium for external fertilization. Gametes are released into the water, where sperm fertilize eggs and produce larvae. This early form of sexual reproduction is still used today by most marine life—most fish, crustaceans, mollusks, and other marine invertebrates. But as animals moved to land, new methods were needed to facilitate the union of sperm and egg, and adaptations were required to protect and nourish the developing embryo.

INTERNAL FERTILIZATION

Part of the game of reproduction on land was finding a safe place for sperm to fertilize eggs and for zygotes to grow to maturity. Most vertebrates, including mammals, reptiles, birds, many amphibians, and quite a few invertebrates, including flatworms, roundworms, some mollusks, and most insects, solved this problem through the use of **internal fertilization**, in which the sperm is placed in the female genital cavity, where it fertilizes the ovum. In internal fertilization, the female herself provides a moist, hospitable, and safe environment for the zygote and developing embryo to thrive.

Sex organs are soft tissues and leave few fossil records, so we have few clues as to when internal fertilization first evolved. Centipedes, those small wormlike creatures with many paired legs and paired antennae, were the first animals to crawl out of the sea and onto land about four hundred million years ago, so modern-day centipedes may provide clues as to how internal fertilization began. Male centipedes package sperm in tiny silk webs, lay the webs near a prospective mate, and often capture the attention of a female with a kind of courtship dance. The female retrieves the sperm-filled web and inserts it into her genitals, where fertilization occurs. This doesn't sound like too much fun, but it seems to work. Centipede eggs are sensitive to drought and vulnerable to mechanical injury, so females lay their fertilized eggs in moist areas under logs and stones, and they sometimes watch over their eggs until the wee centipedes emerge. This is an intermediate form of internal fertilization—internal fertilization without copulation—the kind that might have evolved when animals first walked on land.

Internal fertilization among vertebrates is usually performed with a copulatory organ (a penis or claspers), which facilitates placement of the sperm in the female genital cavity. Copulatory organs are present in all mammals and almost all reptiles but are uncommon among fishes and amphibians and are absent in birds. Although most modern fish do not copulate, ancient armored fish were an exception and may have been the first vertebrates to reproduce with copulation. Fossils of four-hundred-million-year-old placoderms, prehistoric fish with armored plates covering their heads and thoraxes, demonstrated specialized sexual organs—paired bony claspers on the backsides of males with a central groove to facilitate transfer of sperm, and a pair of dermal plates on females with a canal for receiving sperm (Long et al. 2015). Australian paleontologists recently discovered an iconic fossil showing the mineralized remains of a placoderm embryo inside the mother, making the placoderm the oldest known vertebrate to give live birth (Long et al. 2008).

ARRIVAL OF THE AMNIOTES

Amphibians reproduce through both external fertilization (as in frogs), in which fertilization occurs in water, and internal fertilization (as in salamanders and caecilians), in which the fertilized egg is placed in water. In both instances, the fertilized egg develops into a free-swimming larva with gills and fins. The larva matures by shedding its gills and fins, and developing lungs and limbs, as it adapts to life on land. Even in internal fertilization, this reproductive cycle is highly dependent on water, since the larval stage is fully aquatic. As tetrapods moved from moist to drier terrestrial environments, new methods were needed to protect and nourish the developing embryo. Evolving tetrapods achieved independence from water with a number of innovations, but none was more important than the **amniotic egg** (figure 11.1). Appropriately, these new tetrapods were called **amniotes** and today include reptiles, birds, and mammals. The name amniote comes from the amniotic sac, which is located inside the hard shell and serves to provide a protective environment for the developing embryo.

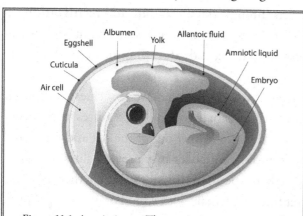

Figure 11.1: Amniotic egg. The amniotic egg protects and nourishes the developing embryo of reptiles and birds and allows the embryo to develop in a hostile terrestrial environment. The amnion and amniotic fluid surround the embryo and provide a moist environment and buffer the embryo from trauma. The yolk provides nutrients. The allantoic fluid and chorion (not shown) transport oxygen through the shell to the embryo and transport metabolic waste from the embryo out through the shell.

Instead of placing a "naked egg" in water, amniotes evolved a hard-shelled amniotic egg that could protect and nourish the developing embryo and could be placed on land. Unlike amphibians, reptiles and birds no longer needed a larval stage. Embryonic development could happen safely within the shelter of the hard shell. The amniotic egg provided nutrition, moisture, and resources for the developing embryo to "breathe," cope with waste, and withstand trauma. When the time was right, the young **amniotes** were ready to hatch and leave their protective environment and face their new terrestrial world.

The Evolution of Placentals

Mammals evolved from mammal-like reptiles (therapsids) at the end of the Triassic period about two hundred million years ago. At that point, they were still laying eggs. This changed during the Jurassic period. A team of scientists at the Carnegie Museum of Natural History described a 160-million-year-old fossil of a small shrewlike mammal that is the earliest known placental mammal—a type of mammal that, like us, gives live birth (Luo et al. 2011). This is the approximate time when the placental mammals (**Eutheria**) diverged from egg-laying mammals (**monotremes**) and mammals who shelter their embryos in pouches (**marsupials**).

Egg-laying mammals and marsupials followed a separate evolutionary path when they were isolated as Australia separated from the southern supercontinent of Gondwana about 160 million years ago. Egg-laying mammals have largely gone extinct, except for modern-day monotremes, which include the platypus and spiny anteaters (echidnas), both of which are endemic to Australia and New Guinea (image 11.2). Marsupials give birth to an immature embryo that finishes development in the mother's pouch. Current-day marsupials include opossums, which are endemic to the Americas, and kangaroos and koalas, both of which are endemic to Australia.

Placental mammals are so named because they have evolved a **placenta**, a unique organ that develops during pregnancy and sits at the interface between the embryo and the uterus. Through the placental attachment, the mother's circulation mixes with the embryo's circulation and supplies the embryo with nutrients and oxygen and disposes of waste. This

Image 11.2: Spiny anteater (echidna). This spiny anteater and the platypus, both of which are endemic to Australia and New Guinea, are the only living monotremes. Monotremes are primitive mammals that have maintained their ancient evolutionary heritage and still lay eggs.

allows the fetus to remain for a long period of time inside the mother, where it is protected from the environment and from predators and can reach an advanced level of maturity and some degree of self-sufficiency by the time of its live birth. The advantages of placental development and live birth have given mammals an adaptive edge over their competition, but there is a downside. The nutritional requirements of the embryo and long gestational periods are very onerous for the mother.

VARIETIES OF SEXUALITY

Image 11.3: Komodo dragon. These monstrous lizards, endemic to Indonesia, reproduce asexually by a process known as parthenogenesis. The mother produces, in a "virgin birth," an unfertilized egg that can mature into an adult male Komodo.

Komodo dragons, native to a few small islands of Indonesia, are monstrous lizards that can reach ten feet in length and sometimes weigh more than three hundred pounds (image 11.3). These exotic creatures will pounce on almost anything: deer, pigs, and even large water buffalo and humans. When a victim ambles by, the dragon springs, using its powerful legs, sharp claws, and serrated sharklike teeth to eviscerate its prey. If the victim survives the initial onslaught, it will succumb to toxic venom that is released into the wounds and induces delayed bleeding and shock.

Komodos are capable of reproducing asexually through an unconventional process known as **parthenogenesis**. Female dragons produce diploid eggs that are capable of developing into embryos without fertilization by a male dragon. However, because of an unusual type of meiosis and duplication, they are not clones of their mother, and all offspring are male (Watts et al. 2006). "Virgin births" through parthenogenesis occur naturally in many plants, in a number of invertebrates, and in several vertebrates, including hammerhead sharks, Amazon mollies, boa constrictors, turkeys, and some reptiles, but they have not been reported in mammals. A number of species that typically use sexual reproduction switch to parthenogenesis when there are no available mates or during conditions that favor rapid population growth.

In the lagoons and bays of Southeast Asia, Persian carpet flatworms attract the attention of local divers with their rich pink stripes and yellow spots as they undulate through the shallow water. Each of these frog-sized **hermaphrodites** has both male and female sex organs. During mating, two Persian flatworms face off like fencers, spraying one another with their two swordlike penises in an attempt to inseminate their opponent (mate) with sperm. The sperm can penetrate the skin and stream through the partner's body, eventually fertilizing eggs nestled in the ovaries. Both worms are capable of becoming pregnant. For these hermaphrodites, every worm is a potential sexual partner, and when times are tough and no mates are available, they can self-fertilize. No need for dating sites here.

We think of separate sexes in separate bodies as a norm, but hermaphrodites are quite common in nature. Most plants and many marine invertebrates—barnacles, snails, starfish, and worms—are hermaphrodites, and about a quarter of the fish we might see when snorkeling in the Caribbean are hermaphrodites. A much smaller number of vertebrates are hermaphrodites.

Because they produce both sperm and eggs, hermaphrodites are capable of self-fertilization, and this may be the best means of reproduction in places where mates are scarce or the transfer of sperm to eggs is difficult. Plants are a good example. Rooted to a spot, plants need help to reproduce sexually. They receive such help in the form of pollinators—insects and birds that transport pollen from a male to a female plant. Since pollinators are not always accessible, the ability to self-pollinate may provide opportunities for reproduction that might not otherwise be available. The majority of flowering plants are hermaphrodites, housing both male and female sex organs (stamen and pistil) in the same plant, and sometimes in the same flower, and this provides opportunities for self-pollination when pollinators are not available (image 11.4). Of course, the downside of self-pollination is a loss of genetic diversity, and hermaphrodites will generally revert back to cross-fertilization when partners are available and when transfer of sperm is feasible.

Hermaphrodites and animals that reproduce through parthenogenesis give us some perspective on the variety of sexuality in the animal kingdom, so it may not be surprising to learn that **homosexuality** and **bisexuality**, in various forms, are also found throughout the animal kingdom and are not rare. Bisexual and homosexual behavior has been documented in nature in close to fifteen hundred species, ranging from primates to gut worms (Jannini et al. 2010). A small percentage of domesticated male sheep (rams), for example, exhibit sexual behavior exclusively with other rams. Homosexuality has been reported in all the great apes. Pigmy chimpanzees (bonobos) are bisexual, notoriously engaging in both overt heterosexual and homosexual behavior. And, of course, we have all seen male dogs mount other male dogs. Perhaps appreciation of the broad sexual diversity in the animal kingdom will make us more tolerant of diverse sexual behavior within our own species.

Image 11.4: Lily. Lilies, like most flowering plants, are hermaphrodites, harboring both male and female sex organs. The male sex organs (stamens) with dark tips surround the taller central female sex organ (the pistil). Stamens produce pollen, which contains sperm, while pistils produce ova.

But doesn't homosexuality seem to contradict Darwinian evolution? Homosexuality does not appear to promote reproductive success. Perhaps not, but there may be a link between homosexuality and other traits that do favor reproductive success. Mothers and aunts of homosexual men have increased fertility rates, suggesting that genes responsible for increased fertility in women may be linked to genes that provide predisposition to homosexuality in male offspring (Camperio et al. 2012). Such genes would be selected if increased reproductive success in women with these genes outweighs decreased reproductive success in their male offspring.

SEXUAL SELECTION

As we discussed in chapter 2, Darwin was initially puzzled by the peacock's extravagant eye-spotted tail, which clearly makes him an easy target for predators (image 11.5). But as he watched the peacock unfurl his tail into a psychedelic-looking curtain and position himself to surround the peahen, Darwin realized the peahen was enamored and strongly attracted by the tail and the ritual. If natural selection were working on its own, peacocks with iridescent tails would have all been eaten long ago—but females might actually favor males with sparkling tails because they signal that the male is robust and nimble. Peacocks with flamboyant tails would be able to attract a mate and have a reproductive advantage.

Based on these observations, Darwin modified his theories of evolution in his second book, *The Descent of Man, and Selection in Relation to Sex*, in which he described a process he called **sexual selection** (Darwin 1871). Just as natural selection favors any heritable trait that increases fitness and survival, sexual selection favors any trait that enhances reproductive success. Reproductive success, along with survival, is a prerequisite for passing genes to the next generation; and sexual selection, like natural selection, guides the course of evolution.

In Darwin's description, animals can achieve reproductive success in one of two ways. Animals, usually males, can make themselves very attractive to the opposite sex. Over millions of years—and millions of generations in which both males and females have chosen mates who seem the most attractive to them—an immense diversity of

Image 11.5: Peacock's tail. The extravagant eye-spotted tail of the peacock is a classic example of a secondary sexual characteristic due to sexual selection. The eye-spotted tail was selected and evolved because it helped the peacock attract the peahen and promoted reproductive success.

secondary sexual characteristics has emerged within species. The peacock's tail is but one example. There are so many more: the manes of male lions, the red-and-blue face and blue buttocks of the male West African baboon, the white legs and white rump of the Banteng bull of Malaysia, the spectacular antlers of the male red deer, and the elaborate feathers, bright colors, and dancing rituals of many birds.

The second way males achieve reproductive success is by making themselves strong and dominant so they can intimidate and defeat same-sex rivals and impose their will over females. The colossal male southern elephant seal found on the beaches of the Falkland Islands and Valdes Peninsula in Argentina provides a striking example. These bull seals are the largest carnivores on our planet, weighing over three tons and reaching lengths of fourteen to sixteen feet—a size more than six times that of females. The dominant bulls, or "harem masters," establish harems of several dozen females with which they alone can mate. Species, such as the elephant seal, that show great physical differences between males and females are said to be highly sexually dimorphic. The striking sexual dimorphism of the elephant seal exceeds that of all other mammals.

When males evolve secondary sexual characteristics that make them attractive to the female, the female has choice and is "in charge," so to speak, but when males evolve secondary sexual characteristics that make them dominant, females are dependent and subjected. We like to think that human sexual selection operates in the first mode, giving the female choice, but this is clearly not always the case. Most advanced Western cultures promote, with varied success, equal rights and opportunities for women, and free choice of mates, but many societies promote and sustain male supremacy, and place women in dependent roles. Biological evolution does not promise justice, morality, or equal opportunity for both sexes. As we will see later in our story, we may soon be in control of our own evolution, transcending natural selection, and may assume responsibility for gender equality.

EVOLUTION OF ROMANTIC LOVE

Charles Darwin proposed that emotions evolved because they were adaptive and helped animals and humans to survive and reproduce (Darwin 1873; Jabr 2010). Feelings of erotic love help humans—and perhaps animals, in a slightly different way—to seek mates and help to facilitate reproduction. While *erotic love* evolved to promote reproduction, *romantic love*, the long-lasting relationship between two people (and in some instances between animal pairs) whose lives are deeply intertwined and who trust and care for one another, must have evolved for some other purpose. Psychologists believe human romantic love evolved as a form of pair-bonding to aid in child rearing (Fletcher et al. 2014). Compare the situation with our closest relatives, the chimpanzees. Chimpanzee males are sexually promiscuous, are often aggressive and sexually abusive toward females, do not engage in long-term pair-bonding with their mates, and contribute little to child rearing. Why are we so different from our cousins?

As we evolved from our hominin ancestors, our brains evolved and grew larger, leading to our dominance over other species on our planet. But as our brains and skulls grew larger, our head size reached a point where it could no longer pass through the birth canal. The only solution to this problem was "premature" birth. So evolution, in its wisdom, arranged for human babies to be born before their brains are fully developed. The brains of newborns increase in size more than 60 percent in the first three months after birth and continue to grow more slowly for another twenty years (Holland et al. 2014).

Because the human brain is so primitive at birth, the young are helpless for years and are heavily dependent on adults for more than a decade. Is your ten-year-old ready to face the world? No other species' newborns are so helpless. This creates a burden of childcare that is too onerous for the mother alone and requires the help of the father. But if the father is to help with child rearing, Darwinian doctrine requires that he be sure the kids are his own and that he is not helping propagate the genes of a rival. He will be faithful to the mother of his kids, but he requires that she be faithful to him as well. To foster pair-bonding between the sexes, nature reduced the size discrepancy between males and females and eliminated obvious signs of ovulation, permitting the female to be receptive throughout the reproductive cycle, increasing the bond between the two partners, and reassuring the male that any offspring are his own. And natural selection reduced (but did not eliminate) man's tendency for sexual promiscuity by reducing testosterone levels in monogamous relationships (Maestripieri 2012). Since a well-developed brain has such enormous evolutionary advantages, big-headed babies, maternal-paternal pair-bonding, cooperative child rearing, and male sexual jealousy all evolved together. Romantic love, which so brightens and gives meaning to our lives, is the legacy of this evolution.

CHAPTER 12

EVOLUTION OF ORGAN SYSTEMS

Without motility, predation as a way of life is barely imaginable. To capture prey and eat it, you must first learn to move, whether like a tiny amoeba, creeping and engulfing, or with the power, speed and grace of a cheetah.

—Nick Lane, in *Life Ascending*

The creatures that suddenly appeared in the Cambrian explosion left few clues as to how they came to be, and there is a large evolutionary gap between the microscopic single-celled organisms that came before and the macroscopic multicellular marine life that seemingly came from out of nowhere. We learned in chapter 9 how single-celled organisms may have come together to form multicellular organisms, and how regulatory genes and transcription factors directed the development of specialized cell types, but until recently knowledge about the development of specialized organ systems has been sparse. The fossil record prior to the Cambrian explosion is very meager and has provided little help.

But now, hundreds of millions of years after these first animals emerged, new technological developments are closing the gaps in our knowledge. Genetic comparisons between modern species allow for the construction of evolutionary trees that reach far back in time. Primitive current-day animals that may have changed little from their ancient ancestors provide additional understanding about transitional life forms. Using this evidence from the living world around us today, we can gain educated insights as to how, from an ocean of single cells, these creatures and their organs came to exist.

To begin, let's take a look at our closest unicellular relatives—the cells that look the most like what our ancestors probably looked like just before they began building bodies.

OUR UNICELLULAR COUSINS

Today, our rivers, lakes, and seas teem with **choanoflagellates**, microscopic single-celled predators that play a critical role in the ocean's food chain. They propel themselves through the water by wiggling a long tail or flagellum and capture bacteria and food particles by filtering the water through a collar of fine filaments called microvilli. These microscopic nanoplankton are where the food chain begins.

Choanoflagellates are our cousins. They are the closest living single-celled relative to all living animals (Carroll 2010). Genetic comparisons have told us that an ancient unicellular ancestor of choanoflagellates gave rise to the lineage leading to the entire animal kingdom (figure 12.1). Choanoflagellates are most closely related to sponges, the oldest and most primitive of all animals, and the similarities between the two are striking.

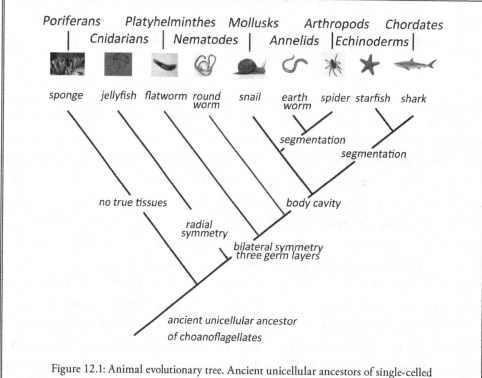

Figure 12.1: Animal evolutionary tree. Ancient unicellular ancestors of single-celled choanoflagellates gave rise to the entire animal kingdom. Poriferans (sponges) branched off the animal tree first and compose the most primitive animal phylum, with no well-defined body plan. Cnidarians (including jellyfish and hydra) branched off next and evolved a body plan with radial symmetry. Subsequent phyla developed three germ layers and body plans with bilateral symmetry.

Sponges are the simplest of animals (image 12.1). They contain no distinct tissue types, as later animals would, but rather consist of amorphous masses of living flesh. They are often brightly colored spectacular specimens, found at rocky ocean shores, on coral reefs, and attached to submerged wood in lakes and streams. You might not recognize them as animals—they look more like plants—but they belong to the animal phylum Porifera (meaning "pore-bearing"), named after the deep pores which cover their surface. The pores lead into a complex network of internal canals lined with feeding cells called choanocytes, which beat their flagella to circulate water and nutrients through the canals and out central openings called oscula at the tip of the sponge.

Image 12.1: Purple stovepipe sponges. Sponges (poriferans) are the simplest form of animal and contain no distinct tissue types.

The choanocytes that line the canals of sponges bear a striking resemblance to choanoflagellates. They both have a characteristic collar of microvilli, which they use to filter water and gather food. It is not hard to imagine how the ancient ancestors of choanoflagellates may have joined together, first in colonies and then in multicellular organisms, and evolved to produce the ancestors of modern-day sponges. Genetic comparisons using molecular clock analyses have shown that current-day choanoflagellates and sponges last shared a unicellular common ancestor in the late Precambrian period, more than six hundred million years ago (see sidebar "Molecular Clocks"). Molecular biologists have identified genes that code for proteins shared by both choanoflagellates and animals that make cells stick to one another (King et al. 2008). Why a single-celled organism like the choanoflagellate would produce proteins that cause cells to stick to one another is not clear—perhaps it is to facilitate the capture of bacterial prey—but the implication is that these proteins may have been used to facilitate the evolution from single-celled ancestors of choanoflagellates to spongelike

MOLECULAR CLOCKS

Comparison of DNA nucleotide codes between species has been used to estimate the relationship between species and to construct phylogenetic trees. DNA comparisons between species have also been used to estimate the time elapsed since species diverged from one another on the evolutionary tree. DNA serves as a molecular clock. It doesn't tick and doesn't tell the time of day, but it does tell time on an epochal scale.

When two species diverge from one another, their DNA codes change over generations as a result of random mutations—the more generations, the greater the change. If a section of DNA in two species differs by forty-two nucleotide letter codes, and if the DNA codes change by one letter every five million generations as a result of mutations, and if the generation times for the two species are one and ten years, then with a little algebra we can calculate the two species diverged 191 million years ago. The rate of change of DNA codes per generation is determined from fossil records which provide the time, and hence the number of generations, since two species diverged. Once the mutation rate is determined, it can be used for other comparisons in which fossil records are not available.

multicellular organisms. Like the common genes we saw with unicellular Chlamydomonas and multicellular Volvox in chapter 8, this is yet another example of how preexisting genes and their associated proteins may be used to serve entirely new purposes—happy accidents that increase the flexibility of life.

LEARNING TO EAT

As single-celled organisms evolved into multicellular creatures, cells differentiated to perform specialized functions, and organs evolved to execute specific jobs for the body. One of the first things multicellular animals needed to survive was a digestive system. Single-celled microorganisms feed themselves simply by engulfing molecules of organic matter through **cytosis** and breaking down the molecules with digestive enzymes into nourishing particles. But a multicellular organism needs a way to access, process, and deliver nutrients to its specialized cells, most of which are located inside of the organism, without easy access to the outside world and the nourishment there.

Sponges, the most primitive modern animals, provide an example of how the earliest body-feeding systems may have worked. Water containing food particles enters the sponge through pores on its surface and flows through internal canals within the sponge. The food particles are engulfed and digested by choanocytes, which line the canals, and digested food simply diffuses passively out of the choanocytes into adjacent cells within the sponge. This is possible because sponges have only two layers of cells, with a soft jellylike layer in between.

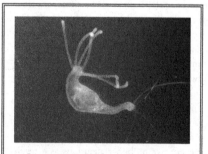

Image 12.2: Hydra with ingested copepod. This tiny freshwater animal has an ingested copepod (small crustacean) visible in its translucent gastric pouch. Hydras have the simplest form of digestive system, consisting of a mouth leading to a blind pouch.

Hydra is a genus of tiny aquatic animals found in ponds, lakes, and streams that are best known for their regenerative ability; they do not appear to age (image 12.2). (We will return to this intriguing detail in chapter 24.) Hydras are members the Cnidarian phylum (the second oldest animal phylum, which includes jellyfish and corals) and may provide insights into early feeding systems. Hydras have a simple mouth leading to a blind pouch. The mouth is surrounded by thin tentacles, each of which contains specialized cells, called cnidocytes, that produce neurotoxins. Hydras capture and paralyze their prey with their tentacles and pull them into their pouch, which is lined with differentiated cells that exude digestive enzymes to dissolve the prey. The organic products are absorbed and diffuse passively to other cells in the body, and after two to three days, the undigested remains are discharged back through the mouth.

Hydras provide an example of external digestion—digestion by enzymes released external to cells. But the hydra's digestive system is very primitive. If we humans had been left with a stomach like the hydra's blind gastric pouch, we might complain. Hydras have that single opening for food to enter and waste to exit: a combined mouth and anus—a truly unappetizing thought. We should be grateful for the evolutionary innovation in digestion that came next in the ancestors of roundworms.

Roundworms today are found almost everywhere—from the poles to the tropics, in oceans and freshwater lakes, and in most terrestrial environments. The roundworms that are best known to us are disease-causing parasites, including those of the genus *Ascaris*, which cause **ascariasis,** and those of the genus *Trichinella,* which cause trichinosis. Roundworms, unlike the more primitive hydra, have an intestinal tract with two openings: a mouth where food enters, and an anus where waste exits. Roundworms' ancient ancestors may have been the first animals to acquire this one-way thoroughfare. The roundworm's circular mouth—which may contain frightening, vampire-like fangs—opens into a pharynx with muscular walls that help it suck down food. Their digestive tracts are lined with glands that ooze food-dissolving enzymes. Their intestines, unlike ours, have no muscles for peristalsis. Food and waste are propelled down the digestive tract by the worm's bodily movement. As in sponges and hydras, nutrients simply diffuse from the digestive tract into the rest of the roundworm's cells. There is no developed circulatory system. With the development of this one-way digestive system, animals were well on their way to developing guts like ours.

LEARNING TO BREATHE

Our simple single-celled ancestors obtained oxygen to run their aerobic metabolisms simply by absorbing it through their cell membranes. As their descendants evolved into multicellular organisms, and as those multicellular organisms grew bigger and bigger, the inside cells got farther and farther away from the outside world and no longer had access to environmental oxygen. With cells no longer able to "breathe" on their own, the organism had to find ways to get oxygen to its cells.

The flatworms had the simplest solution to this problem. Flatworms belong to the animal phylum Platyhelminthes, and their ancestors branched off early in the evolutionary tree of animals—after sponges and cnidarians, but before roundworms (figure 12.1). Like roundworms, the flatworms that are best known to us are those that cause us trouble, such as tapeworms, flukes, and blood flukes. Because the flatworm is flat, all of its cells are near the surface and each cell can obtain oxygen for itself from the outside world. A flatworm has no need for a dedicated circulatory or respiratory system. Everything necessary diffuses from the external environment into the cells.

In order for early marine animals to get thicker and larger, they had to devise new ways to get oxygen to their cells. Octopuses (members of the Mollusk phylum) suck water into their mouths and across gills in their mantle cavities. The gills themselves are feathery filaments of tissue that provide a large surface area for gas exchange. As ocean water flows by, oxygen passively diffuses from the water into capillaries within the octopus's gills and is delivered by the blood to the first of its three hearts before it is pumped to the body tissues. The mollusk's respiratory system is not very efficient, both because of the relatively low concentration of oxygen in the oceans and because of an inefficient circulatory system for delivering oxygen. As a consequence, mollusks have limited stamina and a fairly sedentary lifestyle.

Animal phyla that branched off the evolutionary tree after mollusks—including the tens of thousands of fishes in the Chordate phylum—have evolved improved systems for oxygenation. Gills evolved with greater surface area and thinner membranes to facilitate gas transfer (Morris and Caron 2014). Fish also developed counter-current systems, in which the blood in the gills travels in the opposite direction of ocean water flowing across the gills, allowing greater transfer of oxygen. And as we will see, the evolution of closed circulation systems with blood and hemoglobin let these early marine animals pack their muscles with greater quantities of oxygen. They got bigger and faster, and their muscles and tissues were better able to do their jobs with increased delivery of oxygen.

DELIVERING THE ESSENTIALS

All of the rather primitive animals we have discussed so far in this chapter—sponges, hydras, flatworms, and roundworms—get their oxygen and nutrients directly from the environment using passive cellular diffusion. None of them have evolved circulatory systems. Larger animals that came later needed something more sophisticated, and all animal phyla that branched off the evolutionary tree later than the roundworms, beginning with mollusks, have evolved circulatory systems. While digestive systems digest and absorb nutrients, and respiratory systems take in oxygen from the environment, circulatory systems accomplish the task of getting the nutrients and oxygen to the cells that need them.

The simplest circulatory systems in animals today are **open circulatory systems**, which are found in many invertebrates—arthropods (insects, spiders, crabs, shrimp, and lobsters) and some mollusks (clams, snails, and slugs). These animals have a primitive pumping heart and a limited network of vessels that carry, rather than blood, a surrogate liquid called **hemolymph** to the body tissues. Hemolymph is pumped out of blood vessels and into open body cavities—hemocoels or sinuses, which are the defining feature of open circulatory systems. The hemolymph bathes organs and tissues in a wash of oxygen and nutrients and then drains passively back through the gills, where it picks up a new load of oxygen and returns to the heart.

Hemolymph is not very efficient in transporting oxygen. **Hemocyanin,** a molecular analog to our **hemoglobin** and one of the earliest oxygen-binding proteins, evolved in many invertebrates and greatly enhanced oxygen transport by hemolymph. While hemoglobin is iron based and turns bloodred when bound to oxygen, hemocyanin contains copper atoms and turns hemolymph blue when bound to oxygen.

In many animals with open circulatory systems, the heart is a simple large blood vessel that pumps hemolymph with peristaltic contractions and uses valves to ensure one-way flow. In some animals, the heart has developed further, becoming chambered like our own—with small atria into which the hemolymph drains, and larger ventricles that serve as major pumps to drive hemolymph through the body.

Open circulatory systems have worked successfully for millions of years in these species, but as bodies became larger, a greater efficiency was needed. All vertebrates, as well as annelids (earthworms) and some mollusks (octopuses and squids) have **closed circulatory systems.** Blood is pumped through an intricate network of vessels—arteries, capillaries and veins—and never leaves the circulatory system. Hearts in these closed systems vary between species, from simple contracting blood vessels in earthworms, to two-chambered hearts in fish, to four-chambered hearts like ours in mammals and birds. All of these hearts pump blood first through an oxygenating apparatus (lungs or gills) and then to the rest of the body. We vertebrates have evolved iron-based hemoglobin, which replaced the less efficient hemocyanin, to bind and transport oxygen. We improved oxygen transport further by creating red blood cells to house our oxygen-carrying proteins, which allow us to have higher concentrations of hemoglobin without crowding our blood vessels with sludgy free-floating molecules.

Closed circulatory systems operate at much higher pressures and are able to circulate oxygen and nutrients faster than open systems. Blood coursing through a river of arteries moves much faster than hemolymph sloshing across a lobster's wide-open hemocoels. A closed system can also direct more blood to specific organs in their moments of need—to leg muscles, for instance, when they need oxygen to run. Open circulatory systems work perfectly well in many circumstances, and so they have persisted, but closed circulation is evolution's answer to the problem of a larger, faster animal. A speeding cheetah needs oxygen and nutrients delivered rapidly to its front and hind limbs; its powerful heart and miles of blood vessels can get that job done.

LIFE IN MOTION

All living things must move to stay alive. Even bacteria need to move, both internally and externally. They move to grow, divide, feed, and live. For animals, movement is critical for locating food, finding a mate, searching for suitable habitats, and escaping predators. Enhanced mobility confers great evolutionary advantages.

Modern creatures move however they can. The tiniest of living things—bacteria, unicellular algae, and protozoa—dart to and fro across the surfaces of oceans and lakes, powered by the rhythmic beating of flagella and cilia. Jellyfish, with their umbrella-shaped bodies, propel themselves by contracting muscle-like cells to expel water out behind them—the cnidarian version of jet propulsion. Flatworms glide along the ocean floor, propelled by layers of cilia on their bellies stroking over the stream of mucus they lay down to smooth the way. Earthworms inch along by alternating peristaltic contractions of circular and longitudinal muscles. Crabs and lobsters bottom-walk on multijointed limbs, and eels swim by undulating their flexible bodies, passing a series of waves from head to tail.

The marine animals of the Cambrian period likely used the same methods of locomotion as their modern-day descendants, but the most remarkable feature of all these methods of movement are the muscles that power them. How did these muscles evolve?

FROM CHEMISTRY TO MOTION

In 1954, two scientists, both named Huxley but unrelated and working independently, published landmark papers that first described how muscle cells work (Huxley and Hanson 1954; Huxley and Niedergerke 1954). Using newly developed microscopy techniques, they found that during muscle contraction, thin protein filaments of muscle cells (**actin**) overlapped thicker filaments (**myosin**), resulting in shortening of the muscle (figure 12.2). From this they deduced the sliding filament theory of muscle contraction, which has persisted to the present day. The whole process feels a bit like science fiction. Molecular appendages of myosin filaments, called myosin heads, hook onto actin filaments, forming cross bridges. A change in molecular configuration of the myosin head, driven by energy from ATP, causes the filaments to slide over one another, shortening the muscle units (**sarcomeres**) and causing the muscle fiber to contract. With this process, nature has created a molecular machine that converts chemical energy into mechanical energy.

The actin and myosin proteins in the skeletal muscle of vertebrates and invertebrates are very similar, and it is clear they were inherited from a common ancestor. But actin and myosin go back much further than this. In all eukaryotic cells, actin and myosin are integral parts of the cytoskeleton—the network of protein filaments and tubules in the cytoplasm that determine the cell's shape and provide a means of transport within the cell. The structure of actin and myosin in skeletal muscle is so similar to actin and myosin in the cytoskeleton that there is no doubt our skeletal muscle proteins evolved from the cytoskeleton of an ancient common eukaryote ancestor (Hartman and Spudich 2012; Dominguez and Holmes 2011).

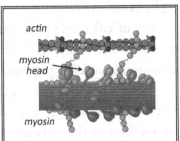

The actin and myosin that make up much of the cytoskeleton of eukaryotic cells are responsible not only for maintenance of the cell's shape but also for movement inside the complex cell. As we saw in chapter 7, molecules are shuttled from place to place within the cell along the course of the cytoskeleton—like a sky rail system in a microscopic city. Actin filaments serve as the rails, and myosin and other motor proteins function as the trains that move along the rails, carrying bundles of proteins as they go. The metaphor collapses when one considers how the myosin actually moves. It doesn't roll like a train on a set of tiny wheels, but rather walks along the actin filaments on two protein legs, powered energetically by ATP. Protein cargo and other molecules, attached to the back of the myosin motors, are transported along the actin rails to where they are needed. This sounds more like futuristic science fiction than reality.

Figure 12.2: Muscle contraction. Muscle contraction occurs by a unique and extraordinary process in which chemical energy is converted to mechanical energy. Actin and myosin protein filaments are essential components of the contractile elements (sarcomeres) of muscle fibers. Appendages of myosin filaments, called myosin heads, are activated ("cocked") using energy from ATP to change the configuration of the molecule. Once activated, the myosin heads bind to actin filaments and, on a signal from a nerve impulse, are deactivated ("uncocked"), pulling the actin filaments over the myosin filaments. This results in shortening of the sarcomeres and shortening of the muscle fibers, causing them to contract. This is one of many examples of molecular machines at work in our bodies.

The cytoskeleton provides for other internal movements within the cell. It is responsible for the separation and clumping of chromosomes during mitosis; for pulling the cell membrane inward at the waist, as if squeezed by a corset, in the process of cell division; and for creating invaginations in the cell membrane to facilitate phagocytosis. These are the most basic motions that enable life on Earth. All biological processes are dependent upon movement generated by this conversion of chemical energy to mechanical energy.

Until recently, it was thought that only eukaryotic cells had cytoskeletons—but recently researchers have found that prokaryotic cells (bacteria and archaea) have cytoskeletal structures of their own (Wickstead and Gull 2011). The bacterial cytoskeleton provides similar functions to that of eukaryotes. Bacterial cytoskeletons are composed of thin protein fibers that are analogous to actin, but their amino acid sequences are not the same, and it was initially thought that the two proteins were completely unrelated. However, even though the molecular composition of the prokaryotic proteins is not the same as actin, their structure—the intricate folds, bends, and contortions that so often define a protein's function—are nearly identical to that of eukaryote actin (van den Ent, Amos, and Lowe 2001). Because of this essential configurational similarity, we can bet that these proteins are homologous, meaning that the genes responsible for producing them evolved from a common ancestor.

That ancestor is LUCA, the last universal common ancestor of all life. LUCA, it appears, had a cytoskeleton. It must have, because all living things must have a cytoskeleton. The cytoskeleton is the cell's source of all movement, and movement is essential to life. Without movement, replicators could not replicate, cells could not divide, proteins and cargo could not be transported, and cells could not capture food. The cytoskeletons present in all the kingdoms of life today—bacteria, archaea, protists, plants, fungi, and animals—were inherited from our last universal common ancestor and are modified to meet our needs. We animals borrowed actin and myosin from the cytoskeleton to build the strong muscles that power our movement. The graceful stride of the cheetah, the soaring flight of the falcon, the undulating speed of the marlin—are all a product of this inheritance. Muscles, and the locomotion they enable, allowed animals to achieve roaring dominion over the Earth.

CHAPTER 13

EVOLUTION OF THE EYE

> It seems that life, at least as we know it on this planet, is almost indecently eager to evolve eyes … And if there is life on other planets around the universe, it is a good bet that there will also be eyes, based on the same range of optical principles as we know on this planet. There are only so many ways to make an eye, and life as we know it may well have found them all.
>
> —Richard Dawkins, in *The Ancestor's Tale*

All living things must be able to detect and respond to changes in their environment in order to survive. We gather information from our environment through our sensory organs using our five senses of sight, hearing, smell, taste, and touch. Other animals, with their own fine-tuned senses, detect signals that we can only imagine. Pigeons feel the Earth's magnetic pulls, insects see ultraviolet light invisible to us, bats see with sound, and sharks sense electrical impulses. Every species has evolved sensory organs allowing it to best adapt to its environment.

Our planet is bathed in light. There is no other stimulus that provides as much information about our environment as light does. The eye, in its role of capturing light, provides our major link to the external world and has become our most important sensory organ. Because the eye has central importance to our being, and because the evolution of the eye has been studied so extensively, it serves as an instructive example for the evolution of our sensory organs.

THE SENSITIVE EYE

The human eye is a magnificent organ that has often been compared to a camera (image 13.1 and figure 13.1). The retina records and captures incoming light like a camera's sensor. It contains millions of **rods** and **cones**—light-sensitive cells that sense incoming light and transmit signals via nerve cells to the brain. The rods and cones are analogous to pixels on the camera's sensor. Rods are sensitive to low light but cannot distinguish color and are used primarily for night vision, while cones are active at a higher intensity of light and can distinguish fine details and a spectrum of colors. The rods and cones utilize photosensitive pigments—**rhodopsins** in rods and **photopsins** in cones—to absorb light. Each rhodopsin and photopsin molecule is composed of a light-sensitive pigment called retinal and an opsin protein. Retinal is a derivative of vitamin A, and opsins are derivatives of proteins used as enzymes to catalyze several biochemical reactions—more examples of the use of existing proteins in the evolution of new structures and new functions.

Each animal species has evolved specific photosensitive pigments that absorb frequencies of light to meet its own needs. Fish have photopsins that sense only blue and green light, because other color frequencies are absorbed by water and are minimally present in deep aquatic environments. We primates have tricolor vision, with three distinct types of cone cells sensitive to red, green, and blue light. Our high color sensitivity may have helped us to distinguish fruits from surrounding foliage when foraging. Nocturnal mammals have little color vision but have well-developed rods, which are able to discern detail in very low light. Conversely, some birds, fish, reptiles, and insects are tetrachromats—they have not three, but *four* types of cones dedicated to seeing differing wavelengths of light—and can therefore discern colors that we cannot. Typically, these animals' fourth cone is dedicated to a shorter wavelength of light, allowing them to see ultraviolet light that is invisible to us humans.

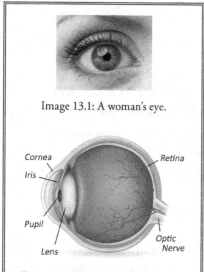

Image 13.1: A woman's eye.

Figure 13.1: Anatomy of the human eye. Light enters the human eye through the cornea, passes through the pupil, is refracted by the lens, and strikes photosensitive cells, rods and cones, in the retina. The rods and cones capture light with photosensitive pigments and transmit signals via the optic nerve to the brain.

STEPWISE EVOLUTION OF THE EYE

Critics of evolutionary theory have argued that if evolution occurs by gradual increments, it could not have created a retina, a lens, an iris, and a cornea, since none of these structures provides any function by itself and therefore could not have developed incrementally over time. How could half an eye evolve, since half an eye has no function and provides no survival advantage? Such a perfect organ as the eye, they argue, must be the product of intelligent design. Darwin himself admitted that the idea of the human eye evolving without a designer seemed "absurd in the highest possible degree," but he later went on to explain how it could be done (Darwin 1859). Let's see how the eye may have evolved in incremental steps slowly over time, and how each step was potentially useful in and of itself in ensuring survival.

The very earliest light-sensing organs—the primitive precursors to eyes—were likely similar to the eyespots found on our unicellular friends the green algae of the genus *Chlamydomonas,* which we met in chapter 9 (Lamb, Collin, and Pugh 2007). Their eyespots consist of light-sensitive pigments embedded in chloroplast membranes within the cell. Light striking the pigments triggers the flagella to beat, and this propels the organism toward the light. For photosynthesizing algae, this ability to sense and respond to light has obvious survival advantages.

The pigments found in eyespots of algae of the genus *Chlamydomonas* are rhodopsin pigments, which are very similar to rhodopsin pigments found in rod cells of human eyes. In fact, rhodopsin pigments are present in all organisms with light-sensitive structures—from microbes to vertebrates. There is little doubt that rhodopsin pigments present in all current-day organisms are homologous—that is, they all evolved from a single common ancestor that already possessed an early form of this photosensitive pigment (Shen et al. 2013).

Flatworms provide clues to the next step in the evolution of the eye. Flatworms have two simple eyespots, each consisting of a single sheet of photosensitive cells layered in a concave cup. The eyespots have no lenses and cannot create an image; they can only detect the presence of light. But the concavity is a new innovation; it allows for the creation of a shadow so the direction of light can be determined. The photosensitive cells contain rhodopsin-like pigments that react to light and send signals to a primitive nervous system, allowing the flatworm to move away from light and into dark hiding spots to avoid predators.

The chambered nautilus is an exotic living example of one of the next transitional stages in the evolution of the eye (image 13.2). The nautilus is a deep-sea mollusk related to the octopus and squid. It has a beautiful spiral shell for protection, more than ninety tentacles with sticky cilia for gripping prey, and a pinhole eye—an unusual type of eye rarely found in nature. Nautiluses have undergone little

Image 13.2: The chambered nautilus (*Nautilus pompilius*). The chambered nautilus, which is in the Mollusca phylum and closely related to the squid and octopus, has a unique pinhole eye. The pinhole opening allows entering light to create an image on a layer of photosensitive cells at the bottom of a cup, providing a great adaptive advantage. The eye has no lens, and functions like an old pinhole camera.

change in over four hundred million years and are considered a kind of living fossil, along with horseshoe crabs, coelacanths, lungfish, and a few other animals that simply haven't had to change over millions of years to survive.

The Nautilus's pinhole eye likely evolved from cupped eyespots similar to the flatworm's. As the concave cup deepened, tissue grew across the top of it to form a smaller opening. As the opening constricted to a pinhole, light entering through the small aperture created an image on the layer of photosensitive cells at the bottom of the cup. This new primitive eye had become a pinhole camera, capable of discerning not just the intensity and direction of light, but actually forming a complex image. Our earliest cameras were pinhole cameras, which used a pinhole to project and capture an image. It's also similar to something you may have done as a kid—creating a pinhole in a piece of cardboard to watch the solar eclipse. Being able to see an image and identify objects has huge adaptive advantages over merely sensing changes in levels of light. The nautilus's pinhole eyes are not very sensitive, because a pinhole allows very little light to reach the photosensitive cells, but they have served their function and have allowed the nautilus to survive for over four hundred million years.

Evolution found a solution to the limitations of the pinhole eye with the invention of the lens. The clear, flexible lens bends incoming light and allows images of objects at various distances to be focused on the retina. No longer is a pinhole needed. Greater amounts of light can now enter the eye through a much larger hole. Lenses of all vertebrates are made of special proteins known as alpha-crystallins, which have unique properties. They are transparent and able to bend light, and they are one of the most durable proteins known—able to remain clear and functional for decades (until clumping of the proteins finally results in cataracts). Like so many proteins used for new inventions of evolution, alpha-crystallin proteins were co-opted from similar proteins originally used for a different function. They originated as heat-shock proteins that acted as molecular chaperones, protecting and repairing other proteins from heat damage (De Jong, Leunissen, and Voorter 1993; Slingsby, Wistow and Clark 2013). At some point, a mutation must have occurred that caused alpha-crystallins to be produced during embryonic development in the wrong place—on the surface of the eye. Through this accidental placement, they provided an unanticipated advantage by refracting and focusing light onto the eye's photosensitive cells. Thus, the vertebrate lens was born.

Each of these small modifications in the evolution of the eye was selected and passed to the next generation, because each one provided an incremental survival advantage, however slight. More fine-tuning came in time. Eyeballs settled into sockets, with extraocular muscles that grabbed and rotated the eye to focus directionally. The tiny muscles of the iris evolved to dilate and constrict the pupil, allowing more or less light to enter the eye. A transparent anterior covering of the eye, the cornea, was added to assist the lens in bending and focusing light. Eyes corresponding to every transitional stage in the hypothetical sequence of evolution that we have described have been found in existing living species. In general, the simplest eyes exist in animals that branched early off the evolutionary tree, and more complex eyes are seen in vertebrates who diverged more recently.

The iconic trilobites from the Cambrian period 530 million years ago provide us with the earliest fossil record of sophisticated eyes. Trilobites' eyes were compound eyes with many lenses, as is typical of today's arthropods, such as flies and other insects. Most investigators believe this remarkable organ evolved over a relatively short period of time—a few hundred thousand years. Once one predator was able to image its prey, other predators and prey would be at a huge disadvantage. They too would need to develop similar tools to survive, and this would touch off an arms race that would accelerate the pace of evolution. Such an arms race may have been a major catalyst for the Cambrian explosion (Parker 2003). Eyes are surely one of the most remarkable inventions of evolution and likely have played a major role in shaping the natural history of life.

BUILDING AN EYE: THE *PAX6* GENE

Without genes that instruct our bodies on how to build themselves during development, we would be formless. Our complex and delicate eyes, like every other part of us, are built with instructions from regulator genes. Regulator genes, if you recall, are genes that activate or suppress structural genes that are responsible for building specialized parts of our organs. Regulatory genes thus control when organ parts are formed and where they are placed during embryonic development. Since regulatory genes determine the structure and organization of the body, mutations in these genes have become the driving force for evolutionary innovation. However, if the instructions from our regulatory genes are misprinted or mutated in a deleterious way, the building process can go awry.

Aniridia is a rare genetic disorder of the eye that most commonly causes, as the name implies, partial or complete absence of the iris—the colored part of the eye that constricts and dilates the pupil. Aniridia can also cause cataracts, defects in extraocular muscles, and abnormalities of the retina and optic nerve that can impair vision. Aniridia is caused by a rare mutation in the **PAX6 gene**, the regulatory gene responsible for controlling development of the eye. *PAX6* genes have been identified in a broad spectrum of animals ranging from the fruit fly to all vertebrates, and their integrity is essential to flawless development of the eye. In classic experiments in 1994, Swiss biologists inserted *PAX6* genes from mice into the genomes of fruit flies and created bizarre fly monsters with eyes on their legs and antennae (Gehring

2005). The *PAX6* gene is the master regulator of eye development, and defects can bring about strange and devastating consequences.

The *PAX6* gene has a remarkable similarity over a wide range of species, despite large differences in the types of species and the structure of their eyes. In fact, it turns out that regulator proteins (transcription factors) produced by *PAX6* genes in all animals with a right and left side—all creatures with bilateral symmetry, from flatworms to vertebrates—share 90 percent of their amino acid sequences (Friedman 1998). It is nearly impossible that such a wide diversity of organisms could have independently evolved *PAX6*-like genes and their regulator proteins with such similar sequences. The *PAX6* gene must have already been present in the common ancestor of all bilaterians before the Cambrian explosion more than 540 million years ago. In other words, the genetic code that directed the development of the cup-shaped eyespots of flatworms before the Cambrian explosion has persisted to today—into *you*. That code directed the development of your fetal eyes, the fetal eyes of every other animal you've ever met, and even the eyes of every modern flatworm. It's a nice reminder that all of us are distant cousins.

Since the emergence of *PAX6* genes in early bilaterians, the eye has evolved in myriad beautiful directions, giving rise to the stunning diversity in living eyes we see today. The results are truly remarkable. Let's look at one fierce and specialized eye—that of the bald eagle (image 13.3). The bald eagle's wide yellow eye has an especially large retina with a high concentration

Image 13.3: Bald eagle. The bald eagle's wide eyes have a large retina with a high concentration of cones and represent the equivalent of a twenty-megapixel camera, as compared to a four-megapixel camera for human eyes.

of cones—one million cones per square millimeter, compared with the two hundred thousand cones per square millimeter in our eyes. This density of cones provides extraordinary optic resolution, allowing the hunting eagle to see tiny mice and fish on the ground and in streams from several hundred feet above. The eagle's eyes are equivalent to a twenty-megapixel camera, compared to four-megapixel human eyes. But eyes get even wilder, stranger, and more extraordinary. The humble, burrowing mantis shrimp arguably has the world's best, most specialized set of eyes. These colorful marine crustaceans have beautiful multifaceted eyes set on flexible rotating stalks, which contain not two or three but *twelve* different types of cones, including six cones for the detection of ultraviolet light. The advantages of such a sophisticated, broad spectrum of light detection are unclear, but these creatures exhibit exotic courtship dances, and the ability to detect many bright colors may be important for sensing and attracting a mate and may have evolved through sexual selection. The eyes of the proud eagle and the mantis shrimp are just two examples of a broad diversity of eyes in the animal kingdom, each adapted to meet the needs of its particular animal in its own environment.

SQUID AND MAN: CONVERGENT EVOLUTION OF THE EYE

The giant squid is among the largest invertebrates on Earth, reaching lengths of over thirteen meters (forty feet), and it has the largest eyes in the animal kingdom—as large as soccer balls. The remarkable thing about the giant squid's eye, apart from its size, is its similarity to the human eye. Both of us have camera-like eyes with irises, lenses, and retinas, but the squid eye and the human eye took different evolutionary pathways to achieve a similar goal. These eyes evolved through **convergent evolution.**

After mollusks and vertebrates diverged from a common ancestor more than five hundred million years ago, the evolution of their eyes proceeded independently, subject to different environmental forces. Despite their long and separate evolutionary pathways, cephalopods and vertebrates both developed camera-style eyes with very similar anatomy. But there is one subtle difference that provides a clue to their separate heritages—the vertebrate retina is wired backward. In our eyes, nerves and blood vessels lie *in front* of the photoreceptor cells on the retina, so light has to pass through these structures before it strikes the photoreceptor cells. The nerve cells lying on top of the retina converge toward a single spot to form the optic nerve, which penetrates the retina on its way to the brain, leaving a blind spot in your eye where no image is received. In contrast, the photoreceptor cells of the squid and other cephalopods lie in front of the nerve cells, so there is no obstruction to light and no blind spot. This and several other differences between the eyes of cephalopods and vertebrates confirm that, though our eyes may be similar, we came about them in different ways over millions of years. Furthermore, comparison of the *PAX6* genes of cephalopods and vertebrates show variations indicating that they evolved independently from the time when these two groups diverged.

Image 13.4: Squid. This small yellow-white squid (not the giant squid mentioned in the text) has camera-like eyes similar to human eyes. Squid and human eyes evolved through separate and independent pathways over millions of years through convergent evolution.

Evolutionary convergence is a remarkable testament to the fine-tuned power of evolution. Two creatures so completely different in history and environment created similar complex eyes through separate and independent evolutionary pathways over millions of years. This was not a highly improbable coincidence. We *needed* sophisticated vision to survive, and apparently the squid's ancestors did too. And so eyes formed in both cases—nearly identical eyes, no less. Either the stepwise fashion of evolution dictated that camera-like eyes would evolve *this* way, and no other—or maybe the common structure of the squid and vertebrate eyes means that our evolutions found the *best* possible design for a camera eye—twice. Would evolution sculpt the same tool, on some distant planet, for an alien creature that needed the power of vision? Until we meet that alien, there's no way to know for sure. But convergent evolution makes it seem possible.

PROCESSING VISUAL INFORMATION

All sensory information from our sensory organs must be transmitted and processed in order to produce a response. What value is information without a response? It is the nature of the response that determines survival of the species. Our eyes are our windows to the world, but our nervous systems are responsible for processing the information we receive from our eyes and other sensory organs, and for initiating a response. Let's explore how the nervous system and brain evolved to provide these functions.

CHAPTER 14

EVOLUTION OF THE NERVOUS SYSTEM

> The brain was built like an ice cream cone (and you are the top scoop): Through evolutionary time, as higher functions were added, a new scoop was placed on top, but the lower scoops were left largely unchanged.
>
> —David J. Linden, in *The Accidental Mind*

The success of metazoan animals has been due in large part to the evolution of the nervous system. Nervous systems have provided the means to transmit sensory input from the environment, process that information, and produce directed responses that help the organism adapt to changing environmental conditions. Nervous systems have evolved the capacity for rapid communication over large distances, facilitating rapid responses essential to survival.

Current-day nervous systems are staggeringly diverse, ranging from the simple nerve net of jellyfish, which has no brain, to the advanced nervous systems of vertebrates with centralized brains containing millions to billions of neurons. Vertebrate nervous systems have what has been classified as peripheral and central nervous systems. The peripheral nervous system is that portion of the nervous system outside the brain and spinal cord that connects the central nervous system with organs, blood vessels, glands and muscles. The central nervous system includes the brain and spinal cord. The spinal cord is a bundle of nerve fibers enclosed within the spine that connects the brain to nearly all parts of the body. The brain is the central processor and the command and control center for the organism. The human brain has thus far been the pinnacle of biological evolution.

Growing a human brain is a daunting task. From a single cell, an undifferentiated zygote, grows a brain with over one hundred billion glial cells and over one hundred billion neurons, each with tens of thousands of connections (Goldman 2010; von Bartheld, Bahney, and Herculano-Houzel 2016). This network of neurons allows us to process reams of sensory input, store memories, stir emotions, learn, think, and finally—most mysteriously of all—develop consciousness and become aware of ourselves. Much of our brain's structure and function is programmed by our genes—it is part of our "nature." But a large part of our brain's structure and function develop over time as new connections and new extensions of our neurons develop in response to stimuli from our environment. Our brains change physically over the course of our lifetimes in response to the daily pressures of our lives. The development of the awesome human brain, the convoluted and ever-changing computer of the mind, is one of the most staggering "miracles" of evolution. Let's see how all this got started.

BEFORE THERE WERE NERVOUS SYSTEMS

Prior to the evolution of animals, and prior to the evolution of nervous systems, unicellular organisms used surrogates that allowed them to transmit signals and respond to changes in their environment. Members of *Chlamydomonas*, the genus of unicellular green algae we visited in the last chapter, have an eyespot that senses light and transmits photo-induced intracellular electrical signals that trigger flagella to move and propel the organism toward light. The signal transmission, which utilizes membrane depolarization and chemical transmitters, has similarities to signal transmission in nerve cells.

Sponges, members of Porifera, the oldest animal phylum, which evolved some six hundred million years ago, have no true tissue types, no true muscles, and no nervous system. Despite their simplicity and lack of a nervous system, they can respond to stimuli through chemical transmitters (Elwanger and Nickel 2006). In the presence of a food stimulus in adjacent seawater, sponges release chemical transmitters that activate synchronous contractions of contracting cells called pinocytes, propelling seawater, with its nutrients, through the internal canals of the sponge (Leys and Meech 2006).

While sponges have no nervous system, the evolutionary origins of the nervous system may still reside in these primitive animals. Sponges contain about twenty-five genes that are very similar to the human genes responsible for producing proteins that are essential for transmitting nerve signals from one neuron to the next (Sakarya et al. 2007; Renard et al. 2009). And, incredibly, these proteins are also present in unicellular choanoflagellates. Remember: ancient ancestors of choanoflagellates are believed to be the precursor to sponges and to all animal life. Presumably these receptor proteins—and the genes responsible for them—were inherited from this last common unicellular ancestor of all animals. But why should sponges and choanoflagellates—two simple neuronless creatures—possess these proteins that are so vital to transmitting signals through neural synapses in our own complex brains? What good were these proteins to a single-celled organism six hundred million years ago? We don't know, but they probably served some other function at that time. This appears to be another example of **exaptation**—a protein evolving for one purpose and then being co-opted further down the evolutionary lineage to solve another problem entirely.

FIRST NERVE CELLS

All cells, including neurons, have an electrical and chemical gradient across their cell membranes, with a negative charge, a high potassium concentration, and a low sodium concentration on the inside of the cell relative to the outside of the cell. These electrochemical gradients are created by energy-requiring pumps that transport ions across the cell membrane. Probably the most important of these pumps is the sodium–potassium pump (Na+/K+-ATPase), which is universal to all animals. This pump evolved initially to help cells respond to osmotic changes in the environment by regulating the osmolality of the cell to prevent swelling or shrinkage (Kristan 2016). The electrochemical gradient also provided the driving force for transporting glucose, amino acids, and other compounds across the cell membrane. In addition, calcium pumps evolved to control the internal metabolic state of the cell. As so often happens in evolution, the electrochemical gradient created by the sodium–potassium pumps and calcium pumps were co-opted by evolving nerve cells for another purpose—signal transmission. Nerve cells use action potentials—traveling waves of depolarization and repolarization of cell membranes—to rapidly transmit signals down a long axon to near and remote regions of the body (figure 14.1).

Neurons have been with us a long time. We don't know exactly when the first neurons appeared, but we do know that the most primitive animals, the sponges, do not have nervous tissues, and the cnidarians, the next animals on the phylogenetic tree, which include jellyfish and hydras, do. So we suspect the first neurons evolved when the first cnidarians appeared in the fossil record in the Precambrian period, about six hundred million years ago.

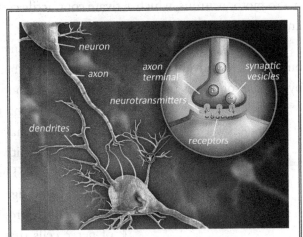

Figure 14.1: Neurons. Neurons are our information highways. They receive signals from their dendrite extensions and transmit these signals in the form of action potentials down their axons to the *axon terminal. Neurotransmitters* housed in *synaptic vesicles* are released from the axon terminal and bind to *neurotransmitter receptors* on dendrites of an adjoining neuron. When neurotransmitters bind to receptors, they create a new action potential that transmits the signal down the adjoining neuron.

Jellyfish may provide us with clues to the origin of the first neurons. In addition to their primitive nervous system, jellyfish have surface epithelial cells joined together by tight junctions, and these cells generate action potentials that transmit electrical signals from one cell to the next (Satterlie 1985). These epithelial cells play a synergistic role with the jellyfish's primitive nervous system in signal transmission that activates contractile cells for swimming. It is a reasonable guess that early neurons may have evolved from these types of epithelial cells, providing more rapid transmission of sensory information (Kristan 2016).

INVERTEBRATE NERVOUS SYSTEMS

Cnidarians, which include hydras and umbrella-shaped gelatinous jellyfish, have the most primitive nervous systems in the animal kingdom, and their nervous systems are likely similar to the first nervous systems to exist on Earth. Jellyfish have a diffuse network of nerves, called a **nerve net**, which lies between their inner and outermost layers of cells (image 14.1) (Satterlie 2011). They have no brains. Sensory neurons generate signals in response to various stimuli, such as light and touch, and these signals are transmitted via intermediate neurons in the nerve net to motor neurons, which activate contractile cells that allow the jellyfish to reflexively move and swim.

Flatworms, the next animal on the phylogenetic tree, inherited the cnidarians' nerve net but made a few advancements to the system. The nerves in the flatworm nerve net are connected by bundles of nerves that run the length of the body, and flatworms have a mass of nerve cells at the head region that qualifies as a primitive brain! The first animal brain may have been similar to this. The flatworms have a few primitive sense organs that communicate with their nervous system. Their eyespots are a single layer of photosensitive cells that send signals to the primitive brain via an optic nerve. Auricles, little projections off the worm's head, contain clusters of nerve cells that are sensitive to touch and chemicals, and enable the flatworm to find food. Remove a flatworm's little primitive nerve-net brain, and most of its complex behaviors—like the ability to find food—will be lost, but simple avoidance reflexes will remain intact.

We are beginning to see a pattern. The evolution of the invertebrate nervous system is one of increasing centralization. As we move up the evolutionary tree, from cnidarians to

Image 14.1: Jellyfish. These saucer-shaped marine animals are an annoyance to us on our ocean beaches but light up the waters of our aquariums with spectacular beauty. They have the most primitive nervous systems in the animal kingdom, consisting of a network of nerves (a nerve net) with no brain.

flatworms to roundworms and onward, the nervous systems of invertebrate phyla become more centralized to the head region. This is the evolution of the brain.

This centralization—at least among invertebrates—reaches its peak in the mollusk phylum with cephalopods (meaning "head-feet"), which include squids and octopuses. The octopus has the largest and most complex brain of all invertebrates, as well as the most advanced cognition. The brain has evolved intricate specialization, with defined regions for specific functions: optic lobes for accepting visual input, motor lobes for control of the eight arms, and about forty additional brain lobes. Octopuses have sophisticated sensory and motor capabilities and an ability to learn quickly, all of which may be adaptations that allowed them to compete with sophisticated vertebrate adversaries.

In recent amazing video footage, marine biologist Julian Finn videotaped an octopus scurrying across the ocean floor carrying two coconut shell halves (Finn, Tregenza, and Norman 2009; Finn 2009). When the octopus reached its destination, it disappeared inside the shells for protection and shelter. This complex behavior of collecting, transporting, and assembling a coconut for future use sets the octopus apart from other invertebrates and puts it in a league with a small number of animals that have demonstrated tool use.

The octopus nervous system shares some surprising similarities with vertebrate nervous systems. Their brains have two lobes dedicated to processing and storing memories. These lobes have evolved smaller neurons and a folded structure to accommodate a higher density of neurons, presumably to meet the increased needs of memory. This pattern is not seen in other invertebrates, but it is a quality typical of the vertebrate hippocampus—the small area in vertebrate brains primarily associated with memory (Hochner, Shomrat, and Fiorito 2006). Perhaps this type of neural wiring is essential to learning and memory, so it has evolved several times—another example of convergent evolution.

The octopus nervous system has another unique feature. Octopus arms contain millions of neurons that seem to function like brains of their own and can independently control basic motions without input from the central brain. An octopus's severed limb can snatch a piece of food and move it toward a phantom mouth (Nuwer 2013). Octopuses are certainly unique and strange creatures.

VERTEBRATE NERVOUS SYSTEMS

Vertebrates inherited many features of their nervous systems from their invertebrate ancestors, but they also acquired many characteristic features of their own. The vertebrate nervous system is larger and contains many more neurons of different types than the invertebrate system, and vertebrates have a central spinal cord, which is enclosed and protected by a bony vertebral column.

Vertebrate brains can be divided into three regions: the hindbrain, the midbrain, and the forebrain (figure 14.2). In fish, the hindbrain is dominant and is involved mostly with simple motor reflexes, while the midbrain and forebrain process visual and olfactory input. In the progression to mammals, the hindbrain becomes less prominent and the forebrain becomes more dominant. The large hemispheric cerebrum of the mammalian forebrain is responsible for the integration of sensory information, learning, and voluntary movement. Mammals have developed a thin layer of neural tissue on the outer surface of the cerebrum called the cerebral cortex, which is the control and information-processing center of the brain. The cerebral cortex is especially enlarged and convoluted in humans and other primates and plays a key role in our abstract thought, language, and consciousness. The convoluted surface provides a large surface area and allows for more neurons, more connections, and more complex patterns of organization, providing a substrate for increased cognitive capacity that has distinguished us from other species.

David Linden, in his book *The Accidental Mind*, describes the evolution of the vertebrate brain as analogous to the building of an ice-cream cone—one scoop added on top of another (Linden 2007, 5–27). As new functions evolved in the vertebrate brain, new structures—such as the cerebrum, and then the cerebral cortex—formed over preexisting brain structures, such as the hindbrain. The older structures remained and retained their previous functions, but new stuff—new "scoops"—were added on top for greater sophistication in thought and computing power. Evolution never has the luxury of starting over with a new design, since each part must continue to function as it is repaired and renovated. The evolution of the vertebrate nervous system provides an illustrative example of this kind of additive evolution. The downside of this process is that the vertebrate brain is simply not as efficient as it could be if it were redesigned from the bottom up. Propagation speeds of neuron action potentials are relatively slow, and there are a limited number of impulses that can be transferred per second—and many impulses don't get through at all. And the organization of the brain's parts and its wiring are not particularly efficient. Despite these limitations, the evolution of the vertebrate brain, and in particular the human brain, has been evolution's greatest invention and has allowed vertebrates and humans to become dominant species.

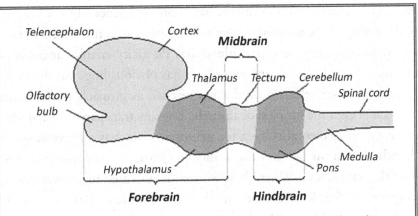

Figure 14.2: Vertebrate brain. The major divisions of the vertebrate brain have remained relatively the same in all vertebrates, from fish to mammals, although the size and shape have varied greatly. The hindbrain includes (a) the medulla, which regulates autonomic functions (breathing and blood pressure) and reflexes (cough and sneezing); (b) the cerebellum, which regulates movement and coordination; and (c) the pons, which serves as a relay station between several areas of the brain. The midbrain includes the tectum, which processes visual and auditory information. The forebrain includes (a) the thalamus, which regulates consciousness and acts as a relay station sending signals to the cerebrum; (b) the hypothalamus, which links the brain to the pituitary gland to control hormone release, (c) the olfactory bulb, which processes sensory input of smell; and (d) the cerebrum, which contains the cerebral cortex—the part of the brain responsible for higher brain functions. Figure courtesy of M. Deric Bownds from *Biology of Mind*.

THE PLASTIC BRAIN

Over millions of years, the animal brain has evolved hardwired circuits that react to sensory input from the environment and dictate basic behavioral responses essential for survival: eating, breathing, locomotion, sexual reproduction, fleeing, predation, and many more. These hardwired circuits function as **algorithms**, analogous to computer algorithms, that define specific responses to a variety of simultaneous sensory inputs. But these algorithms alone—a simple cycle of perceiving, processing, and reacting—is simple robotic stereotypy. Survival in a world full of competitors has required the brain to be capable of much more complex feats. These intrinsic hardwired behavioral traits have been crafted over millions of years and can change only over many generations. Our environment changes much faster than that, and for species to be competitive, they must be able to adapt much quicker.

Evolution has responded to the challenge. Our brains have learned to store memories of our experience, look back and learn from that experience, and change our behavior in response. Our brains have learned to create new algorithms based on experience. This intriguing biological phenomenon, the ability of the nervous system to modify its structure and function in response to input from the environment, is called **plasticity**. Plasticity enables animals to learn from experience and adapt their behavior accordingly. It has obvious huge survival advantages that have favored its evolutionary selection, and it has reached its pinnacle in the human brain.

Plasticity is seen in even the simplest of animals. The tiny transparent nematode roundworm, *C. elegans*, a favorite experimental subject for neuroscientists, has a nervous system with only about three hundred neurons but has the ability to "learn" from experience and respond to stimuli based on that experience. When these worms are nurtured in an environment with food at a certain temperature, they learn to associate that temperature with their food. When they are placed in an environment with a gradient of warmer and colder temperatures, they will migrate toward warmer or colder regions until they reach the region that corresponds to the temperature in which they were nurtured. And once they reach that region, they move isothermally (Kimata 2012). This behavior, called thermotaxis, provides a model showing how nervous systems sense and remember environmental conditions and adapt their behavior accordingly.

HOW DO WE LEARN?

We all learn from experience. When we experience an event, we store a virtual image of that event in our memory and also record our response to that event. When we are faced with a similar situation, we draw on that experience to help us initiate an appropriate response. We have learned from our past experience. To give an example, if the brain experiences associations between illness and various forms of sensory input (place, sound, temperature, and taste), it might "learn" from experience that the strongest association is between illness and taste. It would then use and strengthen those circuits (those memories) and alter behavior to avoid substances (poisons) with characteristic tastes (Bownds 1999, 125).

In 1959, Canadian psychologist Donald Hebb proposed a theory about how we learn that he put as follows: "Neurons that fire together, wire together" (Perlmutter 2013). He postulated that transmission of signals from one neuron to another is enhanced with use. When neurons communicate frequently, the connection between them strengthens; when they communicate infrequently, the connection weakens. Signals that travel the same pathway in the brain over and over transmit faster and faster. If we practice a skill—such as playing tennis, for example—the sequence of movements is very awkward at first, but with repetition and practice, the muscles "learn" how to move to produce precise results. (It is not really muscle learning but rather nerve learning.) The neural circuits become stronger with repeated use, and eventually the motions become almost unconscious. Professional tennis players reach their ultimate goals on the court when they are "in the zone"—when motion flows effortlessly, without any direction from the conscious self. It is the strengthening of synaptic transmission with repeated use that makes this learning possible.

Austrian American neuroscientist Erik Kandel took Hebb's ideas to the next level and investigated how memory and learning occur at the level of the neuron and synapse. In his famous experiments with sea slugs, he introduced the concept of **sensitization** (Mayford, Siegelbaum, and Kandel 2012). When a sea slug is touched, a sensory nerve transmits a signal to a motor nerve, which activates a muscle and causes the sea slug to withdraw from the touch—a simple defensive reflex. If the slug then incurs a separate painful or noxious stimulus to another part of the body, this will create another withdrawal. The slug will "remember" this painful experience, and when the next touch stimulus occurs at the first location, the slug will exhibit an exaggerated withdrawal. The slug "learned" from the noxious stimulus to withdraw quickly to avoid a painful encounter. The response had been "sensitized." Kandel discovered that the noxious stimulus released serotonin at the synapse of the first sensory and motor nerves, which enhanced transmission of the next nerve impulse at this synapse and exaggerated the withdrawal response. His experiments demonstrated the principle that memory and learning can occur by altering the transmission of signals from one neuron to another, and he was awarded the Nobel Prize in 2000 for his work.

Hebb's rule that repeated neural stimuli enhance the strength of connection has been further elucidated at the cellular level (Byrne 1997). When a nerve action potential reaches the axon terminal, it releases neurotransmitters that bind to receptors on dendrites of the adjoining neuron and initiate a partial depolarization of that adjoining neuron. With repeated stimuli, the amount of depolarization for the next stimuli of the same strength results in a greater depolarization. This is termed long-term potentiation (LTP), and we say the strength of the connection is strengthened. When the depolarization reaches a certain level, it triggers an action potential that travels down the neuron.

The enhancement of signal transmission at nerve synapses mediates short-term memory, but this is short-lived, and long-term memory requires something more. Long-term memory is mediated through structural changes in the nervous system with growth of new connections—both new axon terminals and new dendrite branches—in response to repeated stimuli (Lamprecht and LeDoux 2004). This requires the synthesis of new proteins through changes in gene regulation and gene expression. (Remember gene expression from chapter 9?) How this works remains largely unknown, but scientists have begun to identify how repeated stimuli with the release of transmitters are capable of modifying transcription factors that control gene expression. These changes in nerve structure are more permanent and longer-lasting than the short-term changes in transmission at synapses and are essential for long-term memory.

The learning process is much more active and effective in humans in the early years and diminishes but does not go away in adult and later years. A young child can easily learn a first and second language, but once these language pathways are fixed, learning a new language as an adult is very difficult. All of the learning processes (including socialization and personality formation) have a critical time period in which learning is easy and habits are formed. After this critical time period, we are pretty much stuck with most of our habits and our personality.

Research has shown that concentration and focused attention are essential in the learning process (Perlmutter 2013). This seems intuitive. Without focused attention, new nerve connections are not formed, and memory is not stored. Research has also shown that mental activity alone can modify neural pathways and neuroanatomy. Harvard neurologist Alvaro Pascual Leone compared subjects who practiced piano with subjects who only visualized practicing the piano. The group that practiced the piano showed substantial improvement in their skill level, and as assessed by a technique called transcranial magnetic stimulation (TMS), they showed measurable changes in brain structure. The group that used only visualization also showed some improvement in skill level and, astonishingly, changes in brain structure (Begley 2007). This raises the profound new paradigm that focused, directed mental activity can alter brain structure and function.

The remarkable plasticity of the human brain, which has given us an unparalleled ability to learn and remember, has set us apart from the rest of the animal kingdom. We will see how this plays out when we follow the course of human evolution in PART V.

PART IV

FROM SEA TO LAND

CHAPTER 15

THE GREENING OF EARTH

In the Paleozoic era perhaps about 480 million years ago, a grand conquest occurred and yet this victory is not routinely celebrated by any nation on the planet! It was the conquest of the land by green algae …

—Russell Chapman, in *Algae: The World's Most Important "Plants"*

Sometime more than a billion years ago, in what was perhaps an unprecedented onetime event, a lone eukaryotic cell engulfed a cyanobacterium. But rather than simply digesting the bacterium, the eukaryote entered into a symbiotic relationship and transformed the cyanobacterium into the first chloroplast organelle, giving the eukaryote the capacity for photosynthesis (Bhattacharya and Medlin 1998; Howe et al. 2008). The first single-celled green alga was born!

Fast-forward to about 480 million years ago, and tenacious **algae** began venturing out of the water for the first time, attempting to gain a foothold on dry land. These pioneers struggled. Four distinct types of algae originally came ashore: green algae and three types of algae with nonoxygenic photosynthesis. Only one of those four—green algae—managed to evolve enough complexity to survive the harsh conditions of the Precambrian landscape. All the land plants that blanket our lush green planet today—grass, shrubs, trees, flowers, and everything besides and in between—are descendants of green algae. They all contain chloroplasts derived from that first engulfed cyanobacterium, and they all use the pigments chlorophyll A and B in photosynthesis. If this isn't enough proof of the common ancestry of all these land plants, the genetic codes of green algae and all land plants bear a close resemblance to one another. The descendants of other types of ancient algae, including brown and red algae, survive as contemporary algae living in the seas, including seaweeds and the giant kelps, but they never made it on land. Only green algae share a common lineage with land plants. Let's see how this might have happened.

THE SOFTENING OF THE EARTH

Earth's landscape in the Cambrian period—about five hundred million years ago—would have looked, to our green-loving eyes, desolate and empty. Earth was like the surface of Mars, covered with barren rock, sand, and gravel. Algae had gained a clinging hold in lakes and at the mouths of rivers, but there were no other evident signs of life. The vast seas teemed with healthy animal life—the Cambrian explosion was in full swing—but the land was desolate.

Life is always eager to exploit new territories. Bacteria were the first to venture out of the seas (Horodyski and Knauth 1994). Fast-breeding and fast-evolving bacteria had populated pools of water on the landscape, using fool's gold, or pyrite (iron sulfite), as food. They inadvertently prepared the way for the arrival of plants, producing organic acids, which break down rock and release minerals that plants can use for nourishment. Colonies of bacteria formed microbial mats, which bind sediment and prevent erosion by runoff. Dying bacterial bodies provided humus, and soil began to accumulate.

Shifting continental plates left many areas of the landscape covered with shallow water, providing a bridge for algae to venture onto the land. The landscape was nearly ready for the invaders from the sea.

LICHENS: THE FIRST PIONEERS

Lichens are odd life forms. They are not quite plants, and in fact they are not a single organism at all. Rather, lichens are a hybrid of two organisms living together in symbiotic harmony— green algae nestled into a structural framework built of fungal filaments. Green algae capture energy from the sun and use it to produce organic products for the fungi to consume. In return, the fungi absorb water and minerals for both organisms to use, and the surrounding fungal filaments protect the algae against water loss and environmental hazards. Working together like two mountain climbers, the fungi and green algae support one another and are able to survive in almost any environment. Today they inhabit such extremes as the arctic tundra, hot deserts, barren beaches, and rocky areas where glaciers have retreated.

Lichens first appeared in the fossil record about four hundred million years ago, although they could have evolved much earlier than that (Taylor et al. 1995). Because lichens are able to survive in such harsh environments, and because they harbor green algae, the ancestors of all land plants, many scientists believe lichens may have played a pivotal role as a vanguard to the colonization of Earth's landscape by land plants. Housed and nurtured by fungal filaments, algae may have been given a chance to survive long enough on land to adapt to the hostile landscape, and to eventually evolve into the land plants that blanket our planet today.

THE FIRST LAND PLANTS

You've probably seen thick, goopy green algae settled in mats around the edges of stagnant ponds. Hundreds of millions of years ago, during the Ordovician period, algae did exactly the same thing. But back then, there were only algae and no plants, and dry land was still virgin territory. During droughts, green algae at the water's edge may have become stranded in drying mud as the water receded. Over time, they adapted to the dryer conditions and inched their way onto land. They evolved into mosslike vegetation, the first true land plants, settling into the moist soil scant meters from the water's edge. The invasion of the landscape by green algae and land plants had begun.

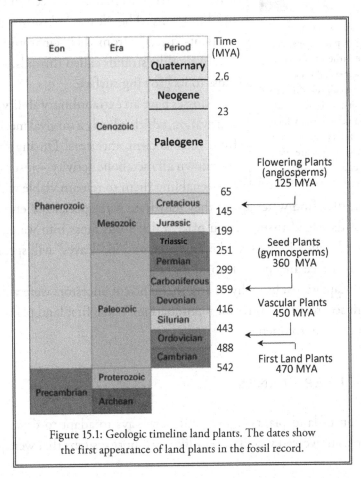

Figure 15.1: Geologic timeline land plants. The dates show the first appearance of land plants in the fossil record.

BRYOPHYTES

Ancient ancestors of current-day **bryophytes** appeared in the fossil record about 470 million years ago in the mid Ordovician period and are believed to be the first plants to colonize the landscape (figure 15.1) (Rubenstein et al. 2010). Bryophytes are a group of primitive nonvascular plants that include contemporary mosses, hornworts, and liverworts. They have provided clues how their ancestors may have adapted to life on land, where water was not always available.

Image 15.1: Moss. Mosses are members of the bryophyte group of nonvascular land plants, the earliest plants to colonize the landscape and the closest living relatives of green algae, which gave rise to all land plants. This moss has a dominant haploid generation, which appears as a mat, and a nondominant diploid generation, which are the stalks with spore capsules at the tips.

Bryophytes are nonvascular, meaning that they do not have specialized channels to transport water and mineral nutrients throughout the plant. Mosses, which we see as lowly green mats on rocks and logs along forest streams, provide an illustrative example of this group (image 15.1). They usually grow in dense, flat mats with leaflike protrusions that are one cell thick and supported by thin stems. The "leaves" absorb water and mineral nutrients directly from their environment. Moss stems end in rootlike strands called rhizoids, which anchor the moss to its growing surface.

Mosses have an extraordinary ability to tolerate water deprivation, which is likely a survival mechanism inherited from their ancient ancestors. During dry periods, they shut down all metabolic activity—just as bacterial spores do—enabling them to remain viable without a source of water. Later, when they find water again, they rehydrate with specialized enzymes that enable the rapid repair of damaged tissue. If you place a dried-out moss into water, you will see an immediate change. It will stiffen as it takes in water, and its "leaves" will spread to expose the greatest amount of surface area possible to sunlight.

Despite these adaptations, bryophytes and their ancient ancestors were vulnerable to water loss and, as we will see, required water for reproduction. These first land plants could not stray far from damp, moist environments.

SEEDLESS VASCULAR PLANTS

The next generation of land plants figured out new ways to adapt to dryer conditions. The seedless vascular plants evolved vascular, vein-like transport systems that were probably similar to those of modern seedless vascular plants, such as today's ferns (figure 15.2). Current-day vascular plants have two types of transport systems: **xylem** and **phloem**. Xylem is made of dead cells stacked together into long tubes. As xylem cells die, their innards are emptied and converted into conduits for water and nutrient transport. Xylem conduits, called veins, function like straws, literally sucking water and nutrients upward.

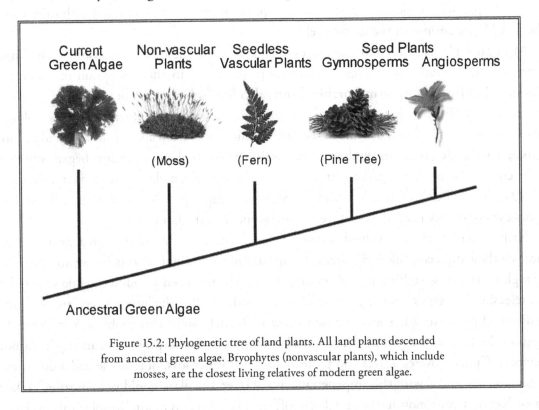

Figure 15.2: Phylogenetic tree of land plants. All land plants descended from ancestral green algae. Bryophytes (nonvascular plants), which include mosses, are the closest living relatives of modern green algae.

Lignin, a stiff complex polymer manufactured by the plant, is incorporated into the xylem cell wall, strengthening the cell wall and, in turn, conferring mechanical strength and support for the upright plant. The wood found in tree trunks and other woody plants is composed of lignin and cellulose fibers. With the development of vascular tissue and lignin for support, plants grew upward and outward, reaching great heights, allowing greater access to sunlight. This is exemplified by the great fern forests of the Carboniferous period.

Phloem, the other transport system in vascular plants, contains cylindrical conducting cells without nuclei, which are stacked end to end, forming long channels called sieve tubes. The conducting cells have perforations at their end walls to allow transport of material from cell to cell. Glucose and other products of photosynthesis are actively transported into the phloem, and water and minerals follow via osmosis, transporting the sugar, water and minerals—what we call sap—throughout the plant.

Plants' next adaptations to living on the landscape were roots, those fingerlike extensions that form an interface between the plant and the soil. They facilitate the absorption of water and minerals, help to anchor the plant to the earth, and sometimes provide a place for food storage. To provide a large surface area for absorption, roots are often covered with thin, hairy, filament-like extensions of the surface cells.

Most land plants today (80 percent or more) have a symbiotic relationship with fungal colonies, which attach themselves to plant roots and assist in the absorption of water and minerals. The fungi in these **mycorrhizal fungal colonies** have long, branching filamentous structures called hyphae that are very efficient at absorbing water and nutrients. Fungal colonies have been observed in fossils of some of the earliest land plants from about four hundred million years ago, indicating that this symbiotic relationship between plants and fungi began very early in the course of evolution—perhaps at the very beginning, when plants invaded land (Remy et al. 1994). It may be that the water- and nutrient-gathering mycorrhizal fungi gave the earliest plants the boost they needed to survive as they adapted to the landscape.

Early vascular plants evolved leaves, which increased the surface area available for photosynthesizing cells, allowing them to capture more sunlight. Plants began to "breathe" through their leaves—taking in carbon dioxide, which they used to build organic molecules, and releasing oxygen as a waste product of photosynthesis. But the large surface area of leaves predisposed plants to water loss and dehydration. To help solve this problem, early vascular plants evolved a thin, waxy film, called **cuticle**, composed of waterproof lipid and hydrocarbon polymers. Cuticle covered leaves and protected against water loss, but cuticle had a downside. If leaves were covered with this impenetrable waxy layer, the plant could not "breathe." Like a plastic bag over your mouth, the cuticle cut off the plant's access to air. To solve this problem, leaves developed openings called **stomata**, through which carbon dioxide could enter and oxygen could be released. Stomata are dynamic, typically opening during the sunny day, when photosynthesis is active and the exchange of gasses is critical, and closing at night to prevent unnecessary water loss.

Vascular plants—both spores and entire plants—were documented in fossils dated to the late Ordovician period about 450 million years ago (Steemans et al. 2009; Edwards, Davies, and Axe 1992). Prior to that time, Earth's vegetation was no more than waist high, but with the evolution of vascularization and lignin to provide structural support, plants grew to stunning new heights. By the end of the Carboniferous period (290 million years ago), those great fern forests that I mentioned earlier dominated the landscape. Height is a great advantage; tall plants can obtain more sunlight by overshadowing their competitors, and they can also distribute their spores—and later, their seeds—farther from on high. The major living descendants of these ancient seedless vascular plants are our modern-day ferns, which grow to great heights and in startling profusion in the cool coastal rainforests of North America (image 15.2). Their graceful arching fronds and shady green foliage characterize those cool, wet forests. They are ancient plants, frozen in time yet vibrantly alive.

Image 15.2: Fern. This seedless vascular plant has multiple brown sori on the undersurface of its leaves, which contain clusters of reproductive organs called sporangia. Sporangia release spores to the ground, which grow into diminutive haploid plants and produce sperm and eggs.

SEEDS CHANGE THE LANDSCAPE

The green armies of nonvascular plants and seedless vascular plants that had invaded the landscape were not able to venture very far from moist habitats. Reproduction required water, and spores and embryos were vulnerable to dehydration in a dry environment. The emergence of the first generation of seed plants (**gymnosperms**), which were documented in the fossil record at the end of the Devonian period about 360 million years ago, changed that (Gillespie, Roghwell, and Scheckler 1981).

Seed plants, as exemplified by contemporary conifer pine trees, produce pollen in male cones and eggs in female cones (image 15.3) Pollen is blown by the wind to the female cones, where fertilization occurs. The newly formed embryo is given a supply of nutrition and enclosed in a hard shell to form a seed, which is released from the cone, falls to earth, and, if conditions are right, germinates into a seedling. Water is not required for reproduction, and the embryo is provided nutrition and protected within the seed against dehydration. Ancient seed plants, equipped with these tools, were ready to march inland.

During the Permian period, which began about three hundred million years ago, the face of the Earth was in furious motion. Major landmasses came together to form the supercontinent Pangaea, which experienced new extremes of climate as a result of its vast size. Southern areas farthest from the equator were cold and frozen under ice, and northern

Image 15.3: Pine cones. The small male cones (seen at the top) of the pine tree produce pollen, which is blown to female woody cones, where eggs are fertilized and then enclosed in hard shells to form seeds. Pine trees are members of the gymnosperm group of seed-producing nonflowering plants.

regions closer to the equator suffered from intense heat, torrential wet periods, and dusty dry periods. This climatic intensity greatly impacted the evolution of plant life. As moist tropical regions began to dry out, ferns and other seedless vascular plants—which had typified the Carboniferous period—surrendered land to more drought-resistant seed plants. The great fern forests lost ground, and piney conifer forests towered in their place. By the end of the Permian period 250 million years ago, conifers and other seed plants dominated the landscape, making up over 60 percent of global flora.

Today, conifers—which include junipers, pines, spruces, and firs—are the most abundant and diverse gymnosperms, and they dominate great swaths of our planet. The northern taiga, the world's largest terrestrial biome, is composed of coniferous pines and spruces and makes up nearly 30 percent of the Earth's modern forests. The largest and tallest living organisms on the planet today—the giant sequoia and coastal redwoods of California—are conifers, which grow taller than the length of a football field. Giant sequoias were the dominant trees in North America and Europe during the Jurassic Period, and sequoia forests must have provided a dramatic backdrop for the giant dinosaurs of the time.

THE FIRST FLOWERS

The emergence of flowering plants and the coevolution of nectar-sipping birds and bees has become an iconic symbol for ecology on our planet. Flowering plants (**angiosperms**) first appeared in the fossil record about 125 million years ago, although some believe they arose much earlier (Hochuli and Feist-Burkardt 2013). They rapidly multiplied, and by the mid-Cretaceous period, about ninety million years ago, they dominated the planet. Christian de Duve, in his book *Vital Dust*, provides a speculative scenario as to how these unique plants may have evolved (de Duve 1995, 183):

> One day in those remote times, a seed plant suffered a mutation that caused the leaves around the sex organs to lack chlorophyll and turn white or, perhaps, yellow or pink if they retained some accessory pigments. The uniformly green landscape of the primeval fields and forests thus became dappled for the first time with bright patches, which acted as beacons for insects that happened to be genetically programmed to move toward light. Because of this fortuitous circumstance, the genetic accident suffered by the plant turned into a benefit. In the course of their visitations, the attracted insects collected pollen grains on their bodies from the male organs of the plant and dropped some grains

again on female organs. The pollination record of the mutated plant increased, and so did its reproductive success. The insects also profited from the plant's mutation, which guided them to nutritious nectar; they proliferated in large amounts. Propelled by the most far-reaching instance of mutually advantageous relationships ever established between plants and animals, a revolution was launched that sprinkled the green expanses of the Earth with countless dabs of color. Flowers were born.

Before the evolution of angiosperms, pollination was not very efficient. Transfer of pollen by gymnosperms from male to female plants was at the mercy of the wind, and only a miniscule proportion of pollen reached its target. The transfer of pollen by pollinators—insects and birds—is much more efficient. Angiosperms also evolved new ways to distribute their seeds. They produced tissues around the seeds, transforming them into fruits or nuts, and this provided additional protection and many new options for distribution. Seeds enclosed in sweet fruit could be eaten and transported in the intestines of animals to be deposited with feces at great distances. Seeds covered with special accoutrement could be blown by the wind (like the "helicopters" of maple trees and the "parachutes" of dandelions), floated on water (like coconuts), and attached to animal fur (like cockleburs). With such an arsenal for reproductive success, angiosperms soon dominated the landscape.

Today angiosperms are the largest group of plants on the planet, with more than three hundred thousand species. They display a huge variety of life-forms, including trees, grasses, herbs, vines, vegetables, legumes, and shrubs. We use angiosperms in so many ways; they supply us with food, timber, pharmaceuticals, fuel, clothing, and ornaments. They are an integral part of our lives, and a mainstay of Earth's ecology.

THE SEX LIFE OF PLANTS

Plants have what seems to be an unusual sex life. They have inherited a complex pattern of reproduction from green algae called **alternation of generations**, which requires two generations

WHY ALTERNATION OF GENERATIONS?

Why would algae and plants evolve such a strange life cycle as alternation of generations? It turns out that this strange cycle is implicit in sexual reproduction. In the beginning, single-celled eukaryotic organisms were almost certainly haploid, with only one set of chromosomes. Sexual reproduction began with the fusion of two haploid cells to form a diploid cell with two sets of chromosomes. This is the diploid generation. The process of meiosis evolved, which allowed the exchange of genetic material between the two sets of chromosomes, followed by duplication of chromosomes and two cell divisions with distribution of chromosomes into four haploid daughter cells. This is the haploid generation. The haploid daughter cells can then fuse to form a diploid cell and repeat the cycle. So, with sexual reproduction, even in this unicellular model, there is a diploid phase and a haploid phase.

to complete the life cycle (see sidebar "Why Alternation of Generations?"). Each subsequent group of land plants evolved modifications of this cycle as they adapted to their new life on land.

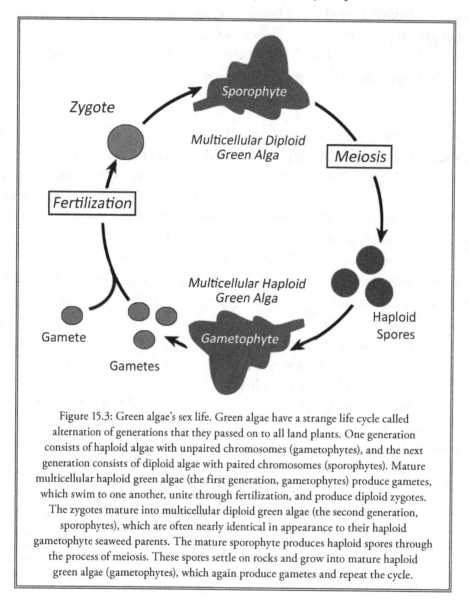

Figure 15.3: Green algae's sex life. Green algae have a strange life cycle called alternation of generations that they passed on to all land plants. One generation consists of haploid algae with unpaired chromosomes (gametophytes), and the next generation consists of diploid algae with paired chromosomes (sporophytes). Mature multicellular haploid green algae (the first generation, gametophytes) produce gametes, which swim to one another, unite through fertilization, and produce diploid zygotes. The zygotes mature into multicellular diploid green algae (the second generation, sporophytes), which are often nearly identical in appearance to their haploid gametophyte seaweed parents. The mature sporophyte produces haploid spores through the process of meiosis. These spores settle on rocks and grow into mature haploid green algae (gametophytes), which again produce gametes and repeat the cycle.

Green algae's sex life is how it got started (figure 15.3). We begin with mature green algae living as seaweed in coastal waters on a rocky ocean shoreline. These algae have only one set of chromosomes and are what we call the haploid generation, or **gametophytes**. They produce gametes, which swim to one another, unite through fertilization, and produce diploid zygotes.

The zygotes mature into multicellular diploid green algae that we call **sporophytes**, which are often identical in appearance to their haploid seaweed parents. The mature diploid algae produce haploid spores through the process of meiosis. These spores settle on rocks and grow into mature haploid green algae, or gametophytes, which again produce gametes and repeat the cycle.

The first generation of land plants, the nonvascular bryophytes, inherited the life cycle of alternating generations from their green algae precursors. Mosses serve as the prototype. The mature moss plant is haploid (a gametophyte), with only one set of unpaired chromosomes. Male haploid mosses produce sperm, which swim to female mosses, where they unite with eggs and produce zygotes, which develop into the diploid generation (sporophytes). The diploid mosses are small dependent structures, stalks with spore capsules at the tips, attached to the haploid moss (image 15.3). They produce **spores** through the process of meiosis, and these are released and grow into mature haploid mosses. So, mosses, according to their aquatic algal origins, need water for their sperm to reach eggs, and therefore they must live in wet habitats. The Bryophytes are the only group of land plants whose dominant mature plants are haploid.

With the next generation of land plants—the seedless vascular plants—the diploid phase of the life cycle became dominant. Ferns are an exemplar of this group. Mature ferns are diploid and have reproductive organs (sporangia) on the undersides of their leaves that produce haploid spores through meiosis (image 15.4). The spores are released into the ferns' forested environment to germinate and grow into the haploid generation—small heart-shaped plants called prothalli, which are less than one millimeter in diameter, only one cell thick, and incredibly delicate. Prothalli must remain wet at all times to survive. Haploid prothalli produce both sperm and eggs. Sperm swim to nearby eggs on the same or neighboring plants, where they unite to form a zygote. The developing embryo attaches to the prothallus and depends upon it for water and nutrition. As the embryo develops into a mature and independent diploid fern, the prothallus atrophies and dies. Both nonvascular mosses and seedless vascular ferns require water for their gametes to find each other, so they can live only in moist environments.

Pine trees provide a good example of how seed plants (gymnosperms) reproduce. The diploid generation is dominant, and the haploid generation is hidden in the cones. Male and female pinecones use meiosis to produce haploid spores, which develop into tiny haploid bodies (rather than mature haploid gametophytes) that produce pollen (sperm) and eggs. As we have seen, the pollen from male cones is blown by the wind to female cones, where fertilization occurs. Since the reproductive cycle does not depend on water, seed plants were able to spread across the landscape.

The life cycle of flowering plants is similar to that of gymnosperms. The mature flowering plant is diploid, and the haploid generation is hidden in the flowers. Within the flowers, haploid spores, created through meiosis, mature into tiny haploid bodies (gametophytes) that produce pollen (sperm) and eggs. Birds and bees bring pollen from male flowers to eggs in female flowers, where fertilization occurs and where seeds are produced and released to germinate. The reproductive cycle does not depend on water, so angiosperms, like seed plants, were able to spread across the landscape.

Unlike plants, we animals do not have a haploid generation. Animals are diploid organisms and produce haploid sperm and eggs through the process of meiosis. But sperm and eggs, unlike spores, are not capable of growing into mature haploid organisms. They can survive only by uniting with each other to form a zygote.

We have seen that diploid organisms have adaptive advantages over haploid organisms because duplicate genes provide protection if one gene is defective. So it is not surprising that plants evolved from haploid dominance in bryophytes to diploid dominance in subsequent generations of plants, and that we animals eliminated the haploid generation entirely.

NITROGEN FIXATION

The universe is composed of over one hundred different atoms, but eight of these—carbon, nitrogen, oxygen, hydrogen, calcium, phosphorous, potassium and sulfur—make up over 98 percent of the mass of living cells (Voet, Voet, and Pratt 2013, 2). Nitrogen is a crucial building block for the synthesis of amino acids, proteins, nucleic acids, and other vital components of living flesh. Nitrogen gas (N_2) is all around us, accounting for 78 percent of our atmosphere—but, ironically, we cannot use it when we breathe it. Nitrogen gas is inert and inactive and must first be converted to ammonia (NH_3) or nitrogen oxides in a process called **nitrogen fixation** to be usable by biologic organisms.

We animals obtain bioavailable nitrogen from our food—from plants or from other animals that eat plants. Plants lower on the food chain obtain their body-building nitrogen from ammonia and nitrogen oxides in the soil, but those supplies are often limited. Fortunately, special bacteria called **diazotrophs** have a synergistic relationship with plants and provide a crucial source of bioavailable nitrogen. Diazotrophs convert nitrogen gas into ammonia, which can be used by plants. In return, these bacteria receive organic molecules from plants to use as food and building materials.

The best known of these relationships is the one between **legumes** and the soil bacteria **rhizobia.** Legumes, which include such plants as alfalfa, peas, beans, lentils and soybeans, develop nodules on their roots where rhizobia live and convert nitrogen gas into usable forms of nitrogen (image 15.4). Because of their relationship with nitrogen-fixing rhizobia, legumes are often used in the practice of crop rotation to increase the nitrogen content of nitrogen-poor soils.

Only very few bacteria and one type of archaeon (methanogens) have the ability to fix nitrogen directly from nitrogen gas in the air. Surprisingly, no eukaryotes possess this ability, despite nitrogen being such an utterly essential material for all lifeforms. Why are we relying on these specialized bacteria to convert nitrogen to a form we can use? How did this tenuous interdependence evolve?

Diazotrophic bacteria "fix" nitrogen by converting nitrogen gas (N_2) and hydrogen ions (H^+) into ammonia (NH_3) in a reaction catalyzed by the enzyme **nitrogenase** and powered by ATP (Kneip et al. 2007). The nitrogenase enzymes in all diazotrophs are strikingly similar, and there is little doubt that these enzymes were inherited from a common ancestor. The patchy distribution of the enzyme in bacteria, and the lack of the enzyme in all eukaryotes, suggest that

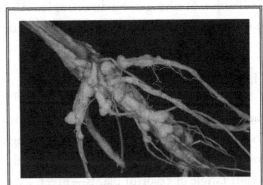

Image 15.4: Root nodules. This plant (a legume) has root nodules, which are colonized by bacteria (known as rhizobia) that convert nitrogen gas into bioavailable ammonia in a process known as nitrogen fixation. Without the help of these bacteria, plants would not have the nitrogen needed to build and run their cells. And we animals depend on plants as our source of nitrogen.

it was not present in the last universal common ancestor (LUCA); otherwise, all living things would have nitrogenase, and maybe all of us would be nitrogen fixers, sipping nitrogen straight from the air. (Boyd 2013) Rather, nitrogenase likely emerged early in the course of evolution and was passed from one bacterium to another by lateral gene transfer.

In prebiotic Earth, atmospheric nitrogen was converted into bioavailable nitrogen oxides by traumatic events, such as lightning, asteroid impacts, and volcanic eruptions. In addition, small amounts of ammonia were produced in deep-sea hydrothermal vents. The Earth was barren of life for hundreds of millions of years, so even though these processes of nitrogen fixation were infrequent and halting, they had time to build a reserve of usable nitrogen by the time life finally emerged.

However, early in the evolution of life on Earth, as the planet's bombardment by meteorites and volcanic eruptions abated and as our biomass grew, bioavailable nitrogen was quickly used up (Boyd and Peters 2013). As nitrogen supplies declined, bacteria looked for ways to adapt. The solution, we now know, was the evolution of nitrogenase. Diazotrophic bacteria "learned" to fix nitrogen into compounds they could use. And the rest was history. Plants eventually teamed up with those bacteria to get the nitrogen they needed, supplying the bacteria with organic food in return. We animals, at the lofty top of the food chain, benefit from these symbiotic relationships. Without those nitrogen-fixing bacteria, we wouldn't have bodies at all. There could hardly be a better illustration of the profound interdependence between organisms than the fixation of nitrogen by diazotrophs. All living things have been deeply dependent on one another since the dawn of evolution of life on our planet.

READYING THE WAY

The greening of Earth was a dramatic transition. Until the Cambrian period, our planet was a barren and colorless rock hurtling through space. With the invasion of land plants, Earth bloomed photosynthetic green. And with the arrival of angiosperms, our home has become a riotous tangle of colorful plants—green leaves mixed with the rich pigments of bug-tempting flowers. With the greening of the Earth, our planet became a breathing oasis in the cold lifelessness of empty space.

Planet Earth was preparing now for a second invasion. The profusion of breathing plants constituted a complete makeover of the ecology of the landscape, and the planet's terra firma was now softer, more hospitable, and ready for the first animals to crawl hesitantly ashore.

CHAPTER 16

INVASION OF THE CONTINENTS

Tiktaalik blurs the boundary between fish and land animals. This animal is both fish and tetrapod; we lovingly call it a "fishapod."

—Neil Shubin, paleontologist

Following the Cambrian explosion, Earth's oceans teemed with marine animals, but the land was still barren. Over the next one hundred million years, animals bided their time in the seas while the continents shifted and reformed. By the Devonian period, four hundred million years ago, our planet's land had coalesced into two supercontinents: Gondwana and Euramerica. They sat close to one another in the Southern Hemisphere, and an endless ocean covered the rest of the globe. The climate had become relatively warm and dry, and land plants, mostly ancient primitive mosses and ferns, had taken root in coastal regions and riverbanks.

The landscape was set for the invasion by animals. But the marine creatures of the Cambrian period had never known life on land, and they would have to overcome a number of obstacles to survive the transition. How does a creature that has spent its entire history in briny water learn how to live on land? They would need to learn to live with limited and intermittent water supplies, to breathe oxygen from the air, to walk on hard ground, and to reproduce without the luxury of water. But life always finds untapped niches, and little by little, animals began exploring the land—first at the coasts and in marshes, where water was plentiful, and gradually moving inland over millions of years.

ARTHROPOD PIONEERS

Arthropods, exemplified by trilobites, dominated the Cambrian seas for millions of years, and their descendants have flourished wildly ever since. Today they are the most diverse phylum in the animal kingdom, accounting for over 80 percent of extant animal species.

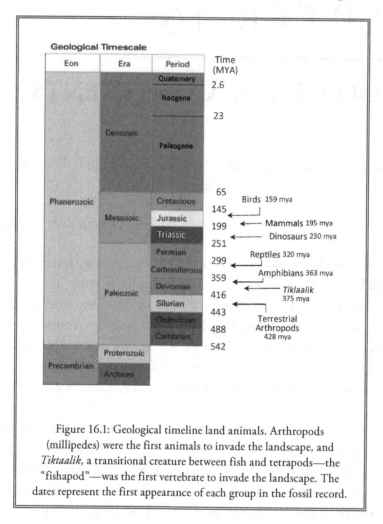

Figure 16.1: Geological timeline land animals. Arthropods (millipedes) were the first animals to invade the landscape, and *Tiktaalik*, a transitional creature between fish and tetrapods—the "fishapod"—was the first vertebrate to invade the landscape. The dates represent the first appearance of each group in the fossil record.

Ancient arthropods were the first members of the animal kingdom to invade the landscape. Fossils of tiny millipedes, those wormlike creatures with innumerable paired legs, were found in the seaside town of Stonehaven, Scotland, and documented the invasion of land by arthropods 428 million years ago, more than 50 million years before the invasion by vertebrates (figure 16.1) (Wilson and Anderson 2004). Small holes in their exoskeletons called spiracles, which fed into the tubular network of tracheae used for breathing, provided clear evidence that the millipedes had adapted to land. Other land-adapted arthropods, such as spiders and insects, began to appear in the fossil record not long after these millipedes.

Arthropods have defining body features that gave them a head start in the transition to land. Think of crabs and lobsters with multiple jointed legs and a characteristic hard-shell exoskeleton. As arthropods moved out of the water, they used their jointed limbs to walk, and their exoskeleton provided a shield of armor for protection and a waterproof covering against dehydration. But every adaptation has its disadvantages. That rigid exoskeleton is the reason that you rarely find an insect bigger than your thumb. Once formed, the exoskeleton limits the size of the arthropod. To grow further, they must shed the hard shell, grow while soft, and form a new, bigger exoskeleton. Arthropods are therefore terribly vulnerable during their soft-skinned molting phases.

As arthropods moved to land, they ditched their gills and developed a dense network of tubes and passages called tracheae, which penetrated the entire body and allowed oxygen to passively diffuse into the tissues and cells. But as they grew larger, their tracheal tubes became so long that oxygen was depleted by the time it reached cells near the center of the organism. This is another reason why arthropods have remained very small.

Like most marine animals, arthropods living in the sea release gametes directly into the water, where sperm swim to eggs to fertilize them. As arthropods moved onto land, they had to devise new methods of reproduction. Millipedes, a current-day arthropod, package large numbers of sperm in airtight sacs called spermatophores, which they leave on the ground for females to pick up and place in their genital cavity, where fertilization can occur. This internal fertilization without copulation may have been how arthropod sex started on land. Most insects today use internal fertilization, with the male depositing the sperm in the female through copulation—no water needed.

FISH OUT OF WATER

In 2006, the *New York Times* published an article with the headline "A 'Missing Link' Between Fishes and Land Animals is Found" (Wilford 2006). A team of scientists at the University of Chicago had discovered in the Canadian Arctic the 375-million-year-old fossil of a creature they named ***Tiktaalik roseae*** (image 16.1) (Shubin, Daeschler and Jenkins 2006). *Tiktaalik* represents one of the most iconic, triumphant images of evolution—the brave fishlike creature crawling out of water and onto dry land. It was a big animal, between five and ten feet long, covered with fins and scales. *Tiktaalik* was still a fish but had evolved many features of a land animal. The big fish skull was attached to an articulate neck, allowing swiveling of its head—a very unfishy adaptation that became characteristic of land animals. Its pectoral fins were built on a framework of bones and joints that had already begun to look like wrists. These must have allowed *Tiktaalik* to pull itself onto shore, later to be used for something akin to walking. It had sharp teeth lining an almost crocodilian mouth. And like the early land-invading millipedes, *Tiktaalik* had spiracles on the top of its head, suggesting that it had evolved primitive lungs for breathing in shallow water. It has become an evolutionary icon: *Tiktaalik roseae*, the "fishapod," the creature that bridged the gap between fishes and **tetrapods**.

Image 16.1: *Tiktaalik. Tiktaalik,* fondly called the "fishapod," is an iconic transitional figure between lobe-finned fishes and tetrapods. *Tiktaalik* and its cousins are believed to be the first vertebrates to venture onto land.

Tiktaalik (or a close relative) is the common ancestor of all tetrapods, which subsequently evolved to include all terrestrial vertebrates, including current-day amphibians, reptiles, birds, and mammals. *Tiktaalik* was a lobe-finned fish whose only living descendants are lungfishes and coelacanths. Lungfishes are considered "living fossils," since they have changed little since the days of *Tiktaalik*, and they provide us with a chance to study the fish–tetrapod transition in a living animal—quite literally in the flesh.

Modern lungfish are restricted to shallow freshwater lakes, ponds, and marshes in South America, Africa, and Australia (image 16.2). They have flattened heads and long eellike bodies sometimes reaching up to five feet in length. As you might expect, they don't breathe using only their gills. They have primitive lungs that evolved from their swim bladders, with honeycomb-like cavities where air enters and flows over tiny blood vessels, allowing gas exchange. They use their protolungs in times of desperate drought. If the lungfish's stream or lake dries up, the lungfish buries itself in the mud, leaving exposed only a small breathing hole through which it can gulp air. In this way, the lungfish can live out of water for months if it needs to, until the rains finally come again. These may be the same type of lungs that enabled their ancient ancestors to adapt to air breathing as they transitioned to land.

Image 16.2: Australian lungfish. The lungfish and its brethren are the closest living relatives of *Tiktaalik*, the "fishapod" that invaded the landscape and gave rise to all tetrapods. The lungfish has been called a living fossil because it has changed little since the days of *Tiktaalik*.

Once *Tiktaalik* and its cousins dragged themselves ashore, land became something more than an emergency backup. It was an opportunity. Earth's landscape was the

New World—rich with plants and insects as potential food sources, and devoid of predators who might eat or compete with the newcomers. Ancient tetrapods showed up in the fossil record in the late Devonian period about 365 million years ago. Faced with a new bounty of resources and empty space, tetrapods diversified in an **adaptive radiation**—the swift cascading evolution of new traits that occurs when species move into many different habitats. One species became many. Tetrapods dragged, crawled, and walked themselves farther inland, found more and better places to live, evolved to fit their new homes on land, and gave rise to amphibians, reptiles, birds and mammals (figure 16.2).

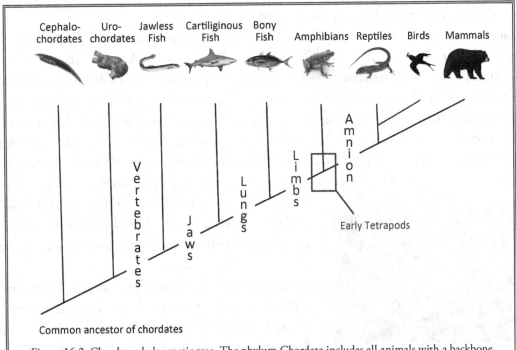

Figure 16.2: Chordate phylogenetic tree. The phylum Chordata includes all animals with a backbone or a precursor to a backbone. Lobe-finned fish, which are a tiny subset of bony fish and include modern lungfish, gave rise to all tetrapods and their descendants: amphibians, reptiles, birds, and mammals.

THE AMPHIBIAN INVASION

During the Carboniferous period, about 350 million years ago, our planet was covered by vast, lush swamps and rainforests choked with mosses and treelike giant ferns. The prodigious biomass of its forests became much of today's coal reserves, which is how this period got its name. The air was warm, humid, and rich with oxygen, reaching levels of 35 percent compared with 21 percent today. Insects, glutted on oxygen, grew massive. Glaciers had long melted, and much of today's land was flooded with extended seas of shallow brackish water. This warm, moist habitat was ideal for amphibians. Bountiful water provided the perfect home, predators had not yet evolved, and food was plentiful. As creatures of the interface between water and land, amphibians began to adapt to their new surroundings.

Today, amphibians include frogs, toads, salamanders, and little-known earthworm-like creatures called caecilians. Amphibians live truly half in and half out of water. They lay their eggs in water, and their larvae literally echo the evolutionary transition from water to land, as their gills and fins turn into lungs and limbs as they mature. Picture, for example, fishlike tadpoles transforming into hopping four-legged frogs.

Sadly, amphibians today are in decline, with decreasing populations of frogs, toads, and salamanders (Strauss 2017). But amphibians once ruled the world! Amphibians evolved from lobe-finned fish and early tetrapods, and first appeared in the fossil record during the Devonian period, about 363 million years ago. *Ichthyostega*, the earliest well-known amphibian genus, was a group of species that grew to about five feet long and appeared to be a cross between a fish and a crocodile (Blom 2005). Some have suggested that *Ichthyostega* was more of a transitional tetrapod rather than a true amphibian. Its well-developed lungs, strong skeletal structure, rigid spine, and strong forelimbs allowed adaptation to the terrestrial environment. Animals in the *Ichthyostega* genus probably spent time on land, basking in the sun to raise their body temperature (amphibians are cold-blooded), and returned to water to cool down, hunt for food, and mate.

Image 16.3: *Mastodonsaurus giganteus.* This giant amphibian was king during the Carboniferous period when amphibians ruled the world. *Mastodonsaurus giganteus* was a reptile-like monster and predator measuring six meters long and resembling a large crocodile.

Amphibians dominated the landscape for tens of millions of years, from the late Carboniferous period through much of the Permian period. By the mid-Permian period, they had achieved much of the diversity they boast today, ranging from small wormy creatures resembling snakes and salamanders to giant reptile-like monsters that resembled large crocodiles—like the six-meter-long tusked monsters and predators of the genus *Mastodonsaurus* (image 16.3).

Amphibians had successfully adapted to life out of water but were still tied to a moist environment—and they still are today. They breathe with their lungs but supplement this by absorbing oxygen through their moist, slimy skin. They reproduce primarily by external fertilization, which requires water, and when they reproduce through internal fertilization, their larvae are placed in an aqueous environment to grow into their adult forms. For all these reasons, amphibians have lived in and around shallow lakes, rivers, and marshes, and have not ventured very far inland.

THE RULING REPTILES

As the Carboniferous period waned and the Permian period began about three hundred million years ago, the two great continents of Gondwana and Euramerica collided, coalescing to form the supercontinent Pangaea (figure 16.3). Pangaea's immense size resulted in a colder, drier climate compared with the moist rain forests of the Carboniferous period. Sea levels fell as the water was frozen in newly formed glaciers. The dry climate caused a collapse of the Carboniferous rain forests and a loss of much of the amphibians' habitat, resulting in a minor extinction event. Amphibians suffered marked declines, and by the end of the Permian period, reptiles had become the dominant tetrapod. The once mighty amphibians had retreated to near obscurity.

The first reptile appeared in the fossil record in the late Carboniferous period about 320 million years ago (Carroll 1964). *Hylonomus* was a genus of twenty-centimeter-long critters that looked distinctly reptilian, with splayfooted posture, sharp teeth, and long tails—resembling the skittering lizards we see on rocks and in pet stores.

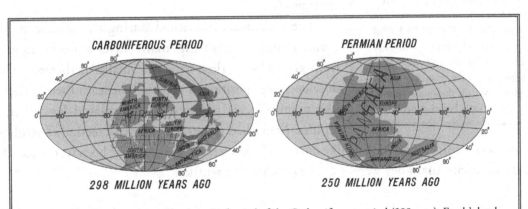

Figure 16.3: Supercontinent Pangaea. At the end of the Carboniferous period (298 mya), Earth's land masses were separated (see left panel), but by the beginning of the Permian period, the land masses began to collide and merge, forming the supercontinent Pangaea (see right panel). The extensive rainforests of the Carboniferous period gave way to a drier climate with vast inland deserts, ending the age of the amphibians and ushering in the age of the reptiles that were better adapted to the drier climate.

Early primitive reptiles evolved scaly, armored skin to protect themselves from dehydration in the open air, powerful limbs for improved mobility, and efficient lungs to improve oxygen absorption. They mated with internal fertilization and began to produce hard-shelled amniotic eggs to protect, house, and nourish developing embryos to self-sufficiency, without the need for a body of water. As we learned in chapter 11, they were the first amniotes. Today the defining characteristics of reptiles are still the amniotic egg, skin covered with scales or scutes, and cold-bloodedness, which we will discuss in the next chapter.

During the late Carboniferous period, about 324 million years ago, two major groups of amniotes emerged. One group, the **sauropsids**, were primitive reptiles whose descendants include dinosaurs, birds, and all modern reptiles. The other group, the **synapsids,** were mammal-like reptiles whose descendants include all modern-day mammals.

Image 16.4: *Dimetrodon.* This large reptilian carnivore genus prevalent in the Permian period had a prominent sail on its back used to warm its body by soaking up the morning sun. *Dimetrodon* was a member of the synapsid reptiles, which were the dominant animals in the early Permian period.

The synapsids became the dominant group by the early Permian period and included the large carnivore genus *Dimetrodon*, which had a characteristic prominent sail on its back used to warm its body by soaking up the morning sun (image 16.4). This may have been the first animal to use solar power. Although it is sometimes mistaken for a dinosaur, members of the *Dimetrodon* genus were neither dinosaurs nor ancestors of the dinosaurs. Rather, their clade gave rise to the lineage of mammal-like reptiles—the **therapsids**.

The therapsids flourished during the middle to late Permian period, evolving many diverse forms ranging from giant-fanged flesh-eaters to smaller herbivores. They developed legs positioned vertically beneath their bodies—like mammals, as opposed to the splayed legs of most reptiles. Many had powerful jaws, with teeth differentiated into incisors, canines, and molars, like ours. And some are believed to have been warm-blooded and covered with fur. They were indeed mammal-like reptiles.

THE END-PERMIAN EXTINCTION

At the end of the Permian period about 260 million years ago, Earth was rocked by tumultuous events that forever changed the course of life on our planet. The End-Permian extinction wiped out more than 90 percent of Earth's species. We will hear more about the extinction and its causes in chapter 18, but for now we are interested in the consequences for the evolution of reptiles and their descendants.

Most reptiles were wiped out in the extinction, but there were a few notable exceptions. One of the most notable surviving clades were the reptilian archosaurs who managed to squeeze through the wall of extinction and quickly became the dominant land vertebrates of the early Triassic period. Archosaurs subsequently gave rise to the mighty dinosaurs.

Another extraordinary survivor was a goofy mammal-like reptile (therapsid) genus named *Lystrosaurus*. Snub-faced and splay-legged, members of this genus looked like a cross between a lizard and a pig. After surviving the horrific extinction, *Lystrosaurus* thrived and dominated the landscape in the early Triassic period, making up more than 95 percent of the fossil record. Man may be the dominant mammal on Earth today, but in the Triassic, the world was overrun by piggish reptiles!

Image 16.5: Cynodont. The cynodont (meaning "dog teeth") group of mammal-like reptiles (therapsids) had many mammalian characteristics: differentiated teeth, warm-bloodedness, and hair. They flourished during the middle to late Permian period, survived the End-Permian extinction, and gave rise to the entire mammalian lineage.

The **cynodonts**, which included the famous "dog-toothed" genus *Cynognathus,* were perhaps the most significant group to survive the "Great Dying" event (image 16.5). Their many mammal-like characteristics—differentiated teeth, warm-bloodedness, and hair—foretold their destiny. These therapsids were the ancestors of all current-day mammals. We are indeed lucky they were able to survive the extinction. But cynodonts and their mammalian descendants had to wait their turn. For the next many millions of years, our diminutive ancestors bided their time in the shadows of the dinosaurs. We'll visit the reign of the dinosaurs in the next chapter.

CHAPTER 17

DINOSAURS REIGN; MAMMALS AND BIRDS EMERGE

Surely some [Mount Everest climbers], staring upward into the pale blue regions where air pressure is even thinner than that which was killing them, watched the serene "V" of majestic geese flocks wending their way over the roof of the world, flying effortlessly thousands of feet above the Everest elevations that readily kill humans.[1]

—Peter D. Ward, in *Out of Thin Air*

Dinosaurs dominated Earth's landscape for some 135 million years before they met an untimely death at the hands of the giant asteroid collision. Because of their size and diversity, which included the monstrous herbivore sauropods and the ferocious theropod predators, and because they were so dominant and ruled the planet for so long, they have captured the imagination of millions and have been the subject of intensive research and the focus of numerous books and movies. And they have left a lasting legacy in the form of avian descendants, numbering over ten thousand species. Their sudden demise in the wake of the asteroid collision has created an added mystery to their epic story and opened the doorway for the emergence and subsequent dominance of mammals. Let's see how the story unfolded.

DINOSAURS REIGN

When the archosaur reptiles slipped through the wall of extinction at the end of the Permian period, they began a lineage that led to the legendary dinosaurs. The archosaurs split into two lineages of dinosaurs distinguished by differing anatomy of their hip bones, which is documented in the fossil record (figure 17.1) (Strauss 2018). One group, the saurischians (meaning "lizard-hipped" dinosaurs), gave rise to the swift, bipedal **theropods** and the giant herbivore **sauropods**. The theropods were carnivores that included such savage predators as *Tyrannosaurus rex* and the velociraptors of *Jurassic Park* fame. *Tyrannosaurus rex* (meaning "king of the tyrant lizards") was one of the largest carnivores to inhabit our planet (image 17.1). This monster stood fourteen feet tall at the hips, extended to a length of forty feet, and weighed up to nine tons. Its thick, muscular thighs, solid tail, large head, and powerful jaws made it a formidable predator, although some have argued it was more of a scavenger. This theropod group of dinosaurs, as we will soon see, subsequently gave rise to the entire lineage of birds.

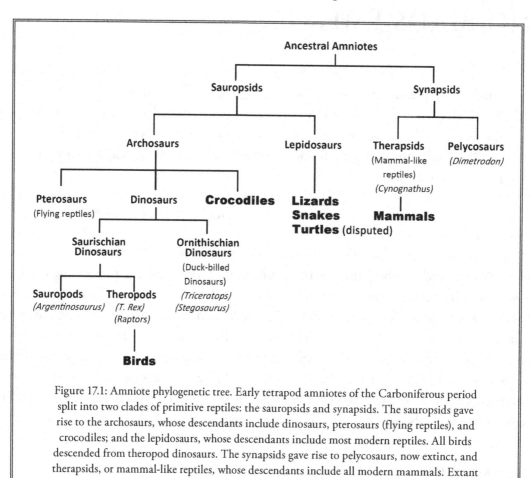

Figure 17.1: Amniote phylogenetic tree. Early tetrapod amniotes of the Carboniferous period split into two clades of primitive reptiles: the sauropsids and synapsids. The sauropsids gave rise to the archosaurs, whose descendants include dinosaurs, pterosaurs (flying reptiles), and crocodiles; and the lepidosaurs, whose descendants include most modern reptiles. All birds descended from theropod dinosaurs. The synapsids gave rise to pelycosaurs, now extinct, and therapsids, or mammal-like reptiles, whose descendants include all modern mammals. Extant groups are shown in bold print. Exemplar genera within groups are shown in parentheses.

The other group of saurischian dinosaurs were the gentle giant sauropods. *Argentinosaurus* (meaning "Argentina's lizard"), the signature genus of this group, was the largest terrestrial creature to ever walk the Earth—reaching lengths of over thirty-six meters and weighing more than eighty tons (image 17.2). The first fossils of this giant were discovered by a rancher in Argentina in 1987, who mistook the leg for a giant piece of petrified wood. A single gigantic fossil of a vertebra was the size of

Image 17.1: *Tyrannosaurus rex* and *Triceratops*

a man (BBC Earth 2018). Sauropod dinosaurs roamed across every continent for over one hundred million years. These long-necked herbivores could pluck foliage from the tops of tree ferns and conifers. They had long tails, rounded bodies, and the smallest brain relative to body size of all the dinosaurs. Most likely they were short on intelligence, but with the abundance of nourishing vegetation produced by their tropical climate—and near-immunity from predators due to their immense size—the sauropods flourished toward the end of the Jurassic period.

The other group that descended from the archosaurs were the "bird-hipped" ornithischians. They were a diverse group of herbivores, including the widely recognized genera *Stegosaurus* and *Triceratops*. Members of the genus *Stegosaurus* were bus-sized quadrupeds distinguished by two rows of bony plates along their back for thermoregulation and two pairs of bony spikes on

their tail for defense (image 17.3). They had a bizarrely small head, and a tiny brain described as being the size of a walnut (Castro 2016). *Triceratops* was a genus of massive herbivore the size of an African elephant, weighing over five tons (image 17.1). Their monstrous heads were as much as eight feet (2.4 meters) long, making up one-third of their body length. They had three protruding horns, two above their eyes and one on their snout, and a spectacular bony frill that framed their head. Members of the genus *Triceratops* roamed the landscape during the Cretaceous period but

Image 17.2: *Argentinosaurus*. This gentle giant vegetarian sauropod dinosaur, weighing over eighty tons, was the largest terrestrial creature to ever walk the Earth.

were confined to North America as they evolved after North America split from the other continents. Paleontologists have uncovered fossil remains of a *Triceratops* skull with partially

Image 17.3: *Stegosaurus.* This bus-sized dinosaur was distinguished by bony plates on its back used for thermoregulation, and a tiny brain the size of a walnut.

healed tooth wounds inflicted by a tyrannosaur, suggesting that the huge herbivore had survived an attack from the ferocious predator.

WHY WERE THE DINOSAURS SO SUCCESSFUL?

The dinosaur dynasty is like no other seen on Earth before or since. These monstrous, ferocious creatures dominated our planet for over 135 million years. In contrast, we, *Homo sapiens*, have been around a mere 300,000 years. What accounts for their spectacular success?

First of all, Earth's climate during the Jurassic and Cretaceous periods was optimally suited for the dinosaurs. Temperatures were much warmer than today, and humidity levels were high because of flooded landmasses. Tropical and subtropical conditions extended well beyond current-day boundaries, and even the polar regions had a temperate climate. Earth was a sauna. These conditions favored lush plant growth, which formed the base of the dinosaur food chain, and allowed the giant sauropods plenty to chew on. The warm climate also allowed these cold-blooded reptiles to lead a more active life.

One trait dinosaurs evolved that set them apart from some of their reptile rivals was the shape and position of their legs (Rogers 2014). Most early reptiles had legs that splayed out to the sides, like those of current-day lizards, and a sprawling gait in which the body twists sideways, back and forth, with ambulation. Dinosaurs evolved legs positioned beneath their bodies, allowing for much more efficient movement and faster locomotion. The mammal-like reptiles (therapsids) also evolved similar long legs positioned under their bodies, providing advantages in locomotion.

Size was also a big advantage. The fossil record is clear. Dinosaurs were the largest land animals that ever lived (figure 17.2). Members of the massive, long-necked herbivore genus *Argentinosaurus* were more than twelve times the size of an African elephant. Other dinosaurs were not so large, but most of them were big. There must have been great adaptive advantages that resulted in the selection of such immense body size.

Large size has obvious advantages in the arms race of predator versus prey. The immense size of *Argentinosaurus* must have made it almost entirely impervious to predators. The large duck-billed hadrosaurids enhanced the protection given by their large size by roaming in herds. And *Tyrannosaurus rex* possessed muscular hind legs, a strong jaw, and slashing teeth, but its immense size is what truly distinguished it as the fiercest carnivore of all time.

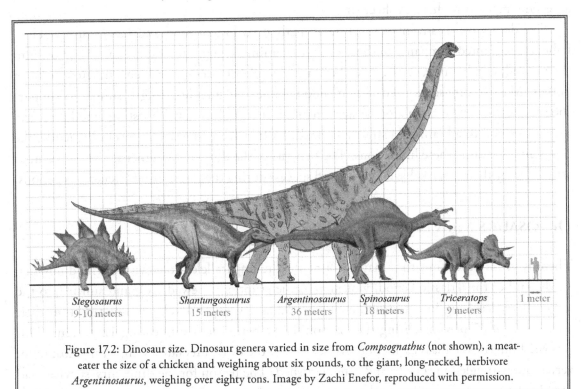

| *Stegosaurus* | *Shantungosaurus* | *Argentinosaurus* | *Spinosaurus* | *Triceratops* | 1 meter |
| 9-10 meters | 15 meters | 36 meters | 18 meters | 9 meters | |

Figure 17.2: Dinosaur size. Dinosaur genera varied in size from *Compsognathus* (not shown), a meat-eater the size of a chicken and weighing about six pounds, to the giant, long-necked, herbivore *Argentinosaurus*, weighing over eighty tons. Image by Zachi Enefor, reproduced with permission.

Large size can also help animals achieve homeothermy—the ability to maintain a relatively constant internal temperature, even as the external environment changes rapidly from hot to cold. This is different from endothermy, or warm-bloodedness, which is the control of internal body temperature by adjustments in metabolism. (We will hear about this later in this chapter.) Large animals retain heat better than smaller animals because, relative to their volume, they have less surface area through which to lose heat. Bus-sized *Stegosaurus*, for example, warmed itself by absorbing heat from the sun during the day using the unique plates on its back, and cooled slowly at night owing to its massive size—heat dissipating off the relatively low surface area of its skin. Its internal temperature was kept fairly constant. A poor, tiny lizard, on the other hand, absorbs heat rapidly during the day—and may even have to get out of the sun to avoid overheating—and loses heat rapidly during the cold night because its surface area is relatively large compared to its volume. *Stegosaurus*, which maintains its body heat during the night because of its size, will remain alert and active in the morning, while the tiny lizard, which loses heat during the night, will be cold and sluggish in the morning. This concept can also help explain why warm-blooded mammals and birds tend not to reach massive dinosaurian sizes. We mammals and birds regulate our temperature metabolically and need not maintain an even temperature through sheer size.

Size does have disadvantages. Large animals need lots of food and lots of space. Their habitats are limited compared to those of smaller animals; giant dinosaurs cannot burrow into the ground, climb a tree, or fly. And there are a number of design problems inherent to huge bodies. Massive dinosaurs needed tremendous bone and muscle strength simply to support their bodies, and their circulatory systems had to generate near vessel-bursting blood pressures to pump blood to a brain held stratospherically aloft at the end of a long neck. Evidently, the dinosaurs overcame these limitations. The benefits of the dinosaurs' immense size won out, as attested to by their remarkable longevity and success during the Jurassic and Cretaceous periods.

DINOSAURS SURVIVE EXTINCTION

There is yet another trait that may account for the unparalleled longevity of the dinosaurs. At the end of the Triassic Period, violent volcanic eruptions spewed CO_2 and sulfur into the atmosphere, resulting in catastrophic climate change. Oxygen levels plummeted from their peak levels of 35 percent during the Carboniferous period to a suffocating 12 percent. This was the End-Triassic mass extinction. Mysteriously, the dinosaurs fared well, while most other vertebrates struggled to survive and often perished. The reason for the dinosaurs' survival had remained a mystery—until recently.

Biologist Peter Ward believes he has the answer, which he explains in his insightful book *Out of Thin Air* (Ward 2006, 180–197). Ward cites fossil evidence documenting air sacs in the bones of saurischian dinosaurs that are nearly identical to air sacs used for respiration in bird bones. Scientists reached the astonishing conclusion that these early dinosaurs had a respiratory system similar to that of current-day birds—a system markedly superior to those of other vertebrates (figure 17.3)! This may explain the dinosaurs' improbable survival during the severe hypoxic period of the End-Triassic mass extinction, which allowed them to make it to the Jurassic and Cretaceous periods, during which they reigned supreme.

Mammalian Lung

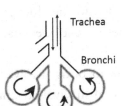

Bird Lung

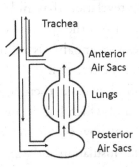

Figure 17.3: Mammalian and bird lungs. Mammals breathe air into dead-end pouches (alveoli) and then out again. Such a dead-end system causes mixing of oxygen-poor air in the alveoli at the end of expiration with new inhaled oxygen-rich air. This is not very efficient. Birds have found a better way. Air is brought into posterior air sacs, pumped through the lungs unidirectionally, and then pumped back out through the trachea by anterior air sacs. The air sacs act as bellows to move the air through the lungs. By pumping air unidirectionally through the lungs, almost all the circulating air that reaches the lungs for gas exchange is oxygen-rich air, and gas exchange can be enhanced by combining the unidirectional flow of air with a cross-current pattern of blood flow (Ritchison 2018).

This very efficient respiratory system has enabled birds to fly long distances at great altitudes. There is evidence that saurischian dinosaurs evolved this type of respiratory system and subsequently passed it onto birds. This may have helped the dinosaurs survive the suffocating low oxygen levels of the End-Triassic extinction.

As discussed below, the theropod dinosaurs gave rise to the bird lineage and gifted their remarkable respiratory system to their avian descendants (Stricherz 2003). Those majestic geese flying thousands of feet above Mount Everest, who were the envy of the Everest climbers described in the epigraph at the beginning of this chapter, owe their place atop the world to their ancient theropod ancestors.

FROM FEATHERED DINOSAURS TO BIRDS

Just as *Tiktaalik* became the iconic evolutionary symbol of the transition from fish to tetrapods, so *Archaeopteryx* became the iconic symbol of the transition from dinosaurs to birds. The 159-million-year-old fossil remains of *Archaeopteryx* were discovered in Germany in 1861 just two years after Darwin published *On The Origin of Species*, and provided prime evidence for Darwin's theories (Erickson et al. 2009). *Archaeopteryx* (meaning "ancient wing") was a jay-sized creature with toothy jaws, a lizard-like tail, and feathers. Although it is often considered the first bird on our planet, it was a transitional figure between a bird and a theropod dinosaur—the dinosaur group that includes *Tyrannosaurus rex* and *Velociraptor*. This discovery and others have led to the astonishing revelation that all birds are the direct descendants of theropods. Every bird you've ever seen—from the sparrow on your feeder, to the chicken laying your eggs, to ostriches on the African savannah—is a dinosaur. They are the ones that survived—a twig on the dinosaur branch of the tree of life.

The complex evolution from theropod dinosaurs to birds involved a number of small stepwise changes, each of which had its own incremental survival advantages. Birds were not "trying" to fly; that's not how evolution works. Hairlike feathers on the bodies of theropods likely evolved to provide insulation and waterproofing. They contain specialized keratin proteins with tight bonding and crosslinks that make them tougher, but lighter, than either hair or nails. Initially theropods had only feathers on their front limbs and tails, but over time feathers would cover more of the body—perhaps to allow nesting mothers to warm their eggs. Later on, the bones of birds became thinner, lighter, and longer, and the bones of the front limbs shortened and fused together to form wings. In the end, these amalgamated changes facilitated flight.

When the giant asteroid collision wiped out the dinosaurs at the end of the Cretaceous period sixty-six million years ago, the birds miraculously survived. How the birds weathered the mass extinction remains a mystery, but some have attributed their perseverance to their small size and ability to build burrows and nest in tree holes to escape environmental hazards. Following the end-Cretaceous extinction—and after millions of years of recovery—birds experienced an adaptive radiation, taking over skies left empty with the extinction of pterosaurs (flying reptiles) that had preceded them. Birds, of course, are still with us and thriving, with more than ten thousand extant species.

MAMMALS EMERGE

Mammals first appeared on our planet during the Jurassic period, at the height of the dinosaur domain. Petrified remains of an ancient tiny furry animal called *Hadrocodium wui* were found in China in 1985 dated 195 million years ago and represent the oldest mammalian fossil (Luo, Crompton, and Sun 2001). The early mammals had descended from cynodonts, the group of mammal-like reptiles (therapsids) that included the famous "dog-toothed" *Cynognathus* that we met in the last chapter. We are fortunate that the cynodonts were lucky enough to survive the end-Permian extinction and continue the lineage that led to us mammals.

No one celebrated the mammals' arrival. They were small, unobtrusive, rodent-like creatures rarely more than a few inches long. These tiny critters could hide in the brushes, scurry up trees, or dig into burrows to avoid the predator dinosaurs. They survived but lived in the shadow of the dinosaurs for millions of years.

When the end-Cretaceous extinction devastated marine and terrestrial life, and wiped out all the dinosaurs, mammals were hit hard too, but a select group survived. Perhaps it was because they were burrowers by nature and avoided the chaos of the asteroid impact by hiding underground.

After millions of years, when the clouds of debris had finally lifted, new habitats opened and provided opportunities for those mammalian species that survived. The giant dinosaur predators were gone, and mammals ushered in a new era with fresh opportunities. Flowering plants provided fruits and berries as new sources of food. A burgeoning insect population provided additional prey. A variety of forests gave rise to alternative habitats for birds and tree-dwelling mammals. In response to all this bounty and space, mammals grew in size. Some migrated back to the oceans to become marine mammals (seals and whales), and some took to the air (bats). The monotremes (egg-laying mammals, such as platypuses and echidnas) and marsupials (pouch-bearers) evolved in the isolated continents of Australia and South America, while placental mammals evolved on other continents. This explosion of diverse mammalian species is one of the most dramatic examples of adaptive radiation in the history of life on Earth.

Mammals have evolved a number of defining traits that have allowed them to survive in new environments and, over time, to become the dominant class of animals on Earth. The placenta, which we discussed in chapter 11, provides the embryo with nutrients and allows the fetus to mature within the mother's womb. Mammary glands provide a source of nutrients for newborns. Vertical limbs facilitate more efficient locomotion, and fur provides protection and insulation against heat loss. But perhaps the most intriguing, and one of the most important, adaptive traits that evolved in mammals was warm-bloodedness.

HOT BLOOD

If you stop eating, you won't survive for two months. You'll die rapidly of starvation. But a crocodile can go comfortably without eating for a year. Why is this? Unlike crocodiles and other reptiles, we generate body heat by revving up our metabolism. We use most of the energy from our food to stay warm. At first glance, this might sound a bit wasteful. The crocodile lives just fine by warming its body basking in the sun and then cooling off in the water, not wasting a bit of food energy on self-heating. Why can't we do the same? It's like turning up your home thermostat to blistering extremes and running up the gas bill. We pay a heavy price to keep our bodies warm in this way. For this trait to have evolved, it must have some great advantages.

AEROBIC CAPACITY

Tour de France champion cyclist Greg LeMond has one of the highest aerobic capacities measured in humans—92.5 ml/kg/min, as compared to that of an average untrained young male of about 45 ml/kg/min. But we humans are nowhere near the elite group of animals. Each animal has evolved an aerobic capacity to meet its needs. The North American pronghorn antelope can consume oxygen at a rate of over 300 ml/kg/min.

Aerobic capacity (VO_2 max) is measured as the maximum rate of oxygen consumption an animal can achieve during maximal exercise. Since oxygen is used in aerobic metabolism to produce ATP, aerobic capacity is a measure of the maximum rate of aerobic energy production.

Oxygen delivery to the body is dependent on a number of systems. Lungs transfer oxygen from the air to the circulatory system. Hemoglobin binds oxygen and increases the oxygen-carrying capacity of blood. The cardiovascular system delivers blood and oxygen through the capillaries to the tissues. And the muscles and their mitochondria accept and utilize oxygen in aerobic metabolism. Animals have evolved each of these components of the oxygen transport system to varying degrees to meet their needs, and each of these systems can be enhanced through exercise training. High aerobic capacity, which provides the capability for sustained physical activity, has provided animals with an important survival advantage.

Mammals keep their body temperature constant within a narrow range, although the temperature varies from species to species. Human body temperatures normally range between 97°F and 99°F (36.1°C and 37.2°C), which is an optimum temperature range for our enzymes to work and our complex metabolisms to run. ATP generation and muscle power, for example, are strongly temperature dependent. We warm-blooded animals, or **endotherms**, can survive in cold habitats and forage during cold nights. In contrast, cold-blooded animals, or **ectotherms,** which rely on heat sucked directly from the environment, have wide swings in their internal temperature. They are sluggish during cool nights and often overheat in the daytime sun. As a consequence, they are unable to live in very cold climates.

Heat is produced as a by-product of metabolism, and animals with high metabolic rates produce the most heat. It is not surprising that warm-blooded animals have higher metabolic rates than cold-blooded animals. But how did warm-blooded animals evolve their higher metabolic rates?

In 1979, researchers Albert Bennett and John Ruben published a classic paper describing their "aerobic capacity hypothesis" (Bennett and Ruben 1979). According to this theory, the high resting metabolic rate responsible for warm-bloodedness evolved as a consequence of the evolution of high **aerobic capacity**. Aerobic capacity is an animal's capacity to consume oxygen during maximal exercise (see sidebar 17.1). Animals able to access and consume more oxygen are able to produce more ATP, and as a consequence, they are able to run faster and longer while chasing prey or fleeing predators. The survival advantages are obvious.

Animals that evolved high aerobic capacity did so by increasing muscle mass, by increasing the number of mitochondria in their muscle cells that produce ATP, and by increasing the efficiency of the cardiorespiratory system for delivering oxygen to the tissues. The consequence of these adaptations was the evolution of a high basal metabolic rate. Metabolism is not totally efficient and generates heat as a by-product, and the heat generated can be used to keep the body warm. Endothermic animals have learned to decouple the process of ATP production, diverting more energy to the production of heat, when needed, and less to the production of ATP (Nedergaard, Ricquier, and Kozak 2005).

Bennett and Ruben believed the need for high aerobic capacity preceded endothermy, but not everyone agrees. Given the significant adaptive advantages of warm-bloodedness, it is certainly possible that warm-bloodedness came first, and that high metabolism and high aerobic capacity followed (Clark and Portner 2010; Wantanabe 2005). Which was the evolutionary driver of change? We will never know for sure, but birds and mammals have certainly benefited from both. Nick Lane, in *Life Ascending*, called hot blood one of the ten greatest inventions of evolution. Armed with eternally temperate bodies and an almost unlimited energy supply, mammals and birds flourished following the end-Cretaceous extinction and the demise of the dinosaurs. They have dominated our land and sky ever since. It seems that the high energy cost of warm-bloodedness has been worth the price.

WARM-BLOODED REPTILES?

Our ancient reptilian ancestors, the mammal-like cynodonts from the late Triassic period, may have been warm-blooded animals. An exemplar from this group that has been well studied is the "dog-toothed" *Cynognathus*. This predator was about a meter long and had a large head and wide jaws with sharp teeth. It laid eggs and had forelimbs that sprawled outward like those of most reptiles but its hind limbs were positioned beneath its body. Like other cynodonts, it had many features suggesting it had high aerobic capacity. *Cynognathus* fossils have a hard palate, like ours, that separates the nasal passages from the mouth, so it could chew and breathe at the same time. They have broad chests and rib cages, and probably had a muscular diaphragm that enhanced breathing and oxygen delivery. Its nasal passages were large, with a delicate latticework of bone called respiratory turbinates to help minimize water loss during breathing. Respiratory turbinates are characteristic of warm-blooded animals and are not seen in reptiles, other than mammal-like reptiles. In other words, there is evidence that cynodonts already had a high aerobic capacity and warm blood, because they had the breathing machinery to enable it. They were well on their way to becoming mammals.

Since dinosaurs are reptiles, it has long been assumed that they are cold-blooded. But theropod dinosaurs, which include *Tyrannosaurus rex* and *Velociraptor*, had high energy rates and led very active vigorous lives, suggesting they may have been warm-blooded. And remember that their descendants, the birds, are warm-blooded. This has kindled a great debate. Recently investigators hoped to resolve this controversy by analyzing growth rates of dinosaurs from fossilized bones (Grady et al. 2014). Bones have growth rings, just like the rings of a tree, which are thought to indicate how fast the animals grew. The results of the analyses are somewhat controversial. Depending on the methodology, dinosaur growth rates are estimated as being either similar to those of mammals or intermediate between those of mammals and reptiles (Geggel 2015; D'Emic 2015). This suggests that dinosaurs were either endothermic or mesothermic (intermediate between endothermic and ectothermic).

DEATH BEGETS LIFE

If there's any lesson to take away from these last two chapters, it's this: destruction makes way for new life. Each great extinction event in the history of animal evolution has led to explosions of new adaptation and new species, as survivors have raced to fill niches left empty by those that died. The formation of the supercontinent Pangaea forced climate change and a minor extinction event that led to the retreat of the amphibians. The end-Permian extinction decimated most reptiles but opened new space for the few survivors. These included the mammal-like reptiles (cynodonts), which gave rise to the mammalian lineage, and archosaurs ("ruling lizards"), which gave rise to the dinosaurs. The dinosaurs survived the end-Triassic extinction, while most of their competitors did not, allowing them to rise to dominance in the Jurassic period. Finally, the giant asteroid collision at Chicxulub ended the unprecedented reign of the dinosaurs, leaving a void and an opportunity for mammals to flourish, diversify, and become the dominant animals on Earth. We will explore these and other great extinction events in the next chapter and see the profound impact they have had on the evolution of life on our planet.

CHAPTER 18

THE MASS EXTINCTIONS

> Mass extinctions are not unswervingly destructive in the history of life. They are a source of creation as well … Mass extinction may be the primary and indispensable seed of major changes and shifts in life's history. Destruction and creation are locked in a dialectic of interaction.
>
> —Stephen Jay Gould, in *The Flamingo's Smile*

We move ever forward in this story of the history of life. The age of the dinosaurs was a time of fertility and growth for our planet. Near the end of the Cretaceous period, about seventy million years ago, the supercontinent Pangaea had completed a tectonic breakup into the seven distinct continents we live upon today. The climate was balmy, seas were high, and the land was covered in newly evolved plant life—trees, grasses, and the extravagance of flowering plants. Bees, dragonflies, beetles, and other insects had evolved to depend on these plants. And the warm seas were teeming with marine invertebrates—nautiluses, barnacles, lobsters, crabs, and sea urchins—as well as ancestral sharks and schools of bony fish. Crocodiles, giant turtles, lizards, and snakes had all appeared for the first time—and the titanic ancestors of birds, the dinosaurs, prowled the Earth on giant clawed feet.

But then one day it came; a rain of fire descended from the sky that would blanket the world in ash.

THE GREAT COLLISION

Sixty-six million years ago, marking the end of the Cretaceous Period, a massive asteroid three times the width of Manhattan entered Earth's gravitational field on a collision course with our planet. The sky lit up red and blue as the asteroid punctured our atmosphere with a crash of hissing static noise and plummeted toward the coast of Mexico. It smashed into the Yucatan Peninsula, and its impact released energy ten thousand times as strong as the detonation of today's entire nuclear arsenal (Ward and Brownlee 2000, 158).

The effects of the asteroid were immediate and global. A wave of heat blew out from the crater and vaporized all life in the region. Rock fragments flew into the stratosphere and then fell back to Earth themselves as flaming meteorites. They set forests ablaze and scorched the entire surface of our planet with firestorms. Huge tidal waves smashed the shorelines of neighboring continents. Dust, billowing smoke, and debris rich in sulfate aerosols filled the skies, obliterating the sun. Over the next days, the world grew dark, and temperatures plummeted. Trees and shrubs, deprived of sunlight, withered and died. Starvation rippled upward through the food chains. Herbivores and carnivores alike succumbed to the resulting famine, and the stench of decay permeated the air. No other great extinction had caused such rapid devastation (image 18.1).

Image 18.1: Dinosaurs' extinction. The giant asteroid that collided with Earth and caused the end-Cretaceous extinction wiped out three-quarters of Earth's species, including the mighty dinosaurs.

This is how the great end-Cretaceous extinction (also known as the K/T boundary extinction) likely happened sixty-six million years ago. The most famous victims, of course, were the dinosaurs, but most terrestrial organisms were hit hard; mammals, birds, lizards, insects, and plants were all affected. No large creatures survived. Oceanic life was less affected but was by no means spared. The extinction devastated marine lizards, fish, sharks, and mollusks. About 75 percent of all species were lost (Raup 1991, 66–74).

Even through all this destruction, there were survivors aplenty. The extinction left a void that provided new evolutionary opportunities, although the changes would take millions of years to transpire. Birds and fish underwent remarkable adaptive radiations to fill the environmental niches left vacant by the extinction. But perhaps the greatest adaptive radiation of all was the proliferation of mammals, which transformed into whales, bats, ungulates, rodents, carnivores, primates, and more, filling voids on land, sea, and air. As with previous extinction events, evolutionary pathways were interrupted and reset to begin anew in diverse directions. Many believe that if the extinction event had not occurred and the dinosaurs had not been destroyed, the mammalian lineage would have remained diminutive and lowly—and we, *Homo sapiens*, may never have evolved.

The cause of the end-Cretaceous extinction was long shrouded in mystery. We knew it happened, but we didn't know why. Then, in 1980, Nobel Prize–winning physicist Luis Alvarez and his son, Walter Alvarez, published a paper in the journal *Science* describing a puzzling piece of evidence buried within Earth's crust that provided a clue to solving the mystery (Alvarez et al. 1980).

The investigators had discovered high concentrations of the silvery-white metal iridium in a specific band of ancient sedimentary bedrock found around the world. The rock layer in question was sixty-six million years old—coinciding exactly with the time that paleontologists had recognized as the end-Cretaceous extinction. The Alvarezes and their investigatory team felt that this simply could not be a coincidence. Iridium is exceedingly rare in Earth's crust, but it is abundant in most asteroids and comets. Such a quantity of iridium, they explained, could only have come from one of these space rocks, leading them to believe that there must have been a colossal asteroid collision at the time of the end-Cretaceous extinction. There were also fine particles of soot in those sedimentary layers, suggesting that the Earth's surface at the time had been swept by forest and brush fires.

Evidence for the asteroid's impact was strengthened by the discovery of its fingerprints. In 1990, an asteroid crater 110 miles in diameter was discovered buried beneath the town of Chicxulub near the coast of Yucatan, Mexico. Radiometric dating placed the time of impact to sixty-six million years ago, the time of the great extinction (Richards et al. 2015). The evidence was compelling.

But some have challenged the Alvarezes' theory. Geologists Paul Renne and Mark Richards believe volcanoes contributed to the end-Cretaceous extinction (Renne et al. 2015). They cite evidence found in the Deccan Traps, a huge region of mile-thick basaltic rock stretching across Western India, resulting from massive volcanic eruptions. Using precise dating techniques, they found the volcanoes had been bubbling along slowly until—you guessed it—sixty-six million years ago, when they suddenly and exuberantly came to life, pouring off voluminous flows of lava. The timing of this abrupt change was too improbable to be a coincidence. The investigators believe the asteroid impact resulted in violent earthquakes, which disrupted magma chambers (large underground pools of liquid rock) and changed the plumbing of the Indian volcanos, unleashing a massive lava flow that persisted for hundreds of thousands of years. These investigators postulated the volcanic eruptions spewed huge quantities of carbon dioxide and sulfur compounds into the atmosphere, resulting in calamitous volcano-induced climate change. The Deccan Trap volcanoes alone, they postulated, could easily have been responsible for the great extinction that destroyed the dinosaurs.

Renne and Richards's theory isn't necessarily in conflict with the Alvarezes' collision theory. All of these investigators are faculty members at the University of California–Berkeley, and all have collaborated with one another in previous studies. Ultimately the consensus seems to be that the end-Cretaceous mass extinction and the demise of the dinosaurs were due to a cascade of events triggered by the massive asteroid impact and culminating in massive volcanic eruptions in India and elsewhere. Case closed and may the dinosaurs rest in peace. (Excluding the birds, of course—the only dinosaurs that are still alive and well.)

WHEN LIFE NEARLY DIED

The end-Cretaceous extinction is the best known, best studied, and best understood of all Earth's mass extinctions—but it was not the worst. There have been five mass extinctions documented in the fossil record over the past five hundred million years defined as events in which 75 percent or more of Earth's species have been lost. The greatest mass extinction of all time occurred 251 million years ago at the end of the Permian period and exacted the greatest single death toll the living world has ever seen, wiping out 90 to 96 percent of Earth's species (Benton 2003, 202). Life on Earth nearly died.

Before the extinction, life in the Permian period was flourishing. Pangaea had emerged as a single supercontinent that presented great extremes of climate and environment because of its vast size. The seas were dominated by bony fishes with fan-shaped fins and thick scales, and by large reef communities harboring squid-like mollusks. A variety of conifers and other drought-resistant seed plants populated the landscape. Insects colonized the lands and quiet waters, and included a variety of forms resembling dragonflies, grasshoppers, cockroaches, and beetles. A diversity of amphibians lived in marshes and freshwater lakes in the early Permian period, but by the late Permian period, amphibians were in sharp decline and reptiles ruled the land. The mammal-like reptiles, or therapsids, which included the piglike *Lystrosaurus* and the doglike cynodonts, dominated the landscape in the late Permian. The sauropsid reptiles were a nondominant group composed mostly of small lizard-like creatures, but they included specialized lizards with winglike appendages that enabled them to glide through the air, as well as the five-hundred-kilogram armored pareiasaurs ("cheek lizards"), the world's first large reptile herbivores. Life was plentiful, diverse, and flourishing. But the Permian period was to be the last gasp for much of early prehistoric life. The era came to a calamitous end 251 million years ago, marking a biological dividing line that few creatures would cross. The end-Permian extinction was the greatest mass extinction in the history of our planet. (Benton 2003).

As with the end-Cretaceous extinction, geologists and paleontologists have struggled to put together pieces of an ancient puzzle to solve the mystery of the causes of this unparalleled extinction. The truth includes multiple scenarios and may seem a bit untidy.

Once again, volcanoes may be responsible. Massive volcanic eruptions, evident in the geologic record at the end of the Permian period, spewed masses of hot lava over much of Siberia (Renne and Basu 1991). These eruptions formed the Siberian Traps, which are Earth's biggest igneous province—a sheet of basalt lava rock almost a mile thick, covering an area in Russia the size of Western Europe. Paleontologist Michael Benton describes the likely consequences of these massive volcanic eruptions in his insightful book *The Day Life Almost Died* (Benton 2003). The eruptions spewed forth sulfur aerosols and volcanic ash that choked the Earth's atmosphere, blocking sunlight, rapidly cooling the planet, and smothering plant life. Massive volumes of carbon dioxide and sulfur dioxide gases were released into the atmosphere, where they combined with water vapor to form toxic carbonic and sulfuric acids. The resulting acid rain destroyed most of the plants that hadn't already succumbed to light deprivation. Herbivores starved, and carnivores followed suit.

The volcanic eruptions began a cycle of climate change and global warming that drove the mass extinctions. Carbon dioxide flowed out of the volcanoes, and as the hot lava spread across the crust of Siberia, it intruded into one of the largest coal basins, cooking the fossil fuels and spewing additional massive quantities of CO_2 into the atmosphere (Brannen 2017; Kaplan 2009). Sound familiar? CO_2 levels reached ten times current-day levels, heating up the globe through its massive **greenhouse effect**. To compound the problem, investigators believe global warming may have triggered the release of methane gas from methane gas hydrates—crystalline structures in the ocean consisting of methane enclosed in a cage of frozen water (Benton 2003, 272–77). And the emergence of methane-producing microorganisms may have also contributed to methane release (Rothman et al. 2014). The effects of this combined glut of methane and carbon dioxide were dramatic and lethal. The gasses created an enormous greenhouse effect, causing ocean surface temperatures in tropical and subtropical latitudes to reach 40°C (104°F)—lethal to most life (Puiu 2012; Sun et al. 2012). By comparison, ocean surface temperatures in similar latitudes today average 25–30°C (77–86°F). The global warming impaired oceanic circulation and resulted in greatly reduced oxygen levels in the seas. This, along with acidification of the oceans due to high carbon dioxide levels, wiped out most marine life.

But even before the volcanic eruptions began, events were conspiring against life in the Permian. Oxygen levels were falling. The late Robert Berner, a giant of geology from Yale University, used carbon cycle modeling to determine that oxygen levels plummeted from a glutted 35 percent during the Carboniferous period to a paltry 12 percent in the early Triassic period (Berner 2006). Oxygen levels were approaching these low levels in the late Permian period before the volcanic eruptions. The reasons are not totally clear but may be related to changes in the carbon cycle with increased oxidation of organic material from dying plants and animals, resulting in increased oxygen consumption and falling atmospheric oxygen levels (see sidebar "Changing Oxygen Levels") (Berner 2007). Today such low oxygen levels are found at elevations over fourteen thousand feet—at the summit of Mt. Rainier in Washington State, for instance. The doomed animals at the end of the Permian period must have huddled at sea level and struggled to breathe even there. The very low oxygen levels at the end of the Permian period and early Triassic period certainly accelerated extinction rates and slowed life's recovery for millions of years. As paleontologist Peter Ward said, "You can go without food for a couple of weeks. You can go without water for a few days. How long can you go without oxygen, a couple of minutes? There's nothing with a greater evolutionary effect than oxygen." (Ward 2006).

Geoscientist Lee Kump believes there may have been one final accomplice in the murderous mass extinction—another microorganism that finished the job (Kump, Pavolv, and Arthur 2005). Global warming stifles oceanic circulation, depleting the oceans of oxygen. This creates ideal conditions for anaerobic bacteria to flourish and proliferate into massive blooms. Kump believes that blooms of anaerobic sulfur bacteria released huge quantities of toxic hydrogen sulfide (with its horrid stench of rotten eggs) into the oceans and the atmosphere, killing marine and terrestrial life.

Knowing all that we know now, close your eyes and imagine yourself at the seashore at the time of the end-Permian mass extinction. The atmosphere is filled with a toxic mix of hydrogen sulfide, methane, and carbon dioxide gases, with very little oxygen. The oceans are sludgy and dark with metal sulfides that precipitate because of the hydrogen sulfide in the water. Marine life is almost gone except for blooms of anaerobic bacteria, with their putrid smell of rotten eggs. The landscape is barren—all the plants have withered and died—and many piglike Lystrosaurus huddle near the shoreline. They are hungry, displaced, and gasping at oxygen-starved air. They, too, are dying. It is a dark time in Earth's history. Life on our planet nearly came to an end.

The end-Permian extinction stands out from other extinctions because of its sheer impact and its prolonged recovery. Diverse ecosystems did not return for many millions of years. Like other mass extinctions, the end-Permian extinction altered the course of evolution by triggering the extinction of dominant species, destroying old ecosystems, and opening new ecosystems. It created the conditions for adaptive radiations—the rapid diversification and evolution of new species. As we learned in the last two chapters, several groups did survive the extinction, and they inevitably formed the lineages that are still with us today. Two groups of mammal-like reptiles escaped the holocaust. The dog-like cynodonts escaped and gave rise to all current-day mammals, and the piglike *Lystrosaurus* squeezed through the extinction and came to dominate the Triassic period. The primitive clade of reptilian sauropsids also survived and gave rise to the lepidosaurs and the archosaurs. The lepidosaurs' descendants include modern-day reptiles, and the archosaurs' descendants include the mighty dinosaurs and birds. Death begets life. The end-Permian extinction greatly altered the course of evolution and shaped the biodiversity of our planet forevermore.

BACKGROUND EXTINCTIONS AND MASS EXTINCTIONS

Extinction is part of evolution and part of the natural history of life. No species survives forever. In the absence of mass extinctions, the fossil record has shown that the average lifespan of a species ranges from about one million to ten million years—tending to be on the shorter side in mammals and the longer side in invertebrates (Wilson 2002, 99). Diversity is a measure of the number of different species that are around at the same time. As species go extinct, new species evolve, and if the extinction rate is about the same as the rate of new-species formation, diversity remains about the same. In the past sixty-five million years, changes in diversity have been small; on average, when one species has gone extinct, another has evolved to take its place. Currently, an estimated over eight million species inhabit our planet, of which about 1.8 million have been identified, classified, and documented in registries (Mora et al. 2011).

As we have seen, normal background extinction rates are punctuated by dramatic extinction events, some of which are categorized as mass extinctions. When more than 75 percent of Earth's species are lost within a relatively short geological time period, we consider this a mass extinction. As we mentioned earlier, there have been five such mass extinctions identified by paleontologists within the last five hundred million years, when fossil records have been available (figure 18.1). We have just discussed the two most famous: the end-Permian extinction and the end-Cretaceous extinction. Here are all five, in chronological order:

- end-Ordovician extinction: 450 million years ago (mya)
- late-Devonian extinction: 375 mya

- end-Permian extinction: 251 mya
- end-Triassic extinction: 200 mya
- end-Cretaceous extinction: 66 mya

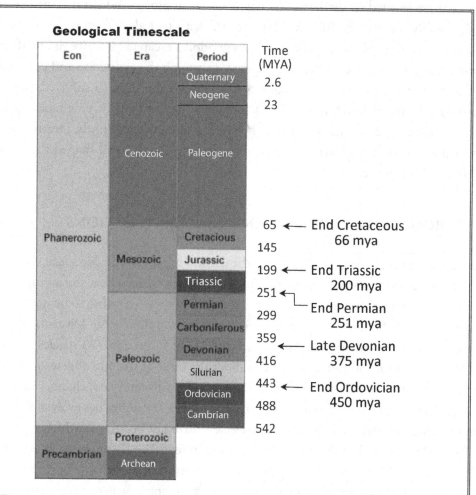

Figure 18.1: Geologic timeline of mass extinctions. There have been five mass extinctions over the past five hundred million years, since fossil records have been available. Mass extinctions are defined as a loss of greater than 75 percent of species over a relatively short geologic time period.

Just a few words about the end-Triassic extinction: Remember from the last chapter that this was the extinction that allowed the dinosaurs to become the dominant land animals on Earth. The cause of the extinction is the subject of considerable debate, but many experts believe volcanic activity, once again, may have triggered the event. At the end of the Triassic period, the supercontinent Pangaea began to break up, and this precipitated massive volcanic activity, releasing CO_2 into the atmosphere and inducing climate change with global warming. At the same time, oxygen levels fell to almost half of current-day levels, suffocating many animals. The low oxygen levels were due to changes in sulfur and carbon cycles (see sidebar 18.1). Massive shifts in tectonic plates with the breakup of Pangaea exposed organic carbon to oxygenation and exposed pyrite to weathering, consuming large quantities of oxygen. An estimated 76 percent of species died. Dinosaurs were luckier than most and were able to survive owing to their efficient respiratory systems.

There is recent evidence that volcanic eruptions may have triggered most of Earth's extinction events. Using updated radioisotopic dating techniques, researchers from the United Kingdom and Canada reviewed twenty extinction events (five mass extinctions and fifteen lesser extinctions) and concluded that at least half of them were driven by large-scale volcanic eruptions, including three of the five mass-extinction events: the end-Ordovician, end-Permian, and end-Triassic mass extinctions (Bond and Grasby 2017). And as we have seen, even the end-Cretaceous extinction that eliminated the dinosaurs was aided by volcanic eruptions. But not all large volcanic eruptions result in mass extinctions. Scientists believe it requires a perfect storm of events. Massive volcanic eruptions release CO_2 into the atmosphere, but probably not enough to cause damaging climate change. However, when volcanic eruptions spew lava across large areas of Earth's crust in regions that contain sequestered fossil fuels, these fuels can be ignited and release enormous quantities of CO_2,

CHANGING OXYGEN LEVELS

Photosynthesis by cyanobacteria, algae, and plants produces the oxygen that we aerobic animals use to breathe. The amount of oxygen produced by photosynthesis and consumed by animals has some effect on the oxygen content of Earth's atmosphere, but atmospheric oxygen levels are determined to a much greater extent by the sulfur and carbon cycles.

Fool's gold, or pyrite (iron disulfide, or FeS_2), is locked in huge quantities within a variety of rocks. When pyrite is exposed to open air by shifting of tectonic plates and geologic upheaval or through normal weathering by rain or wind, it can react with oxygen to produce iron oxide and sulfur dioxide. This process eats up large amounts of oxygen from the atmosphere and can lower atmospheric oxygen concentrations. This process is what we call the **sulfur cycle**.

The carbon cycle is even more important in determining oxygen levels. When animals and plants die and decay, large quantities of organic carbon in their bodies react with and consume atmospheric oxygen. If, on the other hand, the corpse is rapidly buried and sequestered, no oxygen will be consumed. This is what happened in the Carboniferous period when masses of trees and plants were rapidly buried and sequestered in lowland swamps, and their organic material was not exposed to oxygen. (Microorganisms had not yet learned how to break down lignin, which protected the plants from decomposition.) Since the organic carbon in fallen trees did not react with and consume atmospheric oxygen, oxygen levels rose to all-time highs. In contrast, in the late Permian period, with the formation of the supercontinent Pangaea there was a loss of lowland swamps and a consequent reduction in burial and sequestration of organic materials. Oxidation of these organic materials resulted in increased oxygen consumption and falling oxygen levels. (Berner 2007)

During the Triassic period, shifts in tectonic plates exposed buried carbon and pyrite to weathering, and these elements reacted with and consumed large quantities of atmospheric oxygen, causing oxygen levels to plummet to deadly lows near the end of the Triassic period.

The late Yale geologist Robert Berner and his colleagues developed complex computer models to estimate oxygen levels over time based on changes in variables affecting the sulfur and carbon cycles (Berner 2006).

as we saw with the end-Permian extinction (Kaplan 2009). Massive CO_2 release can produce a characteristic sequence of destructive events: global warming from the greenhouse effect, destruction of plant life from acid rain, damage to the ozone layer, acidification of the oceans, and heat-induced interruption of oceanic circulation causing marine hypoxia and suffocation of marine life.

THE GREAT OXYGEN CATASTROPHE

One of the greatest extinctions of all time occurred so early in the course of evolution that it left no fossil record. Microorganisms, after all, have no bones that can fossilize. With no fossil record, this event is not counted among the five mass extinctions—even though it was one of the deadliest moments in the history of life. And this event was curiously different from the five mass extinctions. The cause was biological. This was the great oxygen catastrophe.

Remember: not long after the origin of life, cyanobacteria developed the capacity for photosynthesis and began releasing oxygen as a waste product into the oceans. The gas was quickly consumed as it reacted with iron and other minerals, forming metal oxides, which precipitated to the ocean floor. Microorganisms went about their business, blissfully unaware of the impending disaster. But finally there came a day, about 2.3 billion years ago, when the iron ran out. With the oceans drained of reactive minerals, free oxygen began to accumulate in the atmosphere for the first time in the history of our planet. We humans think of oxygen as a life-sustaining substance; our cells use it through every moment of our lives. But to the microorganisms swimming in the oceans in those ancient times, oxygen was deadly toxic. It destroyed cell membranes on contact. The **great oxygen catastrophe** was the beginning of an ecological disaster that would cause the near extinction of all microbial life on Earth.

Things only got worse. Not only did the emergence of free atmospheric oxygen pose a direct toxic threat to all living things, but it also produced catastrophic global cooling. Prior to the oxygen catastrophe, atmospheric carbon dioxide and methane swaddled the planet like an extra blanket, reducing heat loss and keeping the Earth warm through their greenhouse effect. When cyanobacteria acquired the capacity for photosynthesis, they consumed carbon dioxide, and they released oxygen, which reacted with and removed methane from the atmosphere. With the loss of the two blanketing greenhouse gases, carbon dioxide and methane, Earth became naked to the deep chill of space. Temperatures fell to global averages of -20° to -50°C, and the oceans were covered with ice five hundred to fifteen hundred meters thick from the poles to the equator (Ward and Brownlee 2004, 115) This glacial epoch, known as the Huronian glaciation, was the longest and most severe glacial epoch in Earth's history and earned the name "Snowball Earth."

No matter how icebound the surface of Earth, its core burns hot with molten rock. The only survivors of the great oxygen catastrophe may have been a few colonies of microorganisms that took advantage of this. They clustered at hydrothermal vents on the ocean floor, sheltered in the warmth and richness of the alkaline effluence bubbling from the vents. These are our ancestors—the only survivors of the great oxygen catastrophe. They bided their time at the bottom of the sea for hundreds of millions of years, until the Earth finally began to thaw once again.

MASS EXTINCTIONS AND THE COURSE OF EVOLUTION

Mass extinctions not only reset the clock of evolution; they change its course. Great die-offs force evolution to repeat itself as new lineages adapt to new environments. This can lead to startling new innovations. As David Raup describes in his provocative book *Extinction: Bad Genes or Bad Luck*, without mass extinctions, life can become rather static and unchanging. Over time the natural course of a species' evolution reaches anatomic stability, with smaller and smaller changes in body design occurring. Organisms do not start over from scratch but must build on existing structures, and this limits innovation. As organisms evolve, their basic body structure becomes fundamental, and, after a certain point, mutational changes to that body structure are not compatible with survival. This constrains evolution to minor variations on existing structures. It would be difficult, for example, for mammals to evolve an entirely new circulatory system. Raup calls this **phylogenetic constraint** (Raup 1991, 189).

Mass extinctions provide a bloody but effective solution to the problem of phylogenetic constraint. The fossil record has shown that great adaptive breakthroughs—such as the evolution of entirely new taxonomic orders, with never-before-seen anatomies—tend to occur after mass extinctions. Mass extinctions eliminate species and reduce biodiversity, creating space—both ecologic and geographic—for the design of new species via natural selection. The broad expansion of mammals following the dinosaur extinction is a famous example of this. Mass extinctions are fundamental to the process of evolution. They punctuate the equilibrium of life.

Darwin famously wrote in *On the Origin of Species*, "As natural selection works solely by and for the good of each being, all corporeal and mental endowments will tend to progress towards perfection." (Darwin 1859). Evolution, according to Darwin, was slow, steady, and inevitable—flowing smoothly from simple to complex and toward eventual superiority and perfection. The progression from fish to amphibians to reptiles to mammals and, finally, to us, reflected this orderly process of progress and purpose. This view held much in common with the Aristotelian religious concept of **scala naturae**, or the Great Chain of Being, in which all living things are arranged in a divine order of perfection. Man, the pinnacle of evolution, was closest to perfection among all animals—closest to the angels; closest to God. And the entire history of evolution was the story of our ascent toward this perfection.

However, in recent decades there has been movement away from the ideological paradigm of *scala naturae*. Most scientists now hold a different view. Stephen Jay Gould and fellow evolutionary biologist Niles Eldredge proposed the theory of **punctuated equilibria** in a landmark paper presented in 1972 (Gould and Eldredge 1977). The fossil record, they argued, shows long periods of static equilibrium—with little change in anatomy and very few new species—punctuated by sudden bursts of activity, with the appearance of many new species at once. Bacteria and single-celled organisms, for example, ruled the world for almost three billion years with little change. But the Cambrian explosion upended this equilibrium. Multicellular organisms evolved, and scores of new species suddenly sprang to life within just a few million years. This was followed by another period of stasis with relatively few new species appearing, until mass extinctions triggered a new surge of activity. There were rapid bursts of new species in the fossil record after both the end-Permian and end-Cretaceous extinctions. This is the theory of punctuated equilibria in a nutshell. Evolution happens slowly until some catastrophe happens, and then the world changes overnight.

While there remains debate about the causes of punctuated equilibrium, there is little doubt that mass extinctions contribute greatly to the phenomenon. It is clear that mass extinctions interrupt the natural course of evolution and open new environments and new opportunities for new species to evolve. Gone is the notion of an eternal, methodical march of evolution toward biological perfection. Gone is the concept that human beings are the ordained pinnacle of the natural history of life. Instead, life has been and is a struggle through unpredictable timelines punctuated by catastrophes, which are difficult to predict and impossible to control. Like all other life on Earth, we, *Homo sapiens*, are a beautiful accident—and someday we, too, will be gone.

PART V

HUMAN EVOLUTION

CHAPTER 19

OUT OF AFRICA

> As recently as a hundred thousand years ago that zoologist from Outer Space
> would have viewed us as just one more species of big mammal.
>
> —Jared Diamond, in *The Third Chimpanzee*

One hundred thousand years ago, Neanderthals and *Homo sapiens* were like any other species of large mammal, with nothing in particular to distinguish them. Then Neanderthals drifted away into extinction, leaving *Homo sapiens* as the only remaining human species. About seventy thousand years ago, a great transformation was about to occur. *Homo sapiens* developed a capacity for speech, communication, and cooperation that has surpassed all the creatures that came before us. These new human adaptations were followed by rapid cultural evolution. Jared Diamond, in his insightful book *The Third Chimpanzee*, has called this period the "Great Leap Forward" (Diamond 2006, 32).

The rapidity of our cultural change has far outstripped the slow pace of biological evolution. No one could have foreseen the astonishing adaptive successes that have allowed our species to colonize all corners of the Earth, to build gargantuan steel cities and global empires, to soar across the sky like birds, and to manipulate subatomic particles. We may be the only animals with powers of imagination—and we are certainly the only animals whose technology has allowed us to achieve almost anything that we can dream. Our species doesn't just adapt to the environment; we alter and control our environment to suit us.

And yet we are only animals—Genus *Homo*, species *sapiens*. We are not genetically much different than our closest living relatives, the chimpanzees (Green et al. 2010). How have we come this far? Let's see how we got started.

FROM RODENTS TO MONKEYS TO APES

During the Cretaceous period—when dinosaurs were king before the great asteroid collision at Chicxulub—small rodent-like mammals took up residence in forests and trees, where they found food, shelter, and safety. They were low on the food chain and stayed hidden from view. Somehow many of these lowly creatures survived the great asteroid impact and entered the Paleogene period. With time, some of these little mammals adapted to life in the trees, evolving long limbs, gripping fingers, and prehensile grasping tails. They became the first true primates. Recently the fossil remains of a tiny, big-eyed, and long-tailed mammal were unearthed in China (Ni et al. 2013). This ancient animal resembles a modern tarsier—a small, large-eyed primate endemic today to Southeast Asia. The creature was dubbed *Archicebus*, which translates to "original long-tailed monkey," and the fossil, which was dated to fifty-five million years ago—not long after the extinction of the dinosaurs—is the earliest true primate in the fossil record.

Image 19.1: Lemur of Madagascar. Lemurs and tarsiers are descendants and closest living relatives of the earliest primates.

During the Eocene epoch, fifty-five to thirty-five million years ago, primates resembling the tarsiers and modern lemurs (catlike primates now limited to Madagascar) populated woodlands the world over (image 19.1). These early primates subsequently gave rise to old-world monkeys, new-world monkeys, and apes. One group of these early primates is believed to have crossed a narrow stretch of the Atlantic Ocean about forty million years ago—perhaps riding on a flimsy ark of floating driftwood. Isolated from their ancestors, these primates gave rise to new-world monkeys, which today include spider monkeys, howler monkeys, squirrel monkeys, and capuchins that live in Central and South America.

About twenty-five million years ago, seismic shifts were ripping East Africa apart. Shifting tectonic plates forged the Great Rift Valley and separated many species from their main populations. Some of our primate ancestors were split into two groups that followed separate evolutionary pathways—one giving rise to old-world monkeys, which today include baboons and macaques, and one giving rise to the apes, whose descendants include current-day gibbons, orangutans, gorillas, chimpanzees, and, of course, us (figure 19.1) (Stevens et al. 2013).

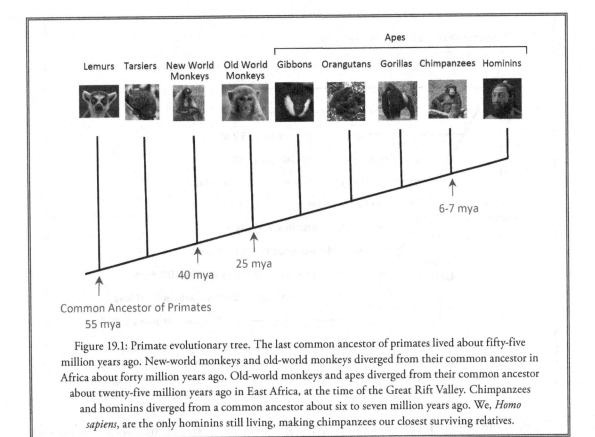

Figure 19.1: Primate evolutionary tree. The last common ancestor of primates lived about fifty-five million years ago. New-world monkeys and old-world monkeys diverged from their common ancestor in Africa about forty million years ago. Old-world monkeys and apes diverged from their common ancestor about twenty-five million years ago in East Africa, at the time of the Great Rift Valley. Chimpanzees and hominins diverged from a common ancestor about six to seven million years ago. We, *Homo sapiens*, are the only hominins still living, making chimpanzees our closest surviving relatives.

OUR FAMILY, THE HOMININS

Ten to twenty million years ago in the Neogene period, our ancient primate ancestors lived in the Great Rift Valley region of East Africa. These apelike creatures walked on all fours, had sharp teeth, were predominantly vegetarian, and were sexually dimorphic, with males much larger than females. About this time, great environmental changes were taking place in East Africa. The climate became much warmer and drier. Tropical forests declined as deserts and grasslands sprawled across the land. Our ancestors suffered overcrowding, food shortages, and dwindling shelter as their forest homes shrank. Faced with extinction, they left their arboreal homes and headed for the open savannas. As our ancestors adapted to their new homeland over the next few million years, they learned to stand erect on their two back legs, freeing their hands for chores and enabling them to sight prey and predators over tall grasses. This was the beginning of the **hominin** lineage.

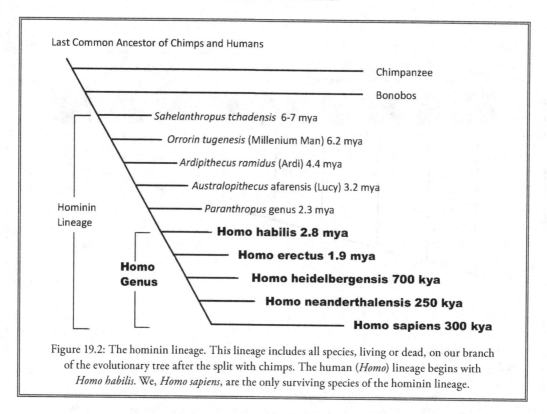

Figure 19.2: The hominin lineage. This lineage includes all species, living or dead, on our branch of the evolutionary tree after the split with chimps. The human (*Homo*) lineage begins with *Homo habilis*. We, *Homo sapiens*, are the only surviving species of the hominin lineage.

The hominin lineage includes all species, living or extinct, on our branch of the evolutionary tree after the split with chimpanzees six to seven million years ago (figure 19.2). As we will see, we, *Homo sapiens*, are the only living species in this group. The fossil record of early hominins shows anatomical changes reflecting an early transition from fully apelike creatures to creatures with some human traits. The most prominent of these traits were adaptations for bipedal walking. The foramen magnum—the hole in the skull through which the spinal cord passes—was already at the base of the skull, suggesting an upright posture with a vertical spine. The pelvis was short and bowl-shaped, with an upper leg or femur that was similar to a human femur—both of which would have facilitated bipedal ambulation. The canine teeth were blunt compared with those of modern apes but were larger and more pointed than those of modern humans. The early hominins were small by our standards—weighing sixty-five to eighty-five pounds—about half the size that we are today—and they had a small, chimpanzee-sized brain of about 450 cubic centimeters. These traits would gradually change as the lineage evolved.

LUCY

There is one fossil, more than any other, that has changed our view of human origins. When paleoanthropologist Donald Johanson and his graduate student Tom Gray discovered the remains of an ancient hominin near the village of Hadar in Ethiopia in 1974, they immediately realized the significance of their find—even though the skeleton, worn by rock and time over millions of years, was smashed to smithereens (Kimbel 2017). The site was a known repository of ancient fossils, and there was no doubt in the researchers' minds that this one was an early human or prehuman species. Johanson and Gray retreated to their camp that night and returned to the site every day for the next two weeks with a team of excavators. They uncovered and sorted through hundreds of preserved bone fragments until finally they had accumulated skeletal parts making up 40 percent of a small hominin skeleton. That night back at the camp, the scientists celebrated their find, drinking, dancing, and singing. It was the 1970s, and the Beatles song "Lucy in the Sky with Diamonds" was played again and again. The name stuck. The new hominin was called Lucy (image 19.2) (Kimbel 2017).

Lucy cemented the belief that humans evolved from an apelike ancestor (image 19.4). She was three feet, seven inches tall, estimated to have weighed just sixty-four pounds, and must have looked a lot like a chimpanzee. Her head was exceedingly apelike, with a flat nose, a protruding jaw, and a small braincase. Her leg bones, though, were like those of modern humans—meaning that Lucy was an upright, bipedal walker. Clearly Lucy was a transitional hominin species somewhere between chimps and us. Her species was dubbed *Australopithecus afarensis*. With Lucy's remains dated to 3.2 million years ago, she was, at the time, the oldest ancestor of all hominid species, and she provided the crucial link between apes and humans that cemented our heritage. Her notoriety spread. She was heralded as the mother of all humanity, and she provided the framework for further discoveries into the origin of human species.

Image 19.2: Lucy (*Australopithecus afarensis*). Lucy was a member of a transitional species between humans and apes. The finding of her fossil remains in 1974 in Ethiopia cemented the belief that humans evolved from apes, and she has been heralded as the mother of all humanity.

The Emergence of Humans: The *Homo* Genus

In the early 1960s, famous fossil hunter Louis Leakey and his wife discovered the 1.75-million-year-old fossilized remains of a unique hominin species at Oldupai Gorge in Northern Tanzania, within the current grounds of Serengeti National Park (Leakey, Tobias, and Napier 1964). The reconstructed specimen stood upright at about four feet tall, weighed an estimated seventy pounds, and had several definitively humanoid features that distinguished him from Australopithecus and the other hominin species that had been found before. The brain was larger, the molar and premolar teeth were smaller, and this was the first species that had an opposable thumb. The Leakeys also found stone tools nearby, indicating that perhaps this species was able to butcher animal carcasses and add meat to his diet—an option unavailable to his forebears. All these features provided justification for designating this hominin as not only a new species but also a new genus. The species had enough similarities to us to be classified as human and accordingly was given the genus name *Homo,* meaning "man," and the species name *habilis*, meaning "handy man." This was *Homo habilis.*

Louis Leakey had been born in Kenya, the son of British missionaries, and grew up living among the native Kikuyu people. According to some sources, he spoke their language from an early age. He gained an early interest in fossils and, after studying at Cambridge, returned to East Africa to continue his work. He and his wife Mary, herself an experienced paleoanthropologist, led numerous fossil-hunting expeditions and became pioneers in the search for human origins. The Leakeys had a longstanding conviction about the African origin for the human species, but this was doubted by many. The discovery of *Homo habilis* came at the end of their long and distinguished careers and served as vindication for their longstanding premise on the African origin of humans.

The Leakeys' find at Oldupai Gorge documented that humans appeared in Africa as early as 1.75 million years ago. However, recent fossil discoveries in Ethiopia of a partial hominin mandible with teeth characteristic of the *Homo* genus have dated the first appearance of humans in Africa to 2.8 million years ago—a much earlier timeline (Villmoare et al. 2015).

There has been some controversy regarding whether *H. habilis* was truly human, because it retained many apelike features, but there is no doubt about the humanity of the next *Homo* species that emerged on our planet (Van Arsdale 2013). **Homo erectus** (meaning "the man who walks upright") was five to six feet tall and had body proportions much like those of modern humans, with long legs and shorter arms, an upright posture, and a major increase in brain size (averaging about nine hundred cubic centimeters—larger than that of *H. habilis* but still smaller than that of modern humans) (image 19.3). *H. erectus* first appeared in the fossil record in East Africa about 1.9 million years ago. This globe-trotting species spread throughout Europe, Asia, and Indonesia, dominating the landscape and leaving many namesakes in the fossil record—Peking Man near Beijing, Java Man in Indonesia, and many others. With the use of tools and fire, *H. erectus* not only adapted to the environment but also began to master the environment and survived 1.8 million years—longer than any other human species. As we will see, we Homo sapiens have been around about three hundred thousand years. As far as longevity goes, we are not likely to be as successful as our *H. erectus* cousins.

The group of *H. erectus* that stayed in Africa may have given rise to another major ancient human species, **Homo heidelbergensis**, although this remains controversial (Manzi 2011). *H. heidelbergensis* populated Africa beginning about seven hundred thousand years ago, and later migrated to Europe. Initially called "archaic *Homo sapiens*," they were later anointed as *Homo heidelbergensis* after Heidelberg, Germany, where the first fossils were discovered. These archaic humans averaged five feet, nine inches tall, weighed about 135 pounds, and had a stout body suited for a cold climate. They may have been a transitional figure between *H. erectus and H. sapiens*, though they still appeared more like *H. erectus* than us. Based on their large brain size (1,200

Image 19.3: *Homo erectus.* *H. erectus* was the longest surviving human species, extant for 1.8 million years.

cc), advanced weaponry, and evidence of group hunting, *H. heidelbergensis* must have had high levels of social cooperation in the way of modern humans. Some of these species moved north to Europe and gave rise to the Neanderthals, while others remained in Africa, where they are believed to have given rise to us, *Homo sapiens* (Manzi 2011).

THE NEANDERTHALS: OUR CLOSEST COUSINS

Of all our human cousins, **Neanderthals** (*Homo neanderthalensis*) are certainly the most familiar. If we know someone who is acting brutish or crass, we might call him a Neanderthal in jest—implying that he is almost, but not quite, human. There's a reason for this familiarity. Neanderthals are the most recently extinct and the most studied of all the non–*Homo sapien* human species. And they are the most closely related to us.

The Neanderthals first appeared in Europe around two hundred thousand to two hundred fifty thousand years ago. They were short, averaging five feet four inches in height, and were put together like bodybuilders, with thick bones and heavy muscles. They had distinct facial features that made them easily differentiable from *Homo sapiens*: a back-sloping forehead, heavy brow ridges, a broad nose, and a protruding jaw (Soper 2007). Their brains, with an average size of 1,400 cc, were actually slightly larger than ours.

What do you think of when you picture a Neanderthal? You probably imagine a hulking cave-dweller wielding a hand-ax or a spear, going out in a hunting party to slay a woolly mammoth. But there was a gentler side to the Neanderthals. Evidence suggests that they took care of their elderly and infirm, and that they buried their dead. Neanderthals were less advanced in art and culture than *Homo sapiens*, but evidence is limited, and we are still learning what Neanderthals were capable of, one discovery at a time.

About forty-four thousand years ago, shortly after *Homo sapiens* arrived in Europe from Africa, Neanderthals disappeared. Their sudden extinction has remained one of the great mysteries of our evolutionary history. To better understand this, we'll first take a look at the origin of our own species: the emergence of *Homo sapiens* in East Africa.

HOMO SAPIENS BORN IN AFRICA

Ancient fossil remains of modern humans were first discovered in 1868, when workmen stumbled across a human skull at the Cro-Magnon rock shelter in southwestern France. These early *Homo sapiens,* given the name Cro-Magnons after their discovery site, inhabited France about twenty-seven thousand years ago. They looked just like us, with only the subtlest of anatomical differences. They have a large brain, a heavy skeletal structure, a nonprotruding jaw, small teeth, and a vertical forehead—rather than the sloping forehead of the Neanderthals— with only very slight brow ridges. They were hunter-gatherers like their hominin ancestors, but the Cro-Magnons were definitively human both in their habits and in their culture. They made tools and weapons, and they left evidence of an artistic culture in the forms of body ornaments and cave paintings.

These early examples of *Homo sapiens* hunted and roamed throughout much of Europe beginning about fifty thousand years ago, but their original home was in Africa. In 1967, Richard Leakey teamed with his venerable fossil-hunter father, Louis (the same man whose team had discovered *Homo habilis* in the early 1960s) in an excavation near the Omo River in Ethiopia. Their team discovered fossils from two individuals they named Omo I and Omo II. These fossils, one hundred ninety-five thousand years old, were identified as *Homo sapiens* and represented the oldest known specimens of our species at the time (McDougall, Brown, and Fleagle 2005). With the discovery of these fossils, Ethiopia was anointed as the "cradle of humankind," but new discoveries have raised a host of new questions. German paleoanthropologists recently announced they had discovered fossils of *Homo sapiens* in a cave in Jebel Irhoud, Morocco, that predate the fossils found in Ethiopia by one hundred thousand years (Hublin et al. 2017)! This revised our thinking about the origin of our species. Does this mean that the "cradle of humankind" needs to be relocated to Morocco? Not necessarily. These findings, combined with reevaluation of other fossil discoveries, suggest that early *Homo sapiens* may have dispersed across the African continent much earlier than previously thought. And the true birthplace of our species may remain a mystery still buried in the fossil record of Africa.

Ancient *Homo sapiens* in Africa had not yet evolved advanced cultural traits, such as artistry, religion, or sophisticated toolmaking. According to the fossil record, these cultural traits emerged in modern humans about fifty thousand to seventy thousand years ago, marking a major behavioral shift. This cultural change differentiated *Homo sapiens* from the Neanderthals by the time the two collided in Europe around forty-four thousand years ago, with existential consequences for both species.

THE MYSTERIOUS EXTINCTION OF THE NEANDERTHALS

So what happened to the Neanderthals? Why did they vanish? And did something about our own species cause their demise? Did our ancestors annihilate them? Or—a more intriguing possibility—did we merge with them?

There is now compelling DNA evidence that our two species did interbreed—even if it was just for a few one-night stands (image 19.4). Sequencing of the Neanderthal genome obtained from fossil DNA has shown that current-day humans are very closely related to Neanderthals, sharing 99.7 percent of their DNA (NIH Research News 2010; Green et al. 2010). By comparison, we share 98.8 percent of our DNA with chimps. A more detailed analysis has shown that current-day humans of European and Asian descent have slightly more DNA in common with Neanderthals than do current-day humans of African descent (Green et al. 2010). This suggests there was some interbreeding between the two species shortly after *Homo sapiens* left Africa. But the amount of Neanderthal DNA in us is small, so there does not appear to be a widespread merging of the two species. So what caused the disappearance of Neanderthals, if not a merger with *Homo sapiens*? While some believe their demise was due to climate change, most experts believe the extinction came at the hands of their *Homo sapiens* cousins (Diamond 2006, 51–52; Klein 2003).

Neanderthals were stronger, more muscular, and better adapted to Europe's cold climate than *Homo sapiens*. But by the time *Homo sapiens* reached Europe, there were signs our species had developed an intellect superior to that of their strong-browed cousins. *Homo sapiens* had more advanced hunting tools, such as snares and fishing nets, and their small teeth implied they had learned to process foods by this time. Evidence of tool parts from faraway regions suggests that our ancestors had developed sophisticated trading and social networking.

Image 19.4: Neanderthal and woman. Genetic studies show there was interbreeding between Neanderthals and *Homo sapiens* after *Homo sapiens* left Africa.

Paleoanthropologist Richard Klein has proposed that because *Homo sapiens* was technologically and intellectually superior to the Neanderthals, individuals of the species may have viewed Neanderthals as subhuman (Kein 2003). He argues that the encounter between the two groups was likely marked by bloodshed, starvation, and the rapid extinction of the Neanderthals. This scenario should sound at least a little familiar; human invaders have caused many extinctions throughout our history. Klein's theory, which is sometimes called the Neanderthal holocaust, is supported by recent evidence that Neanderthals perished almost immediately upon contact with modern humans. New radioactive techniques for dating fossils have shown that Neanderthals disappeared about forty-four thousand years ago—rather than thirty thousand years ago, as was previously thought (Pinhasi et al. 2011). Such a rapid disappearance after the arrival of *Homo sapiens* strongly suggests that *Homo sapiens* destroyed the Neanderthals by violence, virulent disease, or some other insidious means. This may have been the first entry to the ever-lengthening list of genocides committed by our species.

OUT OF AFRICA

The **out-of-Africa theory** holds that *Homo sapiens* evolved in Africa and migrated from Africa to populate Asia, Europe, and the rest of the world (Johanson 2001). It was believed for many years that *Homo sapiens* first evolved in East Africa about two hundred thousand years ago, based on the Leakey fossils, but the recent discovery of *Homo sapiens* fossils in Morocco dated three hundred thousand years ago raises new questions about the birthplace of *Homo sapiens*. There is little doubt that that *Homo sapiens* arose in Africa, but the precise birthplace will have to await further fossil discoveries.

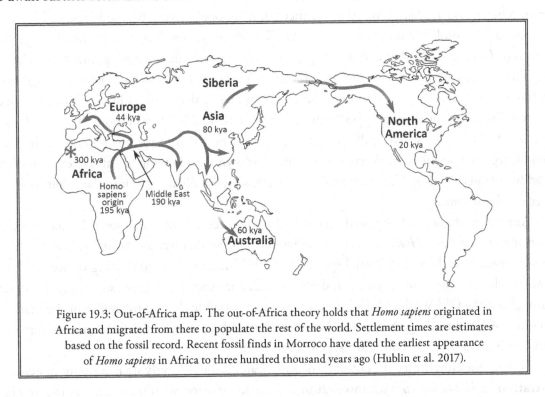

Figure 19.3: Out-of-Africa map. The out-of-Africa theory holds that *Homo sapiens* originated in Africa and migrated from there to populate the rest of the world. Settlement times are estimates based on the fossil record. Recent fossil finds in Morroco have dated the earliest appearance of *Homo sapiens* in Africa to three hundred thousand years ago (Hublin et al. 2017).

Our ancestors first migrated out of Africa and into the Middle East as early as one hundred ninety thousand years ago, taking advantage of low water levels to cross the Red Sea (figure 19.3) (Hershkovitz et al. 2018; Pruitt 2018). The Quaternary ice age, which was ongoing at that time with its subfreezing temperatures and encroaching glaciers, forced *Homo sapiens* to move westward rather than northward. They migrated to Asia about eighty thousand years ago and crossed from southwestern Asia to Australia by boat about sixty thousand years ago, when low ocean levels made the passage more feasible. Finally, when warmer, moister weather finally came to the Middle East and southern Europe, our ancestors migrated from southwestern

Asia and the Middle East to Europe about forty-four thousand years ago, where they met the Neanderthals. As we have seen, the Neanderthals disappeared shortly after the arrival of *Homo sapiens*. The species also moved to Siberia, crossed the Bering Plain, which had surfaced as a result of low ocean levels at the time, and entered the Americas about twenty thousand years ago. By this time *Homo sapiens* had populated all continents of the world, with the exception of Antarctica.

When *H. sapiens* left Africa, there were still a number of ancient human species (genus *Homo*) in Europe and Asia, including *H. erectus* in the Middle East and southeast Asia, the Neanderthals in Europe and western Asia, small groups of cave-dwelling Denisovans holed up in Siberia, and *H. floresiensis*, the one-meter-tall "hobbits" living on the Island of Flores in Indonesia. It is probably no coincidence that all of these species disappeared shortly after the arrival of *H. sapiens*. We out-competed them all. We are the only remaining human species.

The majority of experts now support the out-of-Africa theory about the origin and migration of *Homo sapiens*, but there are dissenters. Several prominent paleoanthropologists have advocated the **multiregional evolution hypothesis** (Wolpoff et al. 1988). They propose that after *Homo erectus* left Africa about two million years ago and dispersed into various regions of the Old World, regional populations slowly evolved on their own—and owing to subsequent extensive worldwide interbreeding, these disparate populations eventually merged into the single modern species *Homo sapiens*.

Advocates of the multiregional scenario cite fossil records of transitional human species with features of both *Homo erectus* and *Homo sapiens*, while proponents of the out-of-Africa theory argue that these transitional species could be due to rare interbreeding between *Homo sapiens*, who migrated out of Africa, and *Homo erectus* or other human species already dispersed throughout the Old World. Out-of-Africa advocates also note the improbability of the massive amount of interbreeding that would have been needed to prevent the formation of separate modern human species in the multiregional scenario.

Perhaps the clincher for the out-of-Africa advocates is the relatively small amount of **genetic variation** in *Homo sapiens* (Johanson 2001). Genetic variation within a species is the average difference in the genomes of individuals within that species, and this increases with time as mutations accumulate. So species that originated a long time ago have greater genetic variation. If the lineage leading to *Homo sapiens* began with *Homo erectus* nearly two million years ago, we would expect *Homo sapiens* to have much more genetic variation. Instead we see surprisingly little genetic variation in *Homo sapiens*, which is consistent with a more recent origin in Africa.

Researchers have also compared genetic variation in African groups with groups from Europe, Asia, Indonesia, and Australia, using data from mitochondrial DNA. The greatest variation was found in the African groups (Cann, Stoneking, and Wilson 1987). Since genetic variation is created by cumulative mutations over time, this evidence suggests that the oldest group, and hence the founding group, of *Homo sapiens* were from Africa. These same investigators used data from mitochondrial DNA to trace the roots of all humanity to a single ancestral mother in Africa two hundred thousand years ago (see sidebar "Mitochondrial Eve") (Cann, Stoneking, and Wilson 1987). Headlines proclaiming the discovery of a **Mitochondrial Eve** erupted in the popular press in 1987, and *Time* magazine heralded Mitochondrial Eve as "Everyone's Genealogical Mother" (Lemonick 1987).

THE NEAR EXTINCTION OF MANKIND

As previously discussed, humans as a species have a relatively small amount of genetic variation. If you compare your genome with that of anyone else in the world, 99.9% of the DNA codes will be identical (new methods suggest the figure may be a little less) (NIH 2017; Levy et al. 2007). In contrast, chimpanzees have a far greater variation in their DNA code—differing by two to three times as much as we do (Bowden et al. 2012). This is surprising, since we split from chimpanzees around seven million years ago and would expect similar genetic variation given the similar time to accumulate genetic mutations. The absence of such diversity suggests our population was drastically reduced at some more recent time, in what is called a **genetic bottleneck**. A genetic bottleneck is a sharp reduction in the size of a population due

---

MITOCHONDRIAL EVE

In 1987 geneticist Rebecca Cann and her colleagues reported evidence that all modern humans are descended from a single *Homo sapiens* woman who lived in Africa about two hundred thousand years ago (Cann, Stoneking, and Wilson 1987). Because their study was based on lineages tracked using mitochondrial DNA (mtDNA), they called this woman "Mitochondrial Eve."

We inherit most of our DNA from DNA housed within the nucleus of the sperm and egg from our father and mother, but we inherent mitochondrial DNA (mtDNA) from our mother's egg alone. The mtDNA in sperm is degraded at conception and is not passed on to offspring. This creates a hereditary line of mtDNA that passes uninterrupted from mother to daughter to daughter to daughter, until the chain is broken when there are no surviving female offspring. Maternal mtDNA passed to male offspring are not passed to the next generation.

Unlike nuclear DNA, mitochondrial DNA passes unchanged from one generation to the next. There is no mixing or recombination between maternal and paternal mtDNA, as there is with nuclear DNA. The only changes that occur are due to random mutations, which are relatively rare. Cann and colleagues compared mtDNA from disparate populations across the planet—from Africans, Asians, Caucasians, aboriginal New Guineans, and aboriginal Australians. They measured the differences in DNA codes within each group, which is considered the genetic variation of that group. When the investigators compared groups, they found the greatest genetic variation in the African group. Since genetic variation is created by cumulative mutations over time, this suggests that the oldest and founding group of *Homo sapiens* began in Africa. Smaller groups that left Africa sometime after the founding of the original group would start out with less genetic variation than the larger founding group, and while their variation would increase over time, it would never reach the level of variation of the founding group. The relative age of the non-African groups could be determined by their genetic variation; the greater the variation, the older the group.

The Mitochondrial Eve hypothesis has reshaped our thinking about human evolution and provided support for the Out-of-Africa theory. Cann and her co-investigators were curious as to when this common maternal ancestor might have lived. By tracking the differences in mtDNA within the African group and estimating the average rate of mutation, they extrapolated that our genealogical mother may have lived in Africa about two hundred thousand years ago. The story of Mitochondrial Eve lives on, and she has not lost her mystery or intrigue.

to environmental events (earthquakes, floods, disease, or famine). Such events, which reduce population size, also reduce the variation in the gene pool. It's like starting over with a new, smaller, much less diverse gene pool. The very low genetic variation in the human population suggests we experienced a genetic bottleneck in our recent geological past.

Geneticists, based on analyses of genomes from diverse populations, believe the bottleneck occurred sometime between fifty thousand and one hundred thousand years ago (Gibbons 2009; Amos and Hoffman 2010). Estimates are that human populations in Africa were decimated, leaving fewer than six thousand surviving individuals, and populations in Asia and Europe may have been hit even harder (Li and Durbin 2011). We may have come close to extinction! But why?

The cause of the bottleneck has been shrouded in mystery. A popular theory advocated by anthropologist Stanley Ambrose and others holds that the bottleneck was caused by a massive volcanic eruption (Ambrose 1998; Ambrose 2003). Geologists have documented that Mount Toba, in Sumatra, produced a colossal volcanic eruption about seventy thousand years ago, depositing six centimeters of ash across southern Asia, the Indian Ocean, and the South China Sea. Proponents of the Toba catastrophe theory argue that the ash filled the sky, dimmed the sun, and caused a precipitous drop in temperature followed by drought and famine. They postulate that the effects extended not only through Indonesia and Southern Asia but also to Africa, which is just across the Indian Ocean from Mount Toba. This "volcanic winter" lasted six years, they argue, and was followed by the coldest ice age (glacial period) in recent record, persisting for one thousand years.

But the Toba catastrophe theory has skeptics. A recent study from Oxford University analyzed seventy-thousand-year-old volcanic ash from the Toba supereruption in sediment in East Africa (Lane, Chorn, and Johnson 2013). The Oxford investigators found no evidence of a significant temperature drop, no evidence of major climate change, and no evidence of large-scale killings of humans and other animals. They concluded there was no evidence of a "volcanic winter" in East Africa.

While there remains great uncertainty about the cause of the human bottleneck, there is little disagreement that it happened. And the consequences are felt today. This bottleneck is likely responsible for the physical differences between our current-day races, through what we call the **founder effect**. When a small group is separated from a larger population, the smaller group will, by chance, have a much lower frequency of some traits and a much higher frequency of other traits. If the group remains isolated, as the population grows, there will be a predominance of the higher-frequency traits and a paucity of the lower-frequency traits. These differences in traits may distinguish one race from another. Experts believe the bottleneck occurring fifty thousand to one hundred thousand years ago created a number of small, isolated *Homo sapiens* groups that subsequently led to our separate races through the founder effect (see sidebar 19.2) (Ambrose 1998).

Harvard geneticist David Reich explains that, contrary to common dogma, there are genetic differences between different races, although those differences are smaller than the differences between individuals within specific races (Reich 2018). The average genetic variation between human individuals is about 0.1 percent. (We are 99.9 percent the same.) About 85 to 90 percent of this variation is within each of the major races, while 10 to 15 percent of the variation is between races (Jorde and Wooding 2004). Reich emphasizes the importance of acknowledging these differences rather than denying them in the name of political correctness, and of appreciating each other's unique traits and strengths. He advocates that we all accept the challenge "to empower all people, regardless of what hand they are dealt from the deck of life" (Reich 2018).

The genetic bottleneck, which decimated the human population and reduced its numbers to a precarious few thousand individuals, nearly ended the life story of *Homo sapiens*. It makes us, once again, appreciate the fragility of life. We almost disappeared from the face of the Earth. This bottleneck, like the mass extinctions we saw in the last chapter, shows how unpredictable catastrophic events can alter the course of evolution. If we rewind the tape of the natural history of life on planet Earth, we would likely see an entirely different outcome. We are lucky to be here.

A RAINBOW OF DIFFERENCES

One of the curious characteristics of modern humans is that we, more than other species, are genetically very similar to one another. We share 99.9 percent of our genetic code (maybe a little less by new methods) (NIH 2017; Levy et al. 2007). And yet we show a number of physical differences between various population groups. We usually call these racial differences. Why don't we all look more alike? And, more particularly, why don't we all look more like Africans?

All of this can best be explained by the genetic bottleneck that occurred fifty thousand to one hundred thousand years ago when a cataclysmic event, be it the volcanic eruption of Mount Toba or some other precipitous event, precipitated a dramatic reduction in the size of the human population. When a bottleneck dramatically reduces the population size of a species, some alleles (genes coding for specific traits) are eliminated altogether, and the species has less genetic diversity as a result. It's like starting the process of evolution over again, beginning with a genetically more homogeneous group, and with less time for new mutations to add to diversity. This explains the low genetic diversity that characterizes our species today.

At the same time, the bottleneck may also have triggered physical differences between our races through what we call the **founder effect.** When a small group is selected and isolated from a larger group, the smaller group will, by chance, have a higher or lower frequency of one or more alleles that code for specific traits when compared with the original population. If this group remains isolated and reproduces to form a larger population, there will be a predominance of those higher-frequency traits compared with populations that exist elsewhere. For example, Native Americans, in contrast to other ethnic and racial groups, rarely have blood type B. This is almost certainly the result of a founder effect. Native Americans all descend from the small group of people who crossed the Bering Strait land bridge from Siberia to the Americas about twenty thousand years ago. This founder population must have had, by chance, a very low proportion of people with genes coding for blood type B. As they reproduced and multiplied, they produced a large Native American population that retained a very low frequency of blood type B.

As anthropologist Stanley Ambrose, champion of the Toba catastrophe theory, has summarized: "When the modern African human diaspora passed through the prism of Toba's volcanic winter, a rainbow of differences appeared" (Ambrose 1998). Geneticists now recognize that there are genetic differences between races and stress the importance of acknowledging these differences rather than denying them in the name of political correctness (Reich 2018). We should appreciate each other's unique traits and strengths and strive to empower all people regardless of the genetic hand they have been dealt.

CHAPTER 20

THE RISE TO DOMINANCE OF *HOMO SAPIENS*

> It's cooperation (or, better said, the synergies produced by cooperation) that has been primarily responsible for the progressive evolution of ever more complex species on Earth, culminating in humankind.
>
> —Peter Corning, in *Synergistic Selection*

Jared Diamond, in his fascinating and provocative book *The Third Chimpanzee*, observes that we share 98.8 per cent of our DNA with our closest relatives, the chimpanzees—and yet we are so different from them (image 20.1) (Green et al. 2010; NIH Research News 2010). We *Homo sapiens* communicate with language, cooperate in large numbers with multiple complex groups, and build enormous societies and civilizations, while chimps—although they show a remarkable degree of behavioral complexity themselves—are confined to the forests and, sadly, to zoos and experimental laboratories. What is it about the small differences in our genetics that make us humans so unique?

Here's the short answer: We, *Homo sapiens*, evolved structures and connections in our brains that gave us the unique ability to communicate through language. We joined together in groups and learned to cooperate in a synergistic way, providing benefits for both the individual and the group. We evolved brains with the capacity for abstract thinking that allowed us to cooperate in vast numbers with people we have never met. As we coalesced into larger groups and formed societies, exchange of information was accelerated, and each society developed its own shared information—its own culture. The ability to cooperate in large numbers and accumulate collective knowledge and skills over time is responsible for the cultural evolution that has outstripped the slow pace of biological evolution and has transformed our planet and led to the dominance of *Homo sapiens*. Let's see how the story unfolded.

Image 20.1: Chimpanzee. Chimps share a whopping 98.8 percent of their DNA with us, yet while we build vast civilizations with advanced technology, chimps are confined to the forests and, sadly, to zoos and experimental laboratories (Green et al. 2010, NIH Research News 2010). Researchers are still struggling to understand why.

THE ORIGIN OF LANGUAGE

Through the simple act of pushing air out of our lungs and through our vocal cords and modulating the sound by changing the shape of our mouth, lips, tongue, cheeks, and palate, we can create an entirely new abstract concept in the mind of another individual who is listening to us speak. This ability is unique in the animal kingdom. Other animals communicate in myriad ways, with excellent accuracy, about a number of things—but no animal is able to communicate with the abstracted accuracy and complexity of human language. Our linguistic capacity, and the unprecedented ability it gives us to communicate and cooperate as social animals, has been cited as the most compelling reason for the evolution of our culture and the consequent unique accomplishments of our species. We know that other animals, like us, use tools, build social networks, solve tough problems, and experience complex emotions. In language and communication, though, we have them all beat.

Understanding the evolution of language is considered one of the toughest challenges in evolutionary biology. There is no fossil record of the first word, and the neural bases for language—as with other cognitive abilities—are highly complex and not fully understood. But we have made some progress.

Charles Darwin wrote in *The Descent of Man*, "man has an instinctive tendency to speak, as we see in the babble of our young children; while no child has an instinctive tendency to brew, bake, or write" (Cziko 1995, 179). Most linguists and evolutionary biologists today support Darwin's original concept: The capacity for language is an innate human trait. It is coded in our genes and hard-wired in our brains. Children learn language with minimal exposure, unlike other skills such as math and music. And all languages share some universal qualities across diverse cultures. All languages have nouns, verbs, adjectives, and prepositions; they have verb tenses that distinguish past, present, and future; and they all have common syntax. Language is a window into how our brains organize the way we see the world as humans.

We also know that chimpanzees, our closest living relatives, do not appear to be capable of learning language in the way we conceive of it. In the 1930s and 1940s, two separate psychologist couples undertook a dramatic scientific venture by adopting baby chimpanzees and raising them alongside their children (Pinker 1994, 343). Like their children, the chimps learned to dress themselves, brush their teeth, and use the bathroom, but they were unable to speak or understand language, even with special instruction. The chimps' brains seemed wired to learn many skills, but not language. (We know from other experiments that chimps can master a few linguistic skills—like learning abstract signifiers—but they seem to be unable to learn most basic language skills the way our children can.)

The human language has many key facets: semantics (meanings of words, phrases, and sentences), grammar (sentence structure), phonetics (production and perception of vocal sounds), and more. And it is open-ended and potentially infinite, in that new words can be formed and combined in never-before-used ways to communicate ideas of ever-greater complexity. Psychologist Steven Pinker believes that each of these facets of language is facilitated by wired neural circuits, and that each one evolved incrementally over the past six to seven million years (Pinker 2007).

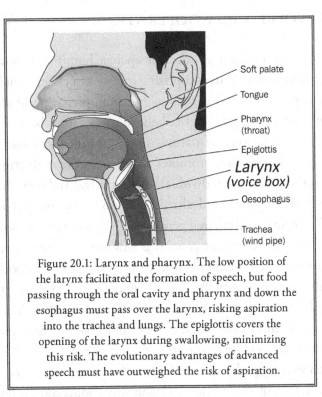

Figure 20.1: Larynx and pharynx. The low position of the larynx facilitated the formation of speech, but food passing through the oral cavity and pharynx and down the esophagus must pass over the larynx, risking aspiration into the trachea and lungs. The epiglottis covers the opening of the larynx during swallowing, minimizing this risk. The evolutionary advantages of advanced speech must have outweighed the risk of aspiration.

Sophisticated language is unique to humans. Animals can communicate using sounds, but their systems of communication don't have all the intricate qualities of our language. Vervet monkeys, for example, have at least ten distinct vocalizations, including separate sounds to alert others to certain predators. If a vervet hears the alarm call for an eagle, say, as opposed to a snake, it will flee accordingly (to the ground for an eagle, or to the treetops for a snake). These monkeys have an impressive system using abstract signifiers, but it is quite primitive compared to human language.

Most mammals produce sounds by passing air from the lungs through vocal cords situated in the larynx. The human larynx has an unusual location very low in the throat, and it has a large cavity, the pharynx, located just above, which facilitates the production of vowel sounds in concert with the tongue and lips (figure 20.1). This has aided the formation of speech, but because of the low position of the larynx, ingested food must pass over the larynx (which leads to the trachea and lungs) on its way to the stomach. This poses the potential risk of choking and aspiration of food into the lungs. Evidently, the benefits of facilitated speech have outweighed the risks of aspiration, because the low larynx has survived and is with us today.

HOW HUMANS EVOLVED TO COOPERATE

Millions of years ago in the Pliocene epoch, our ancient hominin ancestors formed clans of twenty-five to one hundred individuals comprising mostly kin. These small groups cooperated in many essential activities of daily life, such as hunting, preparing food, child rearing, and protecting the group. Evolutionary biologists believe cooperation between members of these clans evolved through a special type of natural selection called **kin selection.** Kin selection is selection of altruistic behavior that promotes survival and reproduction of an individual's relatives, potentially at the cost of the individual's own survival and reproduction. Kin selection arises because relatives share genes, and altruistic behavior by an individual toward family members and kin helps them to pass shared genes to subsequent generations. Since shared genes in relatives also likely carry the altruistic gene, this will promote transmission of these helping genes to the next generation and eventually lead to a population of individuals that all tend to help their families. Renowned twentieth-century biologist J. B. S. Haldane deduced "he was willing to lay down his life to save two brothers or ten cousins." (Smith and Szathmary 1999, 125). This is kin selection, and it is believed to be responsible for the beginnings of pro-social behavior in social animals, including hominin and human groups.

But if kin selection were the only selective force influencing human cooperation, our society would be very different. Clearly there is some reason why cooperation is favorable even outside the family. If it were not, we might not have the large cooperative societies we have today, and we might consider all people outside our families as potentially hostile. Most members of human social groups are not genetically related kin, and yet we humans still direct altruistic helping behavior toward each other. How could helping someone who's not even related to you, potentially at high cost to your own survival, be something that is evolutionarily selected?

The answer lies in **reciprocal altruism**—a social system in which individuals do favors for other nonrelated individuals with the expectation that these favors will be returned. It is easy to see how this might have evolved. If you have some genetic predisposition toward being kind and giving, and you do a favor for someone in your social circle who happens to share your proclivity for prosocial niceness, the likelihood is that they will do the same for you at some point, simply because it's who they are. Since two of you working together are more effective than each working alone, *both* of you benefit, increasing the likelihood that both of you will have long, successful lives and will reproduce to have children who have those same nice-guy genes. Over time this leads to a social group full of nice-guy reciprocators who can expect that the losses they incur by giving will be recouped in return—sometimes many times over.

A society built on reciprocal altruism requires that we have complex psychological machinery. We need to be able to assess costs and benefits, store favors and paybacks in memory, and detect those who might be cheating us and taking advantage of our kindness. For all of this to work, groups initially had to be small enough for individuals to know and trust one another. Groups with a lot of prosocial people—who exhibit traits like empathy, compassion, and sharing—would cooperate better and have a survival advantage over less cooperative rival groups. Such cooperative groups would survive by group selection and pass genetic tendencies toward cooperation on to the next generation. Over time, humankind would become more and more cooperative.

But then, as human groups grew even larger, a new level of cooperation emerged. Reciprocal altruism is a powerful society-organizing force, but it depends upon knowing the people you're giving to—and knowing that they'll give back to you someday in the future. But over time our social networks simply became too big for us to keep track of everyone. And yet we still act with kindness—even toward perfect strangers whom we may never see again. Why bother?

The answer lies in the concept of **indirect reciprocity** (Nowak and Sigmund 2005). I do something nice for you, and although you may not return the favor, someone else eventually will. We do unto others not necessarily expecting payback from them in particular, but knowing that, in the long term, others will do similarly unto us. Obviously, the potential for cheating is still an issue. How can we prevent others from cheating? The answer is reputation. Cheaters are discovered and are either punished or gain a tarnished reputation and are no longer eligible for favors. If the culture of the group rewards cooperators and punishes cheaters, cooperators will have a survival advantage over cheaters, and their prosocial behavior can be propagated by classic Darwinian evolution through **individual selection.** As philosopher Richard Joyce put it,

ALTRUISTIC YEAST?

We have seen how humans in complex societies have evolved altruistic traits to be kind and helpful to strangers. Could simple unicellular organisms with no brain also evolve altruistic behavioral traits? Isn't it too complicated? Simple single-celled brewer's yeast may provide answers.

Within a brewer's yeast cell colony, there are some yeast cells that produce the enzyme invertase, which breaks down complex sugars but at the same time expends energy needed for growth and reproduction. These altruistic yeast cells sacrifice themselves for the good of the group. Other yeast cells, which we label as cheaters, gobble up the simple sugars without spending energy to produce invertase. The group will survive only if enough cooperating cells produce a large enough quantity of food to support the population. There is, of course, a great temptation *not* to be an enzyme producer—to eat the sugars produced by others and survive and reproduce. The survival of these yeast colonies depends on the self-sacrifice of the few. Scientists have expected the cheaters to take over, leading to the eventual extinction of the colony, but surprisingly many yeast colonies survive maintaining a large contingent of altruistic yeast (Wloch-Salamon 2013).

A team of evolutionary biologists and mathematicians from Princeton University and the University of Bath, United Kingdom, has recently provided some answers (Constable et al. 2016). They created mathematical models to predict the evolutionary path of colonies, such as our yeast cells, populated by cooperators (who sacrifice their reproductive success to produce enzymes) and cheats (who use up resources but don't contribute). Unlike prior models, they incorporated changes in population size due to random fluctuations in birth and death rates—a process called demographic stochasticity. Their models showed that if, by chance, there are more cooperators to produce more food, the population grows larger. However, if, by chance, there are more cheats, food will be scarce, and the population will decrease and become extinct. In this way, groups with a preponderance of cooperators will survive and pass their cooperative genes to subsequent generations, while groups with a paucity of cooperators will perish. So altruistic behavior favors survival of the group and, with it, propagation of altruistic genes that lead to a stable altruistic population. Though our ancestors had bigger brains and more complex thought than yeast, group selection of societies with a predominance of altruistic behavior may have made us into the (sometimes) kindhearted, cooperative creatures that we are today.

"Individual selection occurring at the genetic level could now produce psychological traits designed to enhance one's success in this environment where pro-sociality is so heavily rewarded." (Joyce 2007 43).

Indirect reciprocity might also be propagated by **group selection**, in which natural selection acts at the level of the group. Groups that have more cooperators and fewer cheaters have adaptive advantages and triumph over less-cooperative groups, passing prosocial traits of indirect reciprocity to subsequent generations. The concept of group selection was in disfavor in the last part of the twentieth century when genecentric neo-Darwinism and selection at the level of the individual ruled the day. Richard Dawkins propagated the concept of neo-Darwinism and individual selection in his 1989 best-selling book *The Selfish Gene,* which played a major role in shaping public perceptions (Dawkins 1989). But multilevel selection theory gained strong advocates early this century. Eminent biologist E. O. Wilson, in his landmark book *The Social Conquest of Earth,* championed the paradigm that group selection is the major driving force of human evolution (Wilson 2012). More recently, evolutionary biologist Peter Corning has advocated the importance of synergistic cooperation and group selection in the evolution of complex systems in his recent book *Synergistic Selection: How Cooperation Has Shaped Evolution and the Rise of Humankind* (Corning 2018). He explains how the huge and disproportionate advantages that arise from cooperative synergistic behavior between individual organisms—from microorganisms to large mammals to humans—account for the march to higher levels of complexity and organization in the natural history of life (see sidebar "Altruistic Yeast?"). Natural selection acts on these higher levels of organization in living systems, selecting what works. Complex human societies are the pinnacle of this complexity and organization.

ABSTRACT THINKING AND COOPERATION

Historian Yuval Harari argues in his insightful book *Sapiens: A Brief History of Humankind* that the evolution of human's ability to engage in abstract thought is an essential factor in facilitating humans' capacity to cooperate in large numbers (Harari 2015). He calls our abstract thoughts "imagined realities" or "myths." Only humans possess the ability to think in this way. Our imagined realities can be shared by many and provide a common vision and a common purpose that allow for large groups to cooperate and work together toward common causes. Shared laws enable large societies to live with one another in order and peace. Monetary policy and rules of trade foster exchange of goods between strangers. And patriotism and love of country can unite peoples in a march to war.

Our ancient hunter-gatherer ancestors lived in small groups of twenty-five to one hundred foragers. Groups of this size were able to cooperate through reciprocal altruism, but reciprocal altruism comes from knowing and trusting every member of your community and only works in groups of up to about 150 people. Indirect reciprocity allowed cooperation in larger groups but did not allow cooperation at the level of large societies today. Only when we began to cooperate in large numbers—thousands, hundreds of thousands, and millions—did we truly begin to stand apart from our ape cousins. There is no way that a group of just 150 people could have built the first skyscraper, made the first rocket, or flown that rocket to the moon. Think about the number of scientists in diverse disciplines—accumulating knowledge and skills over decades and centuries, building a knowledge base one step at a time, and sharing ideas and cooperating in large numbers—that it took to send a man to the moon. The major differences between us and our closest relatives, the chimps, are our abilities to communicate and cooperate in vast numbers, and our ability to accumulate collective knowledge and skills over time. These abilities depend on abstract thought and common vision—imagined realities that have no basis in the real world. Our ability to have these abstract, imagined realities—and the cooperation they foster—is responsible for the cultural and technological evolution that has so outstripped our biological evolution.

THE UNIQUE HUMAN BRAIN

When we consider differences between humans and chimps—that humans have the capacity for language, the ability to deal in abstract ideas and cooperate in large numbers—the obvious question is that of how our brains differ from those of chimps. Human brains are about three times as large as chimp brains, owing primarily to the evolutionary expansion of the neocortex—the region that supports language, cognitive function, and problem solving. Scientists have also recently found another salient difference between human and chimp brains. Humans have a rare population of neurons in the striatum that produce the neurotransmitter dopamine (Viegas 2017). Dopamine in the midbrain has important roles in cognition and behavior, and its presence in the striatum of humans may have given us important adaptive advantages.

Remember Jared Diamond's question from *The Third Chimpanzee*, which started this whole chapter. Given how awfully similar we are to our chimp cousins from a genetic standpoint, how can we explain the profound differences between us in brain size, mental function, and behavior? The answer may be explained by differences in regulator genes. If you remember from chapter 9, regulator genes are genes that turn on and turn off structural genes—so a small change in one of these regulator genes can have a major effect on the operation of the cell by activating or deactivating many structural genes at once. It may be that our differences in brain size and anatomy, our quintessentially human abilities to communicate with language, and our ability to reason and to deal in abstract concepts are due mostly to mutations in regulator genes, which result in only very small changes in the genome as a whole.

We currently have a very limited understanding of which genes control various mental functions, but we have made progress in understanding genetic control of language. Scientists have identified an inherited defect of speech and language in three generations of a British family known as the KE family. Affected children have problems saying sounds, syllables, and words not because of muscle weakness but rather because of a defect in the brain preventing coordination of the lips, jaw, and tongue in producing speech. Genetic studies have found that all affected members of the KE family have inherited a mutation in a single regulator gene called *FOXP2*, which has been subsequently dubbed the "language gene." (Enard et al. 2002)

The **FOXP2 gene**, which is found in all mammals, produces a *FOXP2* protein, which activates a host of other genes that code for brain functions. The *FOXP2* protein in humans differs from the protein in other primates by just two amino acids—but it is believed that this tiny difference is responsible, in part, for our unique ability to speak. This difference must have come about from a random mutation sometime since our divergence from chimps between six and seven million years ago. The relationship between other genes and other higher brain functions will no doubt be discovered in the not-too-distant future.

CULTURAL EVOLUTION

Humans are a striking anomaly in the animal world. Our abilities to communicate and cooperate are unparalleled and have created an immense aggregate of accumulated shared knowledge and skills. This has triggered a cultural evolution that has created our complex societies and transformed our planet. Biological evolution, which continues to mold human nature, albeit slowly, and cultural evolution, which has so accelerated and transformed our way of life, are inextricably linked, but cultural evolution has become the driver of change. Let's see how this evolved.

CHAPTER 21

CULTURAL EVOLUTION

> A basic element of human nature is that people feel compelled to belong to groups and, having joined, consider them superior to competing groups.
>
> —Edward O. Wilson, in *The Social Conquest of Earth*

About seventy thousand years ago, humans began a transformation that would distinguish them from all other animals. They began to build and use complex and intricate tools— arrows for hunting, needles for sewing, boats for navigating. They began to create art, practice religion, and engage in commerce. They coalesced into larger groups and formed societies. They shared and accumulated information, skills, and technology, and over time each society evolved its own distinctive information, skills, and technology—its own culture. Each society's **culture** is the collective sum of beliefs, behaviors, customs, and characteristics shared by its members. Its culture is its way of life. The many disparities that we see among diverse human societies on our planet today are due to differences in culture, not genetics. We are all *Homo sapiens*, and the genetic and biological differences between societies and ethnic groups are actually quite small and account for very little of the differences in characteristics between societies. We are, to a large extent, defined by our culture.

HOW DID HUMAN CULTURE EVOLVE?

A number of forces came together to drive the evolution of human culture. As seen in the last chapter, our linguistic capacity gave us an unparalleled ability to communicate with one another and pass and store information and knowledge. As we banded together in small tribes, large groups, and then vast societies, we learned how to cooperate synergistically with one another for the benefit of the individual and the group. The ability to have abstract thoughts and virtual reality enabled cooperation in huge numbers with individuals we did not know, and this accelerated the growth of knowledge, skills, and technology. The knowledge, skills, technology, laws, and customs of societies morphed through innovation and shared experience, and this produced evolving, competing cultures.

Since Darwin's time, social scientists have explored analogies between cultural and biological evolution in attempts to better understand how culture evolves. Every society has many cultural elements—language, religion, dress, diet, art, architecture, political structure, and laws, to name a few—and there are huge variations in these cultural elements from society to society. For example, there are over six thousand languages and over ten thousand religions. These cultural elements seem to evolve in ways that behave like accelerated genetic evolution but are actually quite different. Richard Dawkins, in his landmark book *The Selfish Gene*, proposed a model for cultural evolution in which the transmittable units of culture, analogous to genes, are called **memes** (Dawkins 1989). Memes are discrete ideas, beliefs, and behaviors that spread throughout a culture, from brain to brain. Examples of memes are tunes, fashion in shoes, scientific theories, religious beliefs, economic systems, and types of government. But while memes are similar to genes in some ways, they are transmitted quite differently. Genes are transmitted vertically from parents to offspring through union of the sperm and egg, while memes, being abstractions, can in theory be transmitted linguistically to anyone. They can be passed "vertically," like genes, from parent to child, but a child can also receive memes "laterally," from siblings, friends, and teachers, or from radio, TV, books, and newspapers. Genes are transferred as precisely defined information, albeit sometimes with mutations. Memes, on the other hand, are more often altered or modified with transmission.

There is an important caveat in the analogy between biological and cultural evolution. In biological evolution, random mutations provide a variety of offspring, and those best adapted to the environment survive and reproduce and pass their genes to the next generation. The varieties of offspring are due to random-chance mutations and are not self-directed. In contrast, the generation of new memes is not random. The human brain is capable of generating new ideas, inventing new methods, and designing new products, and in doing so we create new memes that shape the evolution of our culture. The search for new sources of renewable energy to decrease the carbon imprint on our planet, for example, is self-directed and has begun to move our culture toward solar and wind-driven energy. Much, if not most, of our culture is driven by ideas and initiatives of individuals and groups. As culture has become the driving force of change, this provides us with an important sliver of hope that we may have some control over the future direction of life on our planet.

THE COEVOLUTION OF BIOLOGY AND CULTURE

All of us living in our given societies must evolve and adapt to ongoing cultural change. And at the same time, as humans evolve biologically, we influence how our cultures evolve. Cultural evolution and biological evolution are intertwined and deeply affect one another.

A classic example of cultural evolution triggering a change in biological evolution occurred about eleven thousand years ago, during the period when our ancestors were first learning to farm. For the first time, domesticated animals introduced milk into the diet of adult humans—and initially we were not at all equipped to handle this. Most of us, when we are born, have an intestinal enzyme called lactase, which helps us digest the lactose sugar that is abundant in our mother's milk. But because breast milk is a temporary food source for us all, this enzyme typically disappears from our guts after weaning—which means that many adults simply cannot digest lactose-rich milk. However, many of us alive today (especially those of us with European ancestry) digest milk just fine—because our forebears, at the time of the agricultural revolution, evolved a lactase enzyme in their guts that persists into adulthood so they could tolerate the milk of domesticated animals in their diets. The Agricultural Revolution and the domestication of animals, which altered our diet, changed the biology of our guts.

While many believe biological evolution has slowed over recent millennia, the opposite is actually true. Human biological evolution has accelerated over the last five thousand years. About 7 percent of our gene pool consists of new genes that have evolved over this time period—a rate about one hundred times the previous rate of evolutionary change (Hawks et al. 2007). The acceleration of biological evolution is certainly related to the need to adapt to the rapid changes in our environment due to the explosion of cultural evolution. Despite this acceleration, the pace of biological evolution still pales in face of the rapid transformation of cultural evolution.

Biologist Peter Richerson and anthropologist Robert Boyd, in their perceptive book *Not by Genes Alone*, have argued that the coevolution of culture and biology is responsible for the high level of cooperation in our societies (Richerson and Boyd 2005; Boyd and Richerson 2009). As groups evolve culture, they establish rules and customs that promote cooperation between individuals and punish and ostracize individuals with antisocial behavior. Individuals with cooperative prosocial behavior survive and reproduce, while individuals with antisocial behavior are shunned and rejected and do not. As a consequence, traits such as cooperativeness, altruism, sharing, and the abilities to feel shame and empathy are propagated to future generatons through natural (individual) selection. Your culture tells you to do good unto others, and through evolution and natural selection, your biology does as well. As individuals evolve more cooperative behavior, they contribute to and enhance the success and survival of the group (through group selection), which further enhances propagation of cooperative genes. The culture of a society begets individual cooperative behavior, and individual cooperative behavior within the group begets survival of the society and its culture. Biology and culture coevolve.

THE AGRICULTURAL REVOLUTION

For most of our time on Earth, we lived as hunter-gatherers. Small groups would venture out to trap and slay wild animals, while others would gather fruits, berries, and nuts. As food supplies became depleted, groups would move from place to place in search of new sources of nutrition. The search for food and shelter was of the highest priority.

About twelve thousand years ago, things began to change. The last ice age had given way to global warming and more plentiful rainfall. Peoples in the Fertile Crescent—the vast verdant land in the Middle East spanning from Egypt to Kuwait, between the Tigris and Euphrates rivers—began to abandon their foraging lifestyle and learned how to cultivate wheat and barley. Growing and harvesting peas, lentils, fruit trees, and flax soon followed. Farming, this brilliant new invention, spread quickly throughout Asia, Egypt, and Europe. Rice was cultivated in Asia, maize and beans in Central America, potatoes in South America, and sugar cane and bananas in New Guinea. Humans soon learned that goats, sheep, cattle, and pigs could be similarly managed and raised to provide a regular supply of meat and milk. With crops and domesticated animals, these early humans no longer needed to venture out in search of food. They could produce what they needed at home. Newly tied to their more fertile lands, they built permanent dwellings for the first time, establishing new communities and villages. The Neolithic Revolution was well underway.

The advent of agriculture was perhaps the greatest single change in modern human history. Agriculture, which was a product of cultural evolution, accelerated the pace of cultural evolution simply by giving people a moment to breathe and think. The settled way of life brought larger and larger groups of people together. As plant cultivation and animal husbandry became more efficient, the many could be fed by the few. Women had more children, and populations soared. Villages became towns, and towns became cities—and the people in them now had time to do things besides search for food, water, and shelter. They learned new skills, becoming toolmakers, carpenters, weavers, artists, and soldiers. The time was ripe for innovation, and the rapid technological progress that began then has accelerated ever since.

The ancient village of Çatalhöyük, in what is now southern Turkey, flourished around 7,000 BC. It provides the earliest and best-studied example of a fully settled Neolithic town. The villagers cultivated wheat and barley, and they bred sheep and cattle for meat, milk, and transport. Çatalhöyük artisans produced pottery and woven textiles. The population grew, reaching about ten thousand at its peak. Curiously, this early Neolithic town had no apparent social classes; all housing was similar, without distinctive features, and women shared equal social status to men (UNESCO 2012). This, of course, would later change.

The Agricultural Revolution was one of the most profound transformations in human history, leading to the rise of modern civilization, but the advent of agriculture has had an underappreciated downside. Societies that were dependent on food from farming became vulnerable to changes in weather that could wipe out crops and result in widespread famine. Population growth and crowding led to epidemics from infectious diseases and parasites. Dietary changes often led to nutritional deficiencies and impaired health. And agriculture led to property ownership that required protection and often led to territorial conflicts and wars.

As villages grew into great cities, hierarchies developed, which benefited the few and often degraded the majority. The excess food and new products produced by artisans stimulated trade, and a new class of merchants emerged. Some of these merchants became enormously wealthy and powerful, while the working class toiled. Although it often goes underappreciated, ancient hunter-gatherer societies were in fact very egalitarian (Ingold, Riches, and Woodburn 1988; Boehm 2001). Men hunted together and shared in equal work. Women stayed home for childbirth and child rearing but shared equal status with the men. In contrast, agricultural societies fostered division of labor, and with division of labor comes division of classes. Life for the elites in agricultural societies may have been better than in hunter-gatherer times, but life for the peasants was assuredly worse. Agricultural societies also aggravated sexual inequalities. Women weakened by more frequent pregnancies—which were simply not possible with the lifestyle of nomadic hunter-gatherers—were relegated to progressively more and more menial labor tasks.

In *The Third Chimpanzee*, Jared Diamond philosophizes about the "choice" we made in transitioning from life as hunter-gatherers to agricultural societies: "Forced to choose between limiting population growth or trying to increase food production, we opted for the latter and ended up with starvation, warfare, and tyranny." (Diamond 1992, 190). But it is hard for us to imagine how this could have been any other way. Farming and domestication of animals arose at similar times in many parts of the world, and it surely would have been difficult for hunter-gatherers to have turned away from its benefits or to have forecast its downside.

THE INDUSTRIAL REVOLUTION

Agrarian societies were the dominant culture across the world for thousands of years—until the idea of the factory was born. The Industrial Revolution, which began in eighteenth-century England, was a transition from farming in rural societies to manufacturing in newly coalescing cities. Hand production gave way to machine-based manufacturing and mass-production in factories. The textile and iron industries led the way, and the transportation industry was transformed with the introduction of the first steam locomotives and steamboats. A new economic system emerged along with this new way of life; people became factory owners, managers, bankers, and financiers. And in 1776 economist Adam Smith published his classic work *The Wealth of Nations,* in which he proposed a system of private ownership and free enterprise that became the foundation for modern capitalism (image 21.1).

Image 21.1: Lower Manhattan (1941). Manhattan is a symbol of our advanced civilization.

As with the Agricultural Revolution, the Industrial Revolution was both a product of cultural evolution and a stimulant to cultural evolution. The introduction of the printing press, the telegraph, the phonograph, and finally the telephone facilitated the exchange of information, rapidly spreading the newly industrialized culture from England to America and to other states in Western Europe.

The Industrial Revolution changed almost every aspect of daily life. Factories produced more and better material goods. Incomes increased, and the standard of living improved for the middle and upper classes. Populations soared to previously unseen heights. But if the egalitarian societies of hunter-gatherers were eroded by the agrarian transition, then the Industrial Revolution took the turn toward inequality much further. The working class had low wages, dangerous and monotonous working conditions, and very little job security. Children initially were part of the workforce and worked long hours under hazardous conditions. Urban life was overcrowded, living conditions were unsanitary, and diseases were rampant. For those at the bottom of the social ladder, the march toward a more perfect life was far from perfect.

COMPLEX SOCIETIES AND HUMAN NATURE

Our innate human instincts have changed little since our hunter-gatherer days and are not always well matched to current complex societies. **Tribal instincts** that we inherited from our hominin ancestors promote cooperative behavior toward those within our small groups who are like us, but they foster hostility toward those outside our group who are different. Xenophobic traits that are part of these tribal instincts may provide a survival advantage for the group in competing with rival groups, but in a large, diverse multicultural society, these innate hostilities toward other ethnic, racial, religious, and political groups must be muted for the large society to function and thrive. Societies have established laws and punishments against antisocial behavior, and in the United States there are special laws with punishments for those committing hate crimes related to a victim's race, ethnicity, religion, or sexual orientation. This type of environment promotes the survival of individuals with prosocial behavior, and at the same time, prosocial behavior promotes survival of the group by group selection, since cooperative societies have competitive advantages. Nevertheless, individuals in complex, diverse societies remain haunted by their latent tribal instincts, fostering hostility toward those not like themselves.

Dominance hierarchies are virtually universal in social animals and are especially well-developed in primates and humans (Rubin 2000; de Waal 1996). The pecking order of hens is a simple example. If an unacquainted group of hens is put in an enclosed area, there will be a rustling commotion and combat until things settle down. After that, if there is a dispute over food or space, it will be settled with a quick peck. Close observation shows that the pecking is not random but that there is a hierarchy among the hens—a pecking order.

Dominance hierarchies evolved in humans—and in most social animals—because they were adaptive in maintaining order within the group. Without leaders and subordinates, groups would have countless unresolved conflicts and would therefore be less competitive in organizing strategies to compete with rival groups. In one classic experiment, boys at a summer camp became more hierarchically organized when they were presented with a competitive challenge from another group (de Waal 1996, 103). Hierarchical behavior evolved through individual selection because it provided a survival advantage to individuals by helping them to avoid excessive violence, and it evolved through group selection because groups with well-ordered hierarchies performed more effectively than their rivals.

Hunter-gather societies untouched by modern civilization have been discovered in remote regions of our planet. These societies have changed little from earlier times and represent vestiges from our ancient past. Two such hunter-gatherer groups are the Kalahari bushmen of the Kalahari Desert in South Africa and the Spinifex indigenous people inhabiting lands in the Great Victoria Desert in Western Australia. Anthropologists studying these and other hunter-gather societies discovered something very surprising; they found these societies to be surprisingly egalitarian (Ingold, Riches, and Woodburn 1988; Boehm 2001; Gavrilets 2012; Gray 2011). They were not immersed in endless squabbles over dominance. There were no kings or elites. All members participated in hunting, gathering, and child-rearing duties. There were gender-specific jobs, but men and women had equal status. There is a stark contrast between these ancient egalitarian societies and today's hierarchal societies, but we have maintained some of our ancient heritage in our resistance to domination and our belief in egalitarian principles (Ingold, Riches, and Woodburn 1988; Boehm 2001). We humans require hierarchy but yearn for our egalitarian heritage.

Cultural evolution has transformed life on our planet and is now the dominant driver of change. Unfortunately, our inherited biological instincts—our human nature—often puts us in conflict with many aspects of our complex societies and modern culture. Whether our evolving human nature can find harmony with our evolving cultures will determine, to a great extent, the course of life on our planet. Cultural evolution is at least partly under our control. Let us hope we have the wisdom and resolve to steer it in the right direction.

CHAPTER 22

EVOLUTION OF MORALITY

And so, to the end of history, murder shall breed murder, always in the name of right and honor and peace, until the gods are tired of blood and create a race that can understand.

—George Bernard Shaw, in *Caesar and Cleopatra*

"In a world without God, all things become permissible," Ivan Karamazov famously cautioned in Dostoyevsky's *The Brothers Karamazov* (Tim 2012). For most of our history, morality has been in the realm of theologians and philosophers. But now times have changed. Religion no longer has a monopoly on our thinking about human good and evil. In 1975, E. O. Wilson suggested in his book *Sociobiology: The New Synthesis* that "the time has come for ethics to be removed temporarily from the hands of the philosophers and biologicized." (Wilson 1975, 562). It was the beginning of a new era in ethics and morality. There followed a groundswell of new ideas and publications that viewed human social behavior through the lens of Darwin's evolutionary theories. These concepts became more mainstream in the 1980s as sociobiology morphed into the field of evolutionary psychology.

There has long been general acceptance of Darwin's theory that physical traits evolved by natural selection, allowing us to adapt to our environments—eyes for seeing, legs for walking, and lungs for breathing. More recently, evolutionary psychologists have made the case that behavioral and psychological traits, like physical traits, evolved by natural selection as adaptations promoting survival and reproductive success, and that these traits have become embedded in our DNA. Today evolutionary psychologists and even many lay people explain human behavior and human nature in Darwinian terms. We look to Darwin for answers to such questions: Are men and women built for monogamy? Why do we feel guilty? Why do we gossip?

Evolutionary psychologists are now asking the question, Is human morality a product of evolution? Have we inherited psychological traits now embedded in our genes that allow us to judge what is moral and what is immoral behavior? Before we try to answer this question, let's explore the nature-versus-nurture debate.

NATURE VERSUS NURTURE

One of the oldest arguments in psychology is the nature-versus-nurture debate. The nurture school has advocated that all (or most) human behavior is learned. Our brains start with a blank slate, and human behavior is acquired by experience and is a product of our environment and culture. Evolution produced our anatomy and physiology, but not our instincts. In 2002 Harvard psychologist Steven Pinker published his best-selling book *The Blank Slate: The Modern Denial of Human Nature*, in which he argued against tabula rasa models and advocated that human behavior and human instincts are a product of evolutionary psychological adaptations (Pinker 2003). This was not entirely new. Darwin had advocated in *The Origin of Species* that instincts evolved by natural selection and was credited as the patron saint of psychology by Nobel Prize–winning Austrian zoologist Konrad Lorenz, a pioneer in animal behavior research. While there is still debate about how much of our behavior is born by nature and how much is due to nurture, there is general agreement that our behavior is profoundly shaped by inherited instincts that are programmed in our DNA.

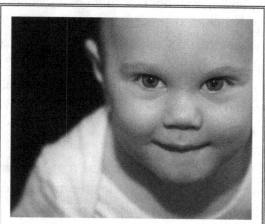

Image 22.1: Human infant. Infants a few months old are capable of empathy and moral judgements, supporting the position that humans have inherited innate morality traits independent of cultural influences.

This premise is supported by observations that many behavioral traits appear in infancy, before cultural influences can shape such traits. Psychologist Paul Bloom and his colleagues at Yale University performed a number of experiments suggesting that babies are capable of identifying with cooperative traits at a very young age (Hamlin, Wynn, and Bloom 2007; Bloom 2010). In one experiment, two-month-old babies watched a puppet show in which a cat is trying to open a big plastic box but just can't seem to get the lid off. A "good" bunny in a green T-shirt comes along and helps the cat open the box. In a second scenario, the cat again is having difficulty opening the box, but is making some progress, when a "bad" bunny in an orange T-shirt comes along and slams the box shut. After watching these puppet shows several times, the babies were presented with the two bunnies, and they almost always reached for the bunny in the green T-shirt—the good guy.

Infants also show empathy to the suffering of others at a very young age. They may show their compassion by offering a stroke or a pat to another infant in distress. And as they become toddlers and enter their third year, they expect individuals to treat others fairly (Hamlin 2013). Of course, young infants and toddlers also exhibit selfish, nonsharing traits at a very young age. These observations, and many others, provide convincing evidence that we are born with innate behavioral traits, both good and bad.

EVOLVED BEHAVIORAL TRAITS

There is little doubt that biological evolution has produced many behavioral traits that produce what we consider moral behavior. We have seen in previous chapters how many of these traits evolved for adaptive advantages. We evolved erotic love to promote reproduction, motherly love to provide for the welfare of offspring, and romantic love and monogamy to facilitate child rearing. As we grouped together into larger societies, we evolved altruistic cooperative behavioral traits through kin selection, reciprocal altruism, and indirect reciprocity that provided adaptive advantages for the individual and the group. We evolved hierarchal behavioral traits to facilitate order and prevent anarchy within the group, while we retained some of our egalitarian traits from our hunter-gatherer days.

All social animals, such as wolves, deer, and monkeys, have evolved behavioral traits that promote cooperation and survival of the group. Dutch primatologist and ethologist Frans de Waal, at the Primate Research Center in Atlanta, Georgia, has shown that monkeys have a sense of fairness, empathy, and cooperation. De Waal performed a famous experiment called the "fairness study," in which capuchin monkeys were placed in adjoining cages where they could see and interact with one another (de Waal 2011). A keeper at the center gave one monkey a piece of cucumber, which he readily

Image 22.2: Capuchin monkey. These monkeys participated in Frans de Waal's famous "fairness study," demonstrating that they have a keen sense of moral fairness.

took and ate. Then the keeper gave the second monkey grapes—a much tastier treat than a cucumber. Of course, the second monkey eagerly took and ate the grapes. But then, when the keeper went back to the first monkey and gave him a piece of cucumber again, the monkey took the cucumber, threw it at the keeper, and violently shook the cage, as if in protest. The monkey sensed the unfairness of its treatment and seemed to be demanding that he, too, should receive a grape.

In another study, an infant rhesus monkey was severely punished and rejected by her elders, and as she cowered in the corner, other infant monkeys approached her to embrace and comfort her. The discomfort of one infant seemed to be shared by her peers (de Waal 2006, 27). This type of empathy, in which animals appear to comfort those in distress, is a common trait among social animals.

Empathy is part of the mammalian heritage and has developed at the highest level in humans, where it helps promote cooperation within the large social group. We have evolved a capacity to connect and to understand others and make their situations and emotional states our own. It is a universal human trait. We are all built with a mind that cannot bear to see the suffering of others (de Waal 2009). Empathy evolved to promote cooperation within the large social group, but as we will see, it is selective and uncommonly applied to those outside the group.

The presence of empathy, cooperation, and an appreciation of fairness in social animals supports their evolutionary origins. No one would argue that these traits in animals are a product of "animal culture." If these prosocial, cooperative, altruistic traits were all we inherited, we could make a strong case that morality is the product of biologic evolution. But we have also evolved traits of selfishness and its coconspirators: fraud, corruption, lust, vindictiveness, ambition, and vanity. These traits have been inherited through natural selection to promote individual survival and are universal in all people in all walks of life. As E. O. Wilson explains, these inherited traits have created a chimeric genotype with a mix of genes for selfish behavior and altruistic behavior. We are all conflicted between heroism and cowardice, between truth and deception. Each of us is part saint and part sinner (Wilson 2012, 289). Our innate behavioral traits include what we consider both moral and immoral behavior. Let's take a look at the darker side of human nature.

TRIBAL INSTINCTS

Our ancient hunter-gatherer hominin ancestors evolved prosocial altruistic traits that enhanced cooperation between individuals within the group because this provided adaptive advantages to both the individuals and the group. But along with altruistic, kind behavior toward those within the group, our ancestors evolved behavioral traits of hostility toward those outside the group, who were seen as enemies. As we discussed in the last chapter, this combination of traits has been referred to as tribal instincts. These tribal instincts promoted survival of the group over rival groups but have left us innately nepotistic, ethnocentric, and xenophobic. These traits promote what we consider today to be both moral and immoral behavior.

Tribal instincts are hardwired in our brains and persist as part of human nature today. We all still belong to tribes. Most of us prefer to associate with people of the same race, religion, politics, and country. We automatically feel more comfortable around people who look, act, speak, and think like we do. In contrast, we often regard those outside our groups in a hostile way, finding them less honest, less competent, and less likable. If racism is the belief that members of another ethnicity possess characteristics that are inferior to ours, then we are all instinctively racist. Fortunately, we are usually able to override some of these hardwired prejudices through cultural learning, but too often our tribal instincts surface. Recent police shootings of unarmed black suspects and the Black Lives Matter protest movement have flared racial tensions and provide an example of the tribal instincts within us all. Understanding tribal instincts helps explain why we have become so politically polarized. We find supporters of the opposing party uniformly less honest, less trustworthy, less intelligent, and less empathetic, while we feel a real kinship with those in the party we support.

THE DARK SIDE OF HUMAN NATURE

> On Monday, September 20, 2004, Islamic militants in Iraq executed an American construction worker named Eugene Armstrong. Four men, masked and clothed in black, tensely clutched their automatic weapons while the bound and blindfolded Armstrong knelt in front of them. "God's soldiers from Tawhid and Jihad were able to abduct three infidels of God's enemies in Baghdad," the leader intoned, "… by the name of God, these three hostages will get nothing from us except their throats slit and necks chopped, so they will serve as an example." The long knife sliced through Armstrong's flesh. He screamed. Blood gushed from his neck. His body shuddered and became limp. The executioner placed the dripping, severed head on the back of Armstrong's lifeless body. (Smith 2007, 1)

This description in David Livingstone Smith's *The Most Dangerous Animal* makes clear the gruesome horror of man's inhumanity to man. We may have strong and possibly universal codes of morality, but even so, no other species organizes violence, plans brutality, and develops weapons of war like we do. Violence against our own kind has been our species birthright dating back to the extermination of the Neanderthals forty-four thousand years ago. In just the last century alone, more than thirty million people have been killed in genocides. Here are the largest of these massacres (Smith 2007, 217–18):

- 1915–16: Muslim Turks slaughtered 1.5 million Armenian Christians.
- 1931–32: The Soviet Union forged a famine that killed 5 million Ukrainians.
- 1940s: Nazi Germany exterminated 11 million Jewish, Romani, Polish, and homosexual people in brutal concentration camps.
- 1970s: The Khmer Rouge massacred 1.7 million Cambodians.
- 1971: The Muslim Pakistani army in East Bengal killed 2.5 million Hindus.
- 1994: The Hutu majority in Rwanda murdered one million Tutsis.
- 2007: The government of Sudan killed two million black Sudanese in Darfur.

While not on the list of massacres in the last century, our American ancestors committed some of the worst acts of genocide. European colonists massacred, raped, and committed numerous atrocities against Native Americans over hundreds of years. Millions were killed, and the spread of infectious diseases wiped out millions more. The native population was reduced from an estimated ten million indigenous peoples in what is now the United States when explorers first arrived in the fifteenth century to less than three hundred thousand in 1900 (United to End Genocide 2018).

Most species engage in violence against their own kind, but it is usually one-on-one violence over mates, food, or territory. However, social animals often exhibit xenophobic behavior toward entire outside groups, showing hostility to those they see as a competitive threat (Smith 2007, 74; Wilson 1975, 249). Our closest evolutionary relatives, the chimpanzees, have become famously known for their xenophobic behavior and vicious intergroup conflicts. Jane Goodall's chimpanzee research center in Gombe, Tanzania, has documented ruthless murders by raiding groups of chimpanzees, in which bands of chimps patrolling neighboring regions will brutally attack, kill, and even dismember lone defenseless individuals of neighboring tribes (Smith 2007, 75–78).

A tendency toward grisly killing of our own kind is part of our hominin heritage. As our hunter-gatherer ancestors learned to use tools, make fire, and cooperate in hunting large carnivores, they took their place at the top of the food chain. But they had to contend with an increasingly intelligent, ruthless, and formidable foe—their own kind. Humans became, as David Livingstone Smith describes, "the most dangerous animal." Early hominins likely participated in bloody raids against neighboring tribes, just as chimpanzees do today. As *Homo sapiens* settled into villages and towns during the Agricultural Revolution, territorial conflicts ensued and pitted one community against another. Agriculture and the domestication of animals provided abundant food and allowed time and resources to be dedicated toward making weapons and establishing armies. With new arsenals, states became increasingly territorial and embarked on wars of conquest. Warfare became commonplace on our planet.

THE MARCH TO WAR

War is sanctioned premeditated violence, pitting one group—usually one nation—against another. The key word is "sanctioned"; war is killing that culture condones. Violence against an outside group is considered a prosocial action by an individual's own group. A person can leave his or her family, go off to war, butcher and murder scores of enemy combatants, and return home as a loyal patriot and hero, welcomed back into society with open arms. We all have inherited behavioral traits that inhibit us from murdering people within our own social groups. So we might expect that we would also have inhibitions against killing those outside our group, since, after all, they're people like us. But this is simply not the case. History has shown that we have been able to overcome our aversion to killing our "enemies" by engaging in a process of self-deception called **dehumanization**. In war, we convince ourselves that the enemy is not quite human and not deserving of our empathy, making it easier for us to commit murder and other atrocities. The Hutu majority in Rwanda characterized the Tutsi minority as "cockroaches" as they set about to exterminate them. That is dehumanization.

David Livingstone Smith argues that the human propensity for war may have been enhanced through the propagation and selection of warrior-like traits in males. Historically, conquering armies, after killing off all the men in an enemy settlement, often engaged in mass rapes of the women that remained. Some of these women survived, delivered offspring, and assimilated into the victors' societies, propagating warrior-like genes in the population. Additionally, victorious warriors returned home as heroes and were sought after by the women of their own society. They became objects of sexual selection and as such had a reproductive advantage in propagating their warrior-like genes. The brutality of human evolution via mass rape is horrifying to contemplate, but these unspeakable acts of violence and hate may be written into our genes through the mechanism of war.

If we are able to better understand the roots of human nature and why human nature makes us predisposed to making war, it may be possible to learn how to constrain these tendencies that have caused so much suffering over the course of human history. Smith believes the best chance we have of deterring war is by eliminating or modifying the process of self-deception that dehumanizes the enemy and makes killing bearable. If left unchecked, these tendencies, combined with a growing arsenal of weapons of mass destruction, may well lead to our extinction. Why would we have evolved such self-destructive behavior as self-deception?

SELF-DECEPTION ENABLES DECEPTION

As humans coalesced into large societies, they evolved traits that enhanced group survival and traits that enhanced individual survival. Deception of others has promoted individual advancement and survival and has become part of our moral—or immoral—fabric. Self-deception, paradoxically, has evolved to enhance deception of others (von Hippel and Trivers 2011). If you do not think you are deceiving, you will not give away cues that might reveal deceptive intent. And if you do not think you are deceiving, you can avoid the cognitive challenge of deceiving and can minimize retribution if the deception is discovered. Self-deception also enhances one's belief in self-righteousness and allows greater self-confidence, which has a host of social advantages. We detest dishonesty and deceit and value truth and honesty above all, yet we all have an innate trait of self-deception to some degree or another. It is not hard to understand that evolution would want to wire our brains with such a tool that would give us an evolutionary advantage over our competitors. It has become part of our human nature.

The implications of self-deception are profound. Evolutionary biologist and anthropologist Robert Trivers states the case:

> If (as Dawkins argues) deceit is fundamental in animal communication, then there must be strong selection to spot deception, and this ought, in turn, to select for a degree of self- deception, rendering some facts and motives unconscious so as to not betray—by the subtle signs of self-knowledge—the deception being practiced. Thus, the conventional view that natural selection favors nervous systems which produce ever more accurate images of the world must be a very naive view of mental evolution. (von Hippel and Trivers 2011)

Natural selection does not care if we have true beliefs; it cares only that our minds evolve to help us survive and reproduce. As the late evolutionary biologist Stephen Jay Gould said, "Nature does not exist for us, had no idea we were coming, and doesn't give a damn about us" (Sage 2017). Self-deception has adaptive advantages for the individual and has promoted individual survival, but most of us consider it immoral, and as we come together in larger and larger groups, it may be counterproductive to the survival of society as a whole. We saw earlier that David Livingstone Smith believes that self-deception, which allows us to dehumanize our enemies, is a major facilitator of war, and that overcoming self-deception may be the best hope of avoiding war.

Unfortunately, self-deception has accelerated in the current political climate in the United States, Europe, and elsewhere. Our politicians and their followers spin their positions with a dogma and confidence born of self-deception. They believe what they say and are entrenched in their positions. There is little room for compromise. Our future, both in the short term and long term, may depend on our ability to overcome this destructive part of our human nature. Overcoming self-deception may be a prerequisite for us to come together on common ground, search for the truth, and find rational solutions to the great problems that confront our world.

IS HUMAN MORALITY A PRODUCT OF EVOLUTION?

Our morality is our conscious belief or conviction regarding what constitutes good, acceptable behavior and what constitutes bad or unacceptable behavior. It is our sense of what is right and what is wrong. Morality varies from culture to culture, but there is a commonality among all cultures. Almost all cultures extoll the virtues of love and respect for human life, honesty, fairness, empathy, compassion, and altruism, while condemning murder, theft, dishonesty, and hypocrisy as immoral. This commonality of belief in what is moral and what is immoral behavior suggests that our moral compass for these traits is inherited and embedded in our genes and is independent of cultural factors.

Why would we evolve such a moral compass? As we saw in chapter 20, as members of *Homo sapiens* came together in large societies, they learned to cooperate and help one another because it benefited both the individuals and the group. The ability to recognize behavior that is beneficial to the group and behavior that is destructive to the group, and to have a conscience to help us make the right choices, would certainly have adaptive advantages for both the individual and the group. It is not difficult to see how such a moral compass would evolve through natural selection.

But our sense of morality is not entirely innate. Morals vary dramatically across time and across cultures, and there is no doubt that contemporary culture also plays an important role in determining what is considered moral and immoral behavior. Consider the words of the Declaration of Independence: "We hold these truths to be self-evident, that all men are created equal, that they are endowed by their Creator with certain unalienable Rights, that among these are Life, Liberty and the pursuit of Happiness." At the time this was written, "all men" did not refer to all men and all women. Slavery was condoned, and women were not given equal rights and did not achieve the right to vote until the twentieth century. Today we interpret these words much more broadly, although African Americans and women are still struggling for equal opportunity.

There are innumerable examples of differences in morality across contemporary cultures. Gay marriage and homosexuality are accepted in many societies and prohibited or punished in others. Subordination of women is common and varies greatly across cultures. Many societies condone polygamy, while it is outlawed in most societies. Female genital mutilation is practiced and considered part of the moral code in a number of African, Middle Eastern, and Asian countries. The list goes on.

As we saw in the last chapter, cultural evolution has coevolved with biological evolution and in many ways has transcended biological evolution. Similarly, morality has evolved through both biological and cultural evolution, and while much of our morality is embedded in our genes, selecting appropriate behavior must often be guided by cultural forces in the context of current societies. We have the advantage that cultural evolution is partly under our control, and we have the opportunity to shape our future morality. Achieving a universal moral code that meets the needs of all humanity may be an essential ingredient to our survival as a species.

IS THERE A MORAL TRUTH?

There is great diversity of morality across cultures, and each culture believes in its own moral code. A number of philosophers have taken the position that there is no inherently right or wrong moral code—a position that has been called the principle of **moral relativism**. Advocates of this principle assert that we should be tolerant of other moral codes even if we disagree with their philosophy. This position has gained popularity because it promotes peace, tolerance, and understanding of other cultures. But it has been criticized because it allows for tolerance of certain behaviors that may be destructive to human lives. Indeed, it seems unacceptable to permit such acts as slavery, beheadings, female genital mutilation, and deprivation of women's rights, to name a few, simply on the premise of having an "open mind" to alternative morally based behaviors.

Many religious groups maintain a code of moral ethics that holds that certain actions are intrinsically right or wrong. Stealing would always be immoral, even if it were stealing food to feed a starving family member. This type of **moral absolutism** begs the question, Who decides what the absolute morals are? The religious would say morals come from God, but what about other religions with other gods? What about nonbelievers?

Neuroscientist Sam Harris, in his provocative book *The Moral Landscape,* has introduced an entirely new way to think about moral truth (Harris 2010). Historically, morality and values were seen as lying outside the realm of science, but Harris has proposed that science can tell us how we ought to think and behave. Just as we use science to figure out how to optimize our physical and mental health, we can use the scientific method to determine which values and behaviors can maximize human and animal happiness while minimizing suffering. Defining and quantifying human happiness and human suffering and identifying moral behavioral correlates of these states is a daunting task, but if we can agree on the goal—maximizing happiness and minimizing suffering for all—perhaps we can make incremental progress toward a universal moral truth.

Here is an example of how this might work. Spanking and other forms of corporal punishment have been commonly used in the United States by parents and teachers to discipline children. Children who misbehave are sometimes beaten with a stick, resulting in bruising, bleeding, and public humiliation. This practice derives directly from a Biblical proverb translated variously as "He who spares his rod hates his son, but he who loves him disciplines him diligently" (Proverbs 13:24). Corporal punishment was thought to encourage healthy emotional development and good behavior, but scientific studies of child development have shown that this is not the case but that, in fact, corporal punishment is detrimental to children's psychological health. From a scientific perspective, therefore, corporal punishment should not be a part of a healthy value system. Many moral questions can be evaluated like this, in an evidence-based, scientific fashion.

Harris contends we should no longer tolerate vast differences in cultures that don't promote human well-being. He believes the development and implementation of a moral system that maximizes human well-being and minimizes human suffering of all mankind—all ethnic, racial, and religious groups—is our best chance of building a global civilization based on shared values. And I would add that such a system will need to respect and value all life, not just human life. A universal moral system adopted by all will be an essential part of the global cooperation that is needed to ensure the future of life on our planet.

CHAPTER 23

THE GREATEST MYSTERY

> What, then, is the mystery? What could be more obvious or certain to each of us than that he or she is a conscious subject of experience, an enjoyer of perceptions and sensations, a sufferer of pain, an entertainer of ideas, and a conscious deliberator? That seems undeniable, but what in the world can consciousness itself be? How can living physical bodies in the physical world produce such phenomena? That is the mystery.
>
> —Daniel Dennett, in *Consciousness Explained*

American neuroscientist Krystof Koch summarized the existential nature of consciousness in his book *Consciousness: Confessions of a Romantic Reductionist*: "Without consciousness, there is nothing. The only way you experience your body and the world of mountains and people, trees and dogs, stars and music is through your subjective experiences, thoughts, and memories. You act and move, see and hear, love and hate, remember the past and imagine the future. But ultimately, you only encounter the world in all of its manifestations via consciousness. And when consciousness ceases, this world ceases as well" (Koch 2012, 23).

Yet consciousness remains one of science's great unsolved mysteries. It has no mass and no measurable energy, and we have little idea about what it is made of. We know brains are required for consciousness, but we are still clueless as to how brain stuff begets mind stuff. And the evolutionary purpose of consciousness remains a subject of great controversy.

CONSCIOUSNESS EXPLAINED

The term "consciousness" is notoriously ambiguous and has many different meanings in diverse contexts. We often use "conscious" or "unconscious" to refer to a creature being awake, asleep, or comatose. In medicine we define these states as "levels" of consciousness. At other times, we use the word to describe awareness or focus on a particular object or space so the subject can respond to that stimulus: "I was conscious of the car approaching on my right."

For our purposes, let's consider consciousness as a state of subjective awareness—either of external objects or internal thoughts and memories. A basic form of consciousness is the simple subjective perception of stimuli from our environment. When we feel pain from touching a hot skillet or perceive vivid red color while looking at an apple or perceive the sweet fragrance of a flower when we smell a gardenia, we are experiencing what neuroscientists and psychologists call **qualia**. Qualia are the subjective components of sensual perceptions. Organisms that experience qualia are considered **sentient** organisms. We think vertebrates and many invertebrates experience pain and other subjective sensations, and as such, we view them as sentient creatures.

This subjective experience of the world—qualia—is what philosopher Ned Block calls **phenomenal consciousness**. He contrasts this sort of consciousness—the raw experience of the world around us—with what he calls **access consciousness** (Block 1995; Van Gulick 2016) Access consciousness is subjective awareness in which information in our brains is accessed and used in thought and reasoning, in communicating to others, and in controlling and modifying behavior. When we recall past experiences, engage in thought and reasoning, and plan for the future, we are utilizing access consciousness—a "higher" or more cognitively sophisticated level of consciousness. Block believes many animals besides humans are capable of this higher level of conscious thought. Chimpanzees, for example, have the foresight to carry stones to nut-cracking sites and to prepare and carry sticks to termite mounds for termite-fishing, suggesting they can plan and anticipate future needs (Boesch 1984).

CARTESIAN DUALISM

French philosopher René Descartes, of "Cogito ergo sum" fame, established himself as the principal advocate of **Cartesian dualism**, the theory that mind and body are completely separate entities acting in parallel and interacting in ways not well understood. Matter is easily described; it has dimensions and weight, and can be seen, touched, smelled, and tasted. Mind is the opposite—invisible, immaterial, and immeasurable. Descartes's dualism had a religious agenda, equating the mind with the immortal soul and concluding that mind–body interactions were facilitated through divine input. Descartes's dualistic philosophy ruled the day in the seventeenth century, and the human mind and consciousness remained in the realm of philosophers and theologians for centuries. There was little appetite for earthly answers.

Times have changed. Inquiries into the mind–body relationship and the nature of consciousness have become one of the hottest fields of research in neuroscience. A slew of technological developments in the twenty-first century—the ability to record from individual neurons with microelectrodes, to acquire neurophysiological data using functional MRIs, and to utilize complex computer modeling—have put the investigation of consciousness in the hands of neuroscientists and psychologists (image 23.1). We are getting closer to understanding how immaterial consciousness can emerge from earthly matter. As evolutionary biologist David Barash proclaimed, "Cartesian dualism is on the run, as well it should be" (Barash 2012).

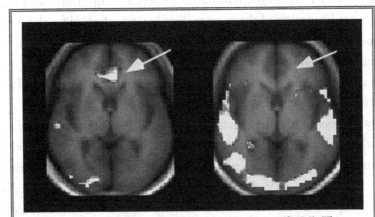

Image 23.1: Functional magnetic resonance image (fMRI). The highlighted regions show areas of increased blood flow and increased neural activity. Functional MRI is one of the new technologies used by psychologists and neuroscientists in the study of consciousness.

THE HARD PROBLEM

Australian philosopher David Chalmers called it the "hard problem." How do we explain the fact that immaterial, subjective experiences arise from the material physical brain? How can ephemeral mind stuff arise from corporeal brain stuff? We have not yet answered that question, but we are making progress.

Neuroscientists and psychologists have acquired a lot of information regarding what anatomical parts of the brain are required for conscious experience. From pathological studies in patients with destructive lesions in the brain, scientists have learned that the reticular activating system (RAS) is essential for the conscious, "awake" state. This evolutionary ancient network of nerve fibers arises in the pons and midbrain, and juts into the hypothalamus and thalamus, where nerve fibers relay projections to all areas of the cerebral cortex. The RAS, sometimes called the "ignition system" of the brain, is what gets your brain started and enables the conscious state. If the RAS is damaged or interrupted, consciousness is lost.

The pons and midbrain in the midline of the brainstem contain clusters of nerve cells that release norepinephrine, serotonin, and dopamine to help modulate the awake state. Damage to this system may also result in loss of consciousness. The thalamus, located in the forebrain just above the brainstem, relays sensory input from the RAS and forwards this information to the cerebral cortex. The thalamus and its connections to the cortex are essential for the consciousness of sensory qualia. If the thalamus is damaged and fails to relay sensory information to the cortex, an individual can suffer a burn to the hand and not feel a thing (Lutkenhoff et al. 2015). The cerebral cortex is the culmination of the mammalian brain and provides the substrate for higher levels of consciousness. However, loss of consciousness requires the disruption of almost all cortical interconnections—injury to the entire cortex.

Neuroscientists at the Max Planck Institute in Germany performed classic experiments using a special technique that enabled them to identify specific areas of brain activity associated with conscious awareness of a visual image (Logothetis, Leopold, and Sheinberg 1996) The investigators simultaneously exposed monkeys to separate images in each eye, which results in an alternating conscious perception of one image and then the other—the brain keeps changing its mind. This phenomenon is called **binocular rivalry.** With this technique, the investigators were able to identify specific patterns of brain activity correlated with conscious perception of each image in a small area of the temporal cortex (figure 23.1). This is surprising, because input from the eye is received in the occipital cortex of the brain—but, it turns out, this is not where the brain becomes "aware" of visual images. Conscious perception of visual images apparently requires integration between various areas of the brain and culminates in the temporal cortex. These specific areas of brain activity correlated with particular conscious experiences are called **neural correlates of consciousness (NCC)** and are the subject of much current research in consciousness.

Identifying NCC is a start toward understanding how neural activity might lead to consciousness, but as we know, correlation does not imply causation. Currently we have no clue what, exactly, thought stuff is made of, or how it arises from the activity of neurons in the brain. Consciousness, as best we know, has no mass or energy and cannot be effectively quantified and studied with current scientific methods. We have a long way to go before we understand how the mind emerges from matter. Best-selling author Robert Wright summarized our limited understanding in *Nonzero*: "Indeed, the more scientific you are in pondering consciousness, the more aware you become of the limits of science; the more inclined you become to approach cosmic questions in general with a touch of humility." (Wright 2001, 305).

This is the "hard problem." But remember: we were unable to explain the behavior of subatomic particles before the introduction of quantum physics and the concept of wave-particle duality. New methods and new theories for evaluating consciousness, possibly from outside the realm of science, will surely emerge in the future, even if we cannot quite imagine them today. For now, the mystery remains.

Figure 23.1: Where does consciousness reside? German neuroscientists sought to answer this age-old question in classic studies using monkeys as experimental subjects (Logothetis 1996). When monkeys look at a banana or a tree, specific brain activity can be detected in the occipital cortex and temporal cortex for each image. When investigators presented monkeys with simultaneous images—a banana viewed with one eye, and a tree viewed with the other—and recorded brain activity, the monkeys were confused. The monkeys' brains apparently processed each image one at a time, and the monkeys were conscious of seeing alternating images first of the banana and then the tree, back and forth—a phenomenon called binocular rivalry. During this process, the investigators taught the monkeys to press one button when they "saw" the banana and another when they "saw" the tree, while the investigators simultaneously recorded brain activity. It turns out that the signals from the occipital cortex were essentially the same whether the monkeys were momentarily "seeing" the banana or the tree. In contrast, signals in a small area of the temporal cortex had one characteristic pattern when the monkey "saw" the bananas and another characteristic pattern when the monkey "saw" the tree. The investigators concluded that neural activity correlating with consciousness of visual images is located in the temporal cortex. These types of correlations provide first steps in solving the mystery of how conscious perception may arise from neural impulses.

EVOLUTION OF CONSCIOUSNESS

Consciousness, like other biological traits, surely has evolved incrementally through natural selection. But thought itself leaves no fossils, and consciousness as we know it has no known genetic markers, so evolutionary biologists have struggled to trace its evolutionary past. There remains great controversy about the origins and evolution of consciousness, and what its adaptive benefits might be.

The brain is bombarded with a vast array of sensory input and must sort out what is important and what is not to initiate an appropriate response. Neuroscientist Michael Graziano believes the brain evolved adaptive mechanisms for selectively processing important signals amid this vast onslaught of sensory data. He believes consciousness evolved as a by-product of that evolutionary process (Graziano 2016).

Graziano believes the **tectum**, a structure in the midbrain of vertebrates responsible for auditory and visual reflexes, selectively processes sensory input, sorting out what is important and coordinating what is called overt attention. For fish and amphibians, the tectum is still the largest and most sophisticated part of the brain. A new enlarged structure, the dorsal pallium, evolved in the forebrains of reptiles and birds between 350 and 180 million years ago. Later, mammals evolved the cerebral cortex, which superseded the dorsal pallium to serve as the substrate for higher levels of consciousness (Feinberg and Mallatt 2013). These structures evolved to focus attention on internal signals from the brain, coordinating what is called covert attention. In the process of creating covert attention, they created a virtual reality of sensory information, stored memories, and thoughts. According to Graziano, this creation of virtual realities morphed into the evolution of consciousness.

Does basic sensory consciousness exist in invertebrates? Sponges, the oldest invertebrates on the phylogenetic tree, do not have nervous systems and therefore surely are not capable of conscious awareness. Likewise, the cnidarians—members of the second-oldest phylum, which includes hydras, corals, and jellyfish—have a nerve net but no brain. These animals can respond reflexively to stimuli but have no substrate for consciousness.

The first animals with primitive brains, ancestors of modern flatworms (platyhelminths), probably had specialized cells that sensed light, as flatworms do today. These worms display phototaxis—a simple movement reflex in response to the presence or absence of light. It is possible that their primitive brains may experience a subjective perception of light or a dim glimmer of shadow, but we do not know. Investigators have tried to determine whether invertebrates have subjective sensory perceptions by assessing their response to pain, but experiments of this sort are often problematic. When an animal receives a noxious stimulus, it will respond by withdrawing its body until the stimulus ends. But this could be a simple reflex, with no signals traveling to the brain and therefore no subjective perception of pain. Is this consciousness? Probably not. Investigators look for responses beyond the reflex—such as nursing a wound, limping, or changes in physiology—to indicate that an organism is experiencing pain.

Many people are horrified at the thought of lobsters being boiled alive, or live crabs having their claws torn off. We hate to see our fellow creatures suffer. But do these and other crustaceans actually feel pain? Recent research suggests that, heartbreakingly, they probably do. In one experiment, acetic acid was brushed on prawns' antennae. Each of the prawns immediately withdrew from the brush and then spent some time grooming their antennae and rubbing them against the side of the tank—just as you might rub at a burned and stinging hand. Furthermore, these responses stopped when the researchers applied a local anesthetic to the prawns' antennae (Barr et al. 2008). In a separate experiment, shore crabs were offered several dark shelters to hide in but were given electric shocks when they entered a specific shelter. After being shocked a few times, most of the crabs moved into a different shelter and avoided the booby-trapped one for the rest of the study (McGee and Elwood 2013). The adaptive function of pain, we believe, is to help organisms avoid similar situations in the future that will also produce pain and cause harm. The fact that shrimp groom their antennae when they are burned, and that crabs avoid electric shocks, strongly suggests that they experience pain. These experiments and others like them may have broad implications as to how we treat prawns, crabs, lobsters, and other crustaceans.

Octopuses (cephalopods in the phylum Mollusca) have the most advanced brains of all invertebrates and have exhibited complex behavior suggestive of an unusually high level of consciousness, similar to that seen in mammals. Like many mammals, octopuses learn rapidly, solve problems, and even make and use tools, such as transporting coconut shells to use in the future for shelter and protection. Octopus brains do not have a tectum, pallium, or cerebral cortex, but they have evolved their own unique brain structures, which are clearly capable of higher-order thought. An advanced level of consciousness appears to have emerged in these disparate branches of the tree of life—cephalopods and vertebrates—by convergent evolution.

Self-awareness, one of the highest levels of consciousness, has been documented in only a few species so far. Classically, self-awareness has been assessed using the mirror test, sometimes called the mark test. This test has been used to assess the self-awareness of young children and a number of animal species. It works like this: A mother marks her young son's nose with rouge and lets him look at himself in the mirror. Until he reaches about fifteen months of age, he will not recognize the image in the mirror with the red spot on its nose as a reflection of himself. But within a few months, the boy will recognize himself and try to wipe away the rouge. Children, on average, achieve this sort of self-awareness between fifteen and twenty-four months.

Self-recognition assessed by the mark test has also been demonstrated in apes (chimpanzees,

Image 23.2: Australian magpie. Surprisingly to most of us, magpies are self-aware; they can recognize themselves in a mirror. Magpies are the only nonmammalian species to show this type of self-awareness. This high level of consciousness arose in magpies and primates by separate evolutionary pathways via convergent evolution.

gorillas, orangutans, and gibbons), dolphins, killer whales, elephants, and, surprisingly, magpies, which are the first nonmammalian species to pass this test (image 23.2) (Prior, Schwarz, and Güntürkün 2008). Before this newest finding, it was believed that self-awareness was correlated with the evolution of the neocortex—the newest part of the mammalian cerebral cortex, with characteristic folds, grooves, and ridges. But the ability of magpies, which do not have a neocortex, to recognize themselves in a mirror challenges this doctrine. It appears that the capacity for self-awareness arose independently in mammals and in birds, providing yet another example of convergent evolution (Prior, Schwarz, and Güntürkün 2008).

WHY DID CONSCIOUSNESS EVOLVE?

Most animal traits exist because they provide an evolutionary advantage. It seems logical that consciousness and complex thought evolved because they provide adaptive advantages, but there is quite a bit of controversy about this idea. Some investigators strongly believe consciousness has evolved because it has survival advantages, but others suspect consciousness may be a complex epiphenomenon arising as a by-product of the way our brains have evolved—like the inconsequential sound of the heartbeat arising as a by-product of the pumping action of the heart. As we have seen, Michael Graziano believes that consciousness may have evolved as a by-product as our brains evolved sophisticated methods of focusing attention (Graziano 2016).

Though it may seem counterintuitive, the position that consciousness is an epiphenomenon without any adaptive advantages has strong advocates. After all, if you think about it, you perform most of your daily activities with little or no conscious awareness. You can brush your teeth, put on clothes, make coffee, and drive to work all without really paying attention to what you are doing. Once motor habits have formed, there is little conscious awareness or direction required to use them in day-to-day life. Psychologists have called our subconscious behavior the zombies within us.

Theoretically it is possible that your friends and neighbors are what theorists call **philosophical zombies**—looking human, acting human, smiling, laughing, and crying like people do, but without consciousness. Poke a philosophical zombie with a pin, and the zombie will pull away and cry out, "Ouch! What did you do that for?"—but will feel no actual pain. Zombies cannot feel a thing. I know that I am conscious because I experience my own thoughts and qualia, but I have no way of knowing if you are conscious. You could be a philosophical zombie. If we can imagine that philosophical zombies exist—and this is becoming easier as we develop sophisticated artificial intelligence with human traits that can perform human tasks— we can also imagine that consciousness might be an epiphenomenon arising from the activity of the brain, with no adaptive advantage in and of itself.

Most evolutionary psychologists, however, believe that consciousness is not just an epiphenomenon. Psychologist Nicholas Humphrey, in his impassioned book *Soul Dust*, argues that by experiencing emotions and feelings in a conscious way, we can better understand the feelings, emotions, and motivations of others, and therefore be in a better position to predict their behavior (Humphrey 2011). This ability to understand the minds of others, called **theory of mind (ToM)**, requires consciousness and is widespread among vertebrate animals (Graziano 2016). A dog is able to look at another dog and figure out whether he is aware of the observing dog and whether he is friendly. We humans are highly social animals, and we form large, complex, ever-fluctuating social groups. Knowing other individuals' intentions, learning whom we can trust, and deciding whom we should mate with or form alliances with, or who will betray us—all of these abilities provide us with major adaptive advantages. Beyond simply allowing us to understand others, having a theory of mind also fosters empathy and promotes mutual respect and cooperation between people, which are qualities that promote group survival and are therefore highly adaptive. Humphrey believes this facilitation of theory of mind may be one of the major adaptive advantages of consciousness responsible for its evolution.

Biophysicist Krystof Koch, in his scientifically thoughtful book *The Quest for Consciousness*, argues that consciousness offers an even more important adaptive advantage than improved social functioning (Koch 2004). In his view, consciousness allows us to plan. Consciousness allows us to extend our sensory experience in time—to give experiences extended duration. We store these experiences in memory, recall them, and use them in forming virtual realities of past experiences. Similarly, consciousness allows us to form virtual realities of imagined future experiences, and we can use these virtual realities to plan and make choices about future actions. Hunter-gatherers used conscious planning to organize hunting parties. Agricultural societies used conscious planning to determine when and how to plant and harvest crops. And industrial societies use conscious planning to create production schedules to meet consumer demand. Today, all of our daily activities are centered on planning. We can't imagine life without it. We have to do things to survive. And to do things effectively, we must plan.

But some even question whether planning requires consciousness. In 1997, when the IBM supercomputer Deep Blue beat world champion Garry Kasparov at chess, Deep Blue was planning its moves—but of course, it was not conscious; it was a computer. Likewise, some behaviorists believe that planning can occur when an organism analyzes potential future scenarios, assesses the existing conditions, and selects the best course of action, all using subconscious algorithms without the mediating help of consciousness. After all, Deep Blue did it in that chess game.

If our conscious brain is to function as the quarterback, coordinating, planning, and directing higher brain functions, it must have the freedom to do so. From somewhere within our brains there must be an executive force, the *self*, that has the freedom to make decisions and to initiate action. There must be free will. Philosophers and theologians have debated the issue of free will for millennia. Does it exist, or doesn't it? Only recently have neuroscientists and psychologists brought the issue into the realm of science. You may be surprised at some of their findings.

FREE TO CHOOSE?

You and I have no doubts about the existence of free will. I certainly have the freedom to decide to get the newspaper in the morning and to read it over a cup of coffee. You have the freedom to arrange a round of golf or a bar crawl with your buddies on the weekend. But psychologists and neuroscientists are not so sure. Many share the belief that free will is actually an illusion.

If you think about it, most of our daily activities are not orchestrated by free will. Brushing your teeth, tying your shoes, driving to work, typing a paper, hitting a tennis ball—most of these things are done at the subconscious level. We might have a conscious moment in which we decide to do any one of these things, but once we've begun tying our shoes, subconscious habitual circuits take over, and consciousness and free will take a backseat. Our fingers do what they've learned to do, and our minds wander to other things.

If we examine any of our "decisions" that we seem to make by free will, we soon realize the innumerable constraints on those decisions. Throughout our entire lives, our genetic makeup and our experiences shape our thoughts and neural pathways, and the unique time, place, and circumstances of any situation add further constraints to our possible choices. If I decide that I want to play tennis with my friend Bert, what are the constraints on that decision? To make this work, at some point in my life I must have learned how to play tennis, and I must be in reasonable health and injury-free. I must have a tennis racket, tennis balls, membership or access to a tennis club, and an available court. I must have a car for transport, and it must not be raining. My friend Bert must live nearby and fulfill all these same conditions. These are just a few of the innumerable constraints on my decision to play tennis with Bert. If any of these conditions are not met, I am not free to play tennis. I am presented with a certain narrow set of choices based on the conditions set by my life. How much room is really left to maneuver? Are we cornered by these many constraints? Are our decisions *really* free?

THE CASE AGAINST FREE WILL

Neuroscientist Benjamin Libet and colleagues at the University of California San Francisco brought free will into the realm of science with their pioneering studies on consciousness (Libet et al. 1983). In a landmark experiment published in 1983, volunteers were asked to push a button and note the exact time when they had consciously decided to push the button. At the same time, the investigators recorded deep brain activity using an electroencephalogram (EEG). They discovered, to the great surprise of most, that deep brain activity, which Libet called a readiness potential, preceded reported conscious intent by four hundred to five hundred milliseconds. Other investigators have repeated these experiments using functional MRI and have shown that neural signals can precede and predict conscious decisions by as much as several seconds. This means that our decisions have already taken place at a subconscious level before we are aware of a conscious decision. The information gathered to make such decisions comes from genes, prior experience, and current sensory input. Once the decision is made, it enters the realm of consciousness and action is taken. Conscious free will appears to be an illusion. If there is free will, it would have to occur at the subconscious level—but how would that work?

Perhaps these arguments against the existence of free will give you pause but you're still not convinced. So let's take this a step further. The conscious mind and free will are ephemeral constructs. They have no matter, no measurable energy, and no physical form. You can tell your friend about a dream you had, but your thoughts themselves cannot be seen, heard, or touched. So how can free will interact with or control our matter-filled, energy-filled, physical computer of a brain? As Krystof Koch explains in *Consciousness: Confessions of a Romantic Reductionist*, "Physics does not permit such ghostly interactions." (Koch 2012, 111).

There is another argument that might be even more convincing—at least for nonbelievers. If we have free will, where do our freely made decisions come from? Our decisions are certainly influenced by past experiences and memories recorded in neural circuits in our brains, but if we truly have free will, some part of these decisions must be free of such influences. It seems they must come from out of nowhere. Maybe free will gets a helping hand from divine intervention. But as scientists, we are left hanging.

If you are *still* not convinced, try this quick exercise suggested by Yuval Noah Harari in his controversial book *Homo Deus* (Harari 2016 285–286). When a thought comes to your mind, ask yourself where that thought came from. Did you will it to come to mind? Stop what you are doing, quiet your mind, and then pay attention to your thoughts over the next minute. Did you control which thoughts came to you? Do you have free will to control your stream of thoughts?

The view that free will is an illusion is counterintuitive to most of us, and it poses a number of ethical problems. The alternative to free will is something like determinism—the idea that even if all events are not entirely predictable, they are predetermined and governed by baseline conditions and the laws of nature. This thought is pretty distasteful to most of us. We want to cling to our free will. We will argue that free will is a prerequisite for morality, for personal responsibility, for goals and achievement, and for the difference between good and evil.

Neuroscientist and author Sam Harris, a strong advocate that free will is an illusion, argues that we can deal with these issues. If we believe that a man who has committed murder did not act out of free will but rather was programed by a series of subconscious mechanisms that are the product of his genes and his environment, we are not necessarily pardoning that man; we simply have a different viewpoint of the crime. Let's look at an example.

A boy is born on the south side of Chicago. His mother, an alcoholic without a job and dependent on welfare, struggles through his entire childhood to make ends meet. At age eleven, he joins a street gang because the gang members are the only group in his community who make him feel accepted and important. The boy enters a culture of guns, drugs, and macho respect. Over the next several years, he integrates fully into the gang and becomes one of its leaders. Then one day, in a fierce argument over a drug deal that stands to put everyone in his gang at risk, the boy kills a rival gang member with three shots to the chest.

The teenager enters the courtroom. He is on trial for murder. He arrives in an orange jumpsuit, his hands cuffed in front of him, his head shaved, and his bare arms covered with tattoos. The judge, jury, attorneys, and most of the spectators in the courtroom all live in his city, but they are from a parallel culture. None of them have ever been in a gang. The jury is there to decide if this boy did indeed commit this murder—and he did. He pulled the trigger three times. But there is a broader question that will not be asked of this jury: Was the boy *responsible* for the act? Did he have a Cartesian free-will choice, or were his actions predetermined by his genetic inheritance and his violent upbringing? Was the boy predestined to kill?

I hope this story gives us some insights about the limitations and perhaps the fallacy of free will. Many of us will agree that this boy did not have an entirely Cartesian choice. This may not have been a willful act. We would agree the teenager still needs to be put in jail to protect society, but we have empathy for the boy, and there would be no need for redemption.

Sam Harris continues his argument. Rather than thinking about how this shift in thinking affects our perception of those who do bad, what about those who do good? How does the lack of free will impact individual responsibility, goals, aspirations, and achievement? Many of us give too much credit to those who are successful. The "self-made" man, like the murderer, is not really self-made; his success is a product of his genes, birthplace, upbringing, and background. His diligence and his willingness to work hard are a product of these things and are not necessarily due to his own volition. He must be lucky to be intelligent and healthy, and to have been at the right place at the right time. This perspective allows us to better understand why people are successful and unsuccessful, and, as Rudyard Kipling counseled, to "treat those two imposters just the same …" (Kipling 1943). Without leaning on free will, we can still value and encourage people to do their best, to become good citizens, and to work hard to be successful—and our actions can still influence their behavior for their benefit and for the benefit of society. We do not need to be taken in by the illusion of totally free will to be good people. In fact, the acknowledgment of the limitations of free will should make us more tolerant and understanding of others.

Philosophers and theologians have struggled with the concepts of free will and consciousness for centuries, and now neuroscientists and psychologists have entered the fray. Still, there has been little resolution, and we will not find that resolution here. The ephemeral nature of consciousness and free will unfortunately put them beyond the realm of scientific investigation—but only for now. Current science can study only those aspects of our universe that are composed of matter and energy in time and space. But just as Einstein discovered the general theory of relativity to describe a new relationship between matter, energy, and space-time, and just as quantum physics described the disparate behaviors of subatomic particles, someday we will develop the tools to more effectively study and understand free will and consciousness. Such an understanding will likely play a major role in shaping the future of our species.

CONSCIOUSNESS AND HUMANITY

While we don't understand how ephemeral consciousness arises out of brain matter, and we still have some doubts about the evolutionary advantages of consciousness, we have little doubt about its importance to our lives. Consciousness is essential to our ethics, morality, and law—all of which have been created with the goal of minimizing conscious human suffering. Without consciousness, life has no meaning. Consciousness transforms our world into an awesome place. We gaze at a flaming fire, listen to the singing of a mockingbird, dangle our feet in a cool stream, and watch the setting sun light up the crimson sky. Our brains transform these sensory experiences into an awesome virtual reality that becomes part of our consciousness. Consciousness allows us to reason, plan, and create visions of the future. Our consciousness embodies our emotions and feelings—the ups and downs that bring drama and meaning to our lives: love and loneliness, hope and despair, courage and fear, agony and ecstasy. Our consciousness embodies our humanity.

CHAPTER 24

AGING, DEATH, AND EVOLUTION

No one wants to die. Even people who want to go to heaven don't want to die to get there. And yet death is the destination we all share. No one has ever escaped it. And that is as it should be, because death is very likely the single best invention of life. It is life's change agent. It clears out the old to make way for the new. Right now, the new is you, but someday not too long from now, you will gradually become the old and be cleared away. Sorry to be so dramatic, but it is quite true.

—Steve Jobs, cofounder of Apple

There were many gods and goddesses in ancient Greece. Among them was Eos, the goddess of the dawn, who had an unusual attraction to mortal men. One morning, as she brought the light of day to Earth, she came across the city of Troy, where she found the handsome Trojan prince Tithonus. It was love at first sight, and it was to last forever. But there was one problem. Eos was immortal, but Tithonus was not. Eos went to her father, Zeus, and pleaded for him to make Tithonus immortal so they could live together forever. After much pleading, Zeus finally gave in and granted her request. Time passed. One morning, as Eos kissed Tithonus on the forehead before leaving for her celestial work, she noticed a gray hair. Then a whole head of white hair. Tithonus got older and older and older. Eos had asked her father to make Tithonus immortal, but she had forgotten to ask for eternal youth. His once powerful limbs shriveled up. His once booming voice faded into a tiny chirp. His mind atrophied, and he declined into a babbling idiot. Finally, Tithonus begged Eos for mercy, and she granted his wish by turning him into a grasshopper (Paleothea 2007).

Today we face a fate similar to that of Tithonus. As medicine prolongs our lives, aging

251

marches on. By late middle age, our skin becomes thinner and wrinkled as it loses its elasticity and tone as a result of the loss of subcutaneous fat and changes in connective tissue. As the female egg supply becomes depleted, ovulation stops, and infertility follows. Diminished production of estrogen and progesterone by the ovaries results in hot flashes and wasting of genital tissues. Men develop testicular atrophy, with decreased testosterone levels and reduced sperm count. Our muscles lose their strength and speed as they slowly atrophy. Our sprint turns into a jog, then a walk, then a shuffle. Our vision dims, and our hearing wanes. The production of digestive enzymes from the pancreas diminishes, and intestinal absorption of nutrients declines. Blood vessels stiffen, absorb fatty plaque, and accumulate blockages. Mental confusion and memory loss are followed by dementia as nerve cells die and the brain withers. "Old age is not for sissies," said Bette Davis, a celebrated actress who died in 1989 at the age of eighty-one after suffering through cancer and several strokes.

IS DEATH INEVITABLE?

Why do we wither and die with age? Historically, the process of aging has been likened to the wear and tear that we see with automobiles or any other piece of equipment after prolonged use. But as we come to better understand the nature of life, we realize that this explanation does not tell the whole story. The universe is subject to the laws of thermodynamics: All systems tend to move from a state of order to a state of disorder. Entropy, the measure of disorder, is always increasing. Life, however, is an exception. Living systems take energy from the environment, grow and build new structures, and systematically repair faulty or damaged parts, unloading entropy as waste. There is no thermodynamic necessity for senescence and death. The rebuilding, machinery-fixing magic of life could just go on and on. Life is not disobeying the laws of thermodynamics; while order is maintained within living systems, disorder is increasing outside living systems, and disorder in the universe as a whole is increasing.

Bacteria, descendants of our most ancient ancestors, provide a good example of the machinery-fixing magic of life. Bacteria reproduce by simply cloning—by splitting in two. When bacteria divide and produce two identical cells, who is the parent and who is the offspring? Who is supposed to get old and die? As it turns out, the answer is neither. When bacteria are grown in a petri dish, provided with an adequate food supply, and "thinned out" to prevent overcrowding, they can divide indefinitely and appear to be biologically immortal. As the eminent German evolutionary biologist August Weismann asked in 1882, "Where is the corpse?" (Clark 2002, 24). Bacteria fall to predation, starvation and overcrowding, but not to old age.

It is not so for us eukaryotes. Remember the eukaryotes? You should—you are one! The eukaryotes are all organisms with complex nucleated cells. Simple single-celled yeasts, used in baking and brewing, are eukaryotes and experience old age and death. Yeast can reproduce asexually by a process called budding. A yeast cell divides, producing a daughter cell that buds from the mother cell. Like bacteria, the daughter cell is a clone of the mother cell, but unlike bacteria, yeast reproduce by forming two distinct cells: a large mother cell and a smaller daughter cell. The mother has a lifespan that encompasses about twenty-five budding events. As she nears the end of this cycle, the budding rate slows, she shows signs of senescence, and then she dies. The daughter cell resets the aging clock to zero as she starts her life anew and begins her own budding cycle. Yeast are primitive eukaryotes, so the genetic programs controlling old age and cell death appeared very early in our evolution and have been conserved over billions of years. The question is why.

Human cells also experience old age and have a limited lifespan. In 1961, Leonard Hayflick, at the Wistar Institute in Philadelphia, discovered that human fibroblasts grown in culture will grow only for a limited time, as if they have a "clock" that counts the number of cell divisions (Hayflick and Moorhead 1961). The cells reach what has been called the **Hayflick limit**; after about fifty-two divisions, the cells stop dividing and enter a phase of replicative senescence. This "winding down" was found to be related to a gradual shortening of **telomeres** that occurs with repeated cell division. Telomeres are short pieces of DNA at the end of chromosomes, similar to the aglets at the ends of shoelaces, that act as buffers to protect the essential information-carrying parts of the chromosomes. Each time DNA replicates prior to cell division, the telomeres shorten. Eventually, when the telomeres have shortened to a critical length, DNA and cells lose their ability to divide. This mechanism probably evolved originally to protect against unregulated cell growth and cancer. As telomeres shorten, and as cells approach the ends of their life cycles, their function is impaired and they enter a phase of senescence, accounting for the effects of aging. The process limiting cell division (and leading to senescence) can potentially be reversed by **telomerase**—an enzyme that facilitates the regeneration of telomeres and thus allows cells to grow indefinitely. Cancer cells have telomerase and can therefore divide indefinitely without regulation.

Death and immortality have different meanings for multicellular and single-celled organisms. Multicellular organisms have evolved two types of cells: Somatic cells, which are cells that have differentiated to perform specialized functions, and germ cells, which remain undifferentiated but pass their genes down from one generation to the next. As Richard Dawkins describes in *The Selfish Gene*, our soma cells constitute a body that is a "survival machine" for our germ cells, allowing our genes to be passed to the next generation (Dawkins 1989). The human germ cell line has been maintained as an unbroken chain of life extending back in time to our earliest ancestors. In this sense, we all have a part in life's immortality, although any individual's contribution to the gene pool is quite small.

Of course, none of us have achieved individual immortality, but this may not be the case for all life forms. There have been several species identified that appear to have escaped aging and may have achieved immortality.

IMMORTAL SPECIES

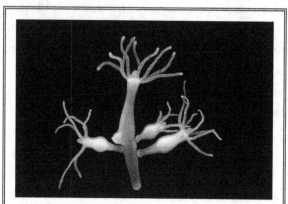

Image 24.1: *Hydra* (*Hydra vulgaris*). These small, predatory freshwater animals, related to jellyfish and corals, appear to be capable of regenerating their tissues indefinitely and appear to be immortal.

The best known and most extensively studied "immortal" species are the hydras—predatory marine animals related to jellyfish and corals (image 24.1). These tiny creatures, a mere centimeter in length, are found in most freshwater ponds and lakes in temperate climates. They have a mouth surrounded by filamentous tentacles, each of which can fire off dart-like threads containing neurotoxins to paralyze their prey.

Hydras are capable of escaping aging by constantly renewing their bodily tissues. They reproduce asexually by budding off germ cells, and they show no decline in reproductive rate, even over many years of observation. Investigators have identified a gene in hydras, the *FoxO* gene, which appears to be responsible for their regenerative capabilities (Boehm et al. 2012). When this gene is inactivated, hydras are no longer able to continually produce the stem cells they need to rebuild themselves. Their immune systems also become suppressed, which may affect aging. Genes similar in structure and evolutionary origin to the *FoxO* gene (homologs) exist in all animals, including humans, and are universally implicated in longevity. We'll hear more about these genes a little later.

CELL SENESCENCE

As we have discussed, repeated cell divisions result in shortening of telomeres, and this leads to DNA damage that triggers a signaling process that initiates cell senescence. Cells can also become senescent prematurely as a result of a variety of stressors that damage DNA (Ben-Porath and Weinberg 2005). These include irradiation, toxins, and damaging genetic mutations. When cancerous genes arise in cells, due to either spontaneous mutations or mutations induced by external factors, they are recognized and can produce signals that induce senescence—a process that protects against the proliferation of cancerous or precancerous cells (Childs et al. 2014).

Animal cells are in a constant cycle of life and death. As cells age, become senescent, and die, new cells are formed from stem cells to take their place. The lifespan of individual human cells varies greatly depending on the cell type: Epidermal skin cells last a mere two weeks, red blood cells last about four months, liver cells turn over every year, and bone cells are replaced every ten years. A few cell types—neurons of the cerebral cortex, and some heart muscle cells—are never replaced at all (Spalding et al. 2005). Once nerve and cardiac cells have formed a mature organ, they stop dividing and persist for the life of the organism. The stem cells that differentiate and replace soma cells ultimately go into replicative senescence and stop dividing. With no stem cells to replace dying soma cells, death of the organism soon follows.

What is potentially a great contributor to cell senescence and the body's aging process had its beginnings long ago in our evolutionary history. After cyanobacteria spilled the first oxygen into Earth's primitive atmosphere, microorganisms, and eventually animals, adapted and used oxygen to drive their powerful aerobic metabolisms. But there is a downside to this system. Aerobic metabolism, which takes place in the mitochondria, produces toxic **reactive oxygen species** (ROS), such as superoxide anions (O_2^-), hydroxyl radicals ($^{\cdot}OH$), and hydrogen peroxide (H_2O_2). These reactive molecules leak into the mitochondrial spaces in cells, and as the name implies, they are highly toxic. They corrode and destroy proteins, membranes, and, most importantly, mitochondrial DNA (mDNA). Remember: mitochondria are remnants of primitive bacteria and have retained some of their own ancestral DNA, which allows local control over cellular energy production. While unproven, damage to mDNA from ROS has been a mainstream hypothesis to explain cell senescence and aging in aerobic animals (Lagouge and Larsson 2013).

The rate of free radical leak in an animal is related to its basal metabolic rate. Larger animals have lower basal metabolic rates per gram of tissue, and smaller animals have higher rates. Because of this, bigger animals have less free radical leak, and this may explain why they live longer than smaller animals. Elephants and whales live longer than mice and men.

Attempts to improve longevity by modulating free radical leak have, unfortunately, not been successful to date. Numerous clinical trials with antioxidant therapies have shown no benefit (although commercial promotion of these supplements is rampant). Antioxidants may not have been effective because, although free radicals are very toxic, they also serve an important function in intercellular signaling (Lagouge and Larsson 2013). These angry little molecules may serve us as they kill us.

CELL SUICIDE

Darwinian evolution teaches us that organisms adapt for self-preservation. So when biologists Valter Longo and Joshua Mitteldorf and others first described how yeast cells sacrifice their lives for the good of their colony, there were many skeptics (Herker 2004; Longo 2005; Buttner 2006). But sacrifice they do. When yeast cells are placed in conditions of potentially lethal food shortage, older or damaged cells commit suicide, removing themselves from the population. Their little unicellular bodies provide food for the rest of the colony and allow survival of a core group that can then replenish the colony. Who could imagine such self-sacrifice in these nonsentient microorganisms? Would we humans ever be so altruistic?

The heroic yeast cells kill themselves in a process we now know as **apoptosis** or programmed cell death. The presence of apoptosis in these unicellular fungi attests to the ancient heritage of this process, which evolved long before the origin of plants and animals. As we all learned from Longo and Middledorf's experiment, cell suicide—while obviously destructive to the individual cell—has definite survival advantages for the colony. In those ancient days, as now, bacteria were at war against a host of invading viruses. Apoptosis must have evolved early on as a defense against invaders. Bacterial cells that become infected with viruses or parasites can destroy themselves to help rid the colony of the foreign threat. And cells with damaged DNA—whether from mutations, ultraviolet light, or free radical damage—can self-destruct before they become cancerous.

When cells are damaged from physical trauma or toxins, they swell and burst their membranes, spilling their contents everywhere. If the cells are part of a multicellular organism, the leakage causes a detrimental inflammatory response. Apoptosis is not so messy. During programmed cell death, the cell undergoes a controlled, rapid degradation of its internal components using powerful enzymes, and waste products are quickly cleaned up before any leaks occur into surrounding tissues. The cell shrinks, becomes gently rounded, and dies with minimal polluting effect on its microscopic neighborhood.

Apoptosis does have a downside. Misguided apoptosis events can destroy normal cells and thereby hasten an organism's death. Unregulated apoptosis is responsible for the destruction of healthy cells in Alzheimer's disease, Parkinson's disease, and Huntington's disease (Favaloro et al. 2012).

WHY DID WE EVOLVE TO AGE AND DIE?

The immortal hydras are a rare exception to the rule of death. For humans and most other animals, mortality is a destiny we all share. But why? Why should it be that evolution, having created a complex sentient human being that is so well-suited for survival in a hostile environment, would then fail to maintain the accomplishments of its own work? Why does natural selection result in such bizarrely self-harming endings as senescence, late-life degenerative diseases, and death instead of eternal youth and immortality? If the body is capable of rebuilding itself over and over again over the course of a lifetime, then why, after a brief race of a few decades, does it slowly fall apart and die?

In 1952, a century after Darwin published his theories of evolution, British biologist Peter Medawar proposed the first modern theory of why we age and die (Gavrilov and Gavrilova 2002). His basic thesis was that the forces of natural selection decline with age.

Here's the logic: Survival of all species decreases with time and age because of predators, infectious diseases, accidents, and starvation, even without the effects of aging. Our reproductive lives are also finite and limited to our younger years. Because of this, genes that have a negative effect on survival later in life, beyond the reproductive period, will have already been passed to the next generation before they inflict damage—and therefore will never be effectively weeded out of the population. In contrast, genes with a negative effect on survival early in life will destroy the organism before it has a chance to reproduce and will not be passed to the next generation; this is the basic mechanism of evolution. Over time, mutations detrimental to survival in old age after the reproductive years will accumulate in a population as they are passed from generation to generation, resulting in diseases of old age, senescence, and mortality in later years. So even as we eliminate mortality due to predators, infectious diseases, and starvation, we will succumb to these diseases of old age. This is our heritage. This is inherent in the process of evolution.

DID AGING AND DEATH EVOLVE FOR THE BENEFIT OF THE GROUP?

In his book *Cracking the Aging Code*, theoretical biologist Joshua Mitteldorf has advocated an alternative hypothesis as to why evolution has left us as aging, mortal souls (Mitteldorf and Sagan 2016). He proposes that aging was programmed into the genetic code very early in the course of evolution for the benefit of the group, despite acting against the survival of the individual. Chapter 20 mentioned group selection—the concept that traits that promote survival of the group will be passed to future generations through group survival, despite some degree of disadvantage to the individual. Mitteldorf believes that genes that control aging and death may have evolved through group selection. Senescence and death, he argues, while fatal to the individual, promote the survival of the group by limiting uncontrolled population growth, which can lead to famine and infectious epidemics that can wipe out entire species. Death frees population resources for the young and promotes genetic diversity, which is essential for the survival of a species in a rapidly changing environment. Steve Jobs, although he was not likely aware of this particular theory, was correct when he said, "[Death] is life's change agent. It clears out the old to make way for the new" (Leadem 2016).

Mitteldorf argues that the neo-Darwinian theory of survival of the fittest, with its emphasis on individual selection alone, is an incomplete paradigm because it neglects the effect cooperative effort has had on evolution. There is always potential conflict between the welfare of the individual and the welfare of the group. This tug-of-war has played out in evolution as a struggle between individual selection of selfish genes and group selection of cooperative genes.

We now know that manipulation of a single gene can dramatically prolong or shorten lifespan in diverse species ranging from worms to mice. Mitteldorf believes that genes that shorten longevity in individuals evolved because they have an adaptive advantage for the group as a whole. This brings us to a discussion of the genetic control of aging and longevity.

GENETIC CONTROL OF AGING

Our understanding about the genetic control of aging took a giant step forward in 1993 when researcher Cynthia Kenyon's team identified a single gene, *daf-2*, responsible for longevity in the nematode roundworm *Caenorhabditis elegans* (Guarente and Kenyon 2000). These worms are favorites of scientists investigating aging and longevity because of their short lifespan (two to three weeks) and the ease with which their genes can be altered. Kenyon found that a mutation in the *daf-2* gene prolonged the lives of these worms twofold. This groundbreaking discovery challenged two widely held beliefs: first, that aging was inevitable because of natural wear and tear on the body and cells over time; and second, that the processes of aging were entirely due to slowly accumulated detrimental mutations in a wide array of genes, as proposed by Medawar. We now have a new possibility: that a single gene or family of genes can control aging and longevity.

Caloric restriction was shown to extend life in rats eighty years ago. Since that time, numerous studies have shown that a low-calorie diet can increase longevity in a wide variety of species, including yeast, worms, flies, and mice (Anderson, Shanmuganayagam, and Weindruch 2009). Initially it was thought that caloric restriction improved longevity by slowing metabolism and thereby decreasing the production of DNA-damaging reactive oxygen species, but this does not appear to be the case. Studies have shown that caloric restriction increases longevity by activating the *SIR2* and *SIRT1* genes. If these genes are inactivated, caloric restriction no longer increases longevity (Canto and Auwerx 2009). The *SIR2* and *SIRT1* genes are now identified as longevity genes, and modification of these genes has been shown to increase longevity in yeast, worms, and mice up to 70 percent (Howitz, Bitterman, and Cohen 2003; Tissenbaum and Guarente 2001; Mitchell et al. 2014).

It may seem bizarre that caloric restriction would enhance longevity, but this may be an adaptation that evolved to improve survival of the group. If longevity can be increased by starvation, longevity would be decreased, on a relative basis, in times of plenty. This may provide a means for population control. Our genes suppress longevity in times of plenty to keep the population down and promote it in times of famine to keep the population up. By this logic, the genes for longevity are a homeostatic control on population size, stabilizing the size of our groups for optimum long-term survival of the group.

The public became aware of do-it-yourself genetic interventions to prolong life when an article headlined "Yes, Red Wine Holds the Answer" was published in the *New York Times* in 2005 (Wade 2006). Geneticist David Sinclair and colleagues at Harvard University had found that resveratrol, a molecule in grape skins and red wine, could prolong the lives of obese mice by more than 50 percent (Baur et al. 2006). Resveratrol has been shown to increase lifespan in worms, fruit flies, and mice through activation of the *SIR2* and *SIRT1* longevity genes (Borra, Smith, and Denu 2005). Scientists are working feverishly to see if resveratrol will have similar beneficial effects in humans (Tome-Carneiro et al. 2013).

HUMAN LONGEVITY GENES

SIRT1 genes in humans are believed to be important for longevity, although this has not been proven. Clinical trials to evaluate the effects of altering *SIRT1* genes on survival in humans are nigh on impossible to perform for ethical reasons. But we do know that these genes are related to *SIR2* genes, which control longevity in a wide variety of nonmammalian species from yeast to worms. And *SIRT1* genes in humans are related to *SIRT1* genes in mice, which also control longevity (Mitchell et al. 2014; Borra, Smith, and Denu 2005). Recent studies have shown that activation of the *SIRT1* gene in mice with an experimental drug, SRT-1720, which is designed to mimic high-dose resveratrol (the chemical in red wine), produces a modest prolongation of lifespan in mice, as well as protection against age-related mental deterioration (Mitchell et al. 2014)

A second human gene, the *FOXO3* gene, may also be important for longevity (Wilcox et al. 2008). The *FOXO3* gene is related to the *FOXO* gene, which has increased longevity in roundworms, fruit flies, and mice by numerous mechanisms, including enhanced DNA repair, modification of the immune system, and breakdown of reactive oxygen species (Morris et al.

2015). As we discussed earlier, the *FOXO* gene also appears to be responsible for the improbable immortality of hydras. In humans, slight variations in the *FOXO3A* gene have been associated with extremely long lifespans in Japanese and German populations, providing evidence these genes may be important in dictating the length of our lives (Flachsbart et al. 2009).

The similarity of these "longevity genes" across a wide variety of species—from yeast, to nematodes, to humans—suggests they are descended from a common ancestor very early in our evolutionary history. They may have provided survival advantages early on in the history of life, repairing cellular and DNA damage in primordial creatures, only to be turned off later in evolution to provide group-benefitting population control. But now, as our technologies advance, we may have the opportunity to reactivate these genes, and in doing so, we may help postpone our senescence and greatly extend our lifespans. But will this help the group? Will this benefit our society?

GENETIC ENGINEERING

With the identification of longevity genes in humans, new treatments to prolong life are likely to be developed in the next few decades. Initially these are not likely to be gene replacement therapies—there are too many cells in an adult human body to "fix" in this way—but rather nongenetic therapies, such as the use of small molecules that could activate or deactivate the longevity genes in the entire body and thereby slow the aging process. Although there will be many ethical and political hurdles to overcome, editing the DNA of embryos created through in vitro fertilization using CRISPR editing technology will surely follow in the not-too-distant future, and this could include editing of longevity genes. Genetic engineering of animal embryos to create transgenic creatures is already well established.

Another potential approach to improving longevity is the prevention of cellular old age by maintaining telomere length with the enzyme telomerase. In humans and most mammals, telomerase is produced only during embryonic development, and during this phase telomeres grow to lengths that will last a lifetime—but only a finite lifetime. If we could produce telomerase throughout life, we may be able to prevent the shortening of the protective telomere caps on our DNA and prevent the DNA degradation that triggers cell senescence and cell death. Unfortunately, studies to date indicate that the potential benefits of telomere lengthening are offset by risks and tradeoffs, and enthusiasm for this approach has waned (Mitteldorf 2018).

But is increasing longevity what we really want? Most members of society are against human genetic modifications, including any extension of life by "artificial" means. Some feel that it interferes with God's plan. Others fear that extending lifespans could worsen a troubling current trend: The survival of more and more seniors who suffer in their decline, who are kept alive too long by expensive drugs and medical devices, who fill up hospitals and nursing homes, and who exhaust our limited resources for Social Security and Medicare. The suffering of the old is already profound. Would prolonging lives not make everything worse? Would we not be creating a new population of Tithonus-like seniors?

It is worth considering that our society has already produced dramatic increases in life expectancy by improving public health and developing treatments for infectious diseases, cardiovascular disease, and cancer. Life expectancy in the United States increased from forty-nine years in 1900 to seventy-nine years in 2011 (Leonhardt 2006; Arias 2015). This is a wild increase, and as a society, we have adapted. Few people, if any, would wish for shorter lives to be the norm again, and future generations that will live even longer will likely feel the same way.

Genetic therapies targeting longevity genes—rather than targeting specific diseases of aging, as is the current practice—could result in a different paradigm. Genetic alterations that postpone senescence and increase longevity should at the same time, according to most experts, push back the ravages of age-related diseases like Alzheimer's disease, Parkinson's disease, cancer, cardiovascular disease, and old age itself. In the future, senior citizens with advanced longevity could be capable of active, energetic, and productive lives. In this ideal future, our aging population would still be capable of contributing in many ways, and our societies would need to adapt to the changing demographics.

However, if Joshua Mitteldorf is correct, there is a very good reason for us to age and die. Aging and mortality, as we have learned, might be an evolutionary device to limit population growth, increase diversity, and promote survival of the group as a whole. It's easy to see that the growing population curve of our own species is not sustainable even now, before we have begun manipulation of our longevity genes. If we begin to reverse the genetic programs for aging and death, we will have to face the consequences. Let's hope we are up to the challenge.

PART VI

THE PRESENT AND THE FUTURE

CHAPTER 25

ARE WE ALONE?

> Our planet is not in a special place in the solar system, our Sun is not in a special
> place in our galaxy, and our galaxy is not in a special place in the universe.
>
> —Marcello Gleiser, in *The Dancing Universe*

Are we alone in the universe? This is the question we have asked ever since we realized there were worlds besides ours. It belies a deep human need for some frame of orientation—some clearer picture of our place in the cosmos. Until recently, this question has largely been answered with wild speculation and fantasies of science fiction. But as our technology advances, questions about the existence of extraterrestrial life have firmly entered the realm of science.

We are here because our Earth is a rare planet with conditions suitable for life, and because we have navigated through a number of improbable biologic and geological events leading to the evolution of complex intelligent life. Statistically, the universe is so massive, with so many opportunities for life to evolve, that even an extremely unlikely occurrence (intelligent life) becomes highly probable. As cosmologists learn more and more about the nature of our galaxy and the universe beyond, and as biochemists and biologists gain a better understanding of the origins and evolution of life on our planet, experts, in increasing numbers, are becoming convinced that we are (probably) not alone.

ONE OUT OF MANY

The number of stars in the universe is astronomical—literally. The Milky Way has between one hundred billion and two hundred billion stars, and our galaxy is just one of more than ten trillion galaxies in the observable universe (Howell 2017). And that's just the observable universe. Who knows how many more galaxies might be out there? And, furthermore, who knows how many planets circle all of those stars, both seen and unseen? The number is staggering. We don't know how many of those planets might be habitable or suitable to support life, but we would have to be radically anthropocentric to think with any certainty that we are the only intelligent civilization in the universe.

There is a growing consensus among scientists and astronomers that our planet is not particularly special or unique. Copernicus pioneered this realization in the sixteenth century, when he placed the sun—not the Earth—at the center of our solar system, where it belongs, with our planet in its orbit. The broader concept, that the Earth is not unique among planets, is now widely called the Copernican principle, and data are emerging that there are likely countless Earth-like planets scattered throughout the cosmos.

In 2009 NASA launched the Kepler space observatory to search for such Earth-like planets (Petigura, Howard, and Marcy 2013). Faraway planets are notoriously difficult to find, since they don't emit any light and their small, dark bodies are dwarfed by the bright stars that they orbit. But clever young scientists have devised new and ingenious techniques to detect lurking faraway planets. Using data from the Kepler telescope, scientists monitored the brightness of more than forty thousand sun-like stars in the Milky Way. If an orbiting planet crossed between its star and Earth, investigators detected a slight dimming of the star as the planet passed by. By measuring the amount of dimming and the time it took for the planet to pass across its star, scientists could determine the size of the planet, the period of its orbit, and its distance from its star. From this they were able to estimate how much each planet is warmed by its star—and therefore whether it is in a habitable zone, able to sustain liquid water. Amazingly, the investigators found that one in five sun-like stars in our galaxy host Earth-sized planets with moderate temperatures, making them potentially hospitable to life (Petigura, Howard, and Marcy 2013). Based on these data, our Milky Way galaxy alone has an estimated eleven *billion* Earth-like planets in habitable zones, circling sun-like stars (Khan 2013)! The universe has plenty of potential real estate for life, and some of it may reside on planets circling stars only twelve light-years away—stars that are visible to the naked eye.

One such planet, Kepler-452b, which lies in our galaxy about fourteen hundred light-years from Earth, is the most Earthlike planet found so far by the Kepler mission and has therefore attracted great scientific interest (see image 25.1) (Jenkins et al. 2015; Chou and Johnson 2015). Its star is about 1.5 billion years older than our sun, and about 20 percent more luminous. Kepler-452b has a diameter about one and a half times that of Earth and orbits its star in 385 days, just slightly longer than our own year. Its distance from its star puts it in a Goldilocks zone, where liquid water can exist on the planet's surface. We are not yet sure whether this cousin planet of ours is a rocky planet like Earth, with a thick atmosphere and active volcanos suitable for life, or a gaseous planet like Neptune, which would render it uninhabitable. Since Kepler-452b is 1.5 billion years older than Earth, it arguably has had more time for life to evolve, and if life exists there, it could be much more advanced than we are—or, paradoxically, it may have already destroyed itself, as we may still do. But of course, we know none of this yet.

Image 25.1: Kepler-452b. Artist's concept of the exoplanet
Kepler-452b, a nearly Earth-sized planet orbiting in the
habitable zone of the sun-like star Kepler-452, fourteen hundred
light-years from Earth in the constellation Cygnus.

Perhaps even more exciting is the recent discovery by a team of NASA astronomers of a cool red dwarf star circled by seven rocky terrestrial worlds (Gillon et al. 2017). The red dwarf star named TRAPPIST-1 is only one-tenth the diameter of our sun and has only a fraction of its brightness. It is therefore not visible to the naked eye, despite its relative closeness in the constellation Aquarius, only about forty light-years away. Seven Earth-sized planets circle this star, and three of them appear to be in a temperate habitable zone. Never before have so many rocky Earth-like planets been found circling one star. It appears that in general, smaller red dwarf stars, which vastly outnumber larger stars, may tend to have more planets orbiting them. The potential for habitable planets may be greater than we thought.

The planets around TRAPPIST-1 are nearby, which means that we can make detailed studies of their atmospheres. This could provide clues to the presence of life. The James Webb Space Telescope, which at the time of this writing is scheduled to launch in 2021, will enhance this effort (ESA 2018). The telescope will be able to determine atmospheric compositions of exoplanets by examining the wavelengths of light emitted by their stars as the light is transmitted through the exoplanets' atmospheres. The presence of atmospheric gasses such as carbon dioxide, water vapor, and ozone could provide us with the first clues of extraterrestrial life.

TALKING TO ALIENS

If there is life besides us in our universe, and if that life is intelligent like us, shouldn't we be able communicate with it? Our best chance for this may be with radio waves. Radio waves travel at the speed of light and could reach our nearest star, Proxima Centauri, in 4.3 years. All of the radio signals we broadcast on Earth flow out into space, and in theory anyone who is listening could hear us. And if there's anyone out there, maybe they're using radio signals too. The Phoenix program, which was developed by NASA in 1992 and supported by private funds through the SETI (Search for Extraterrestrial Intelligence) Institute, was initiated to search for extraterrestrial life using highly sensitive radio telescopes (the Allen Telescope Array) capable of detecting radio wave signals from nearby stars (Project Phoenix 2017; Tarter 2017). But this is a difficult task. Radio signals are subject to interference—first by the celestial bodies that they encounter on their way, and then by our own manmade radio interference once they reach Earth. To date, the Phoenix team has searched for signals from eight hundred nearby stars, and no credible radio signals have been received, despite measures to minimize interference. If any aliens are sending us messages, we still have not heard from them.

Leaders of SETI have proposed searching for extraterrestrial life by sending radio signals to parts of our galaxy where Earth-like planets have been found. But there is a conundrum. The late renowned cosmologist Stephen Hawking was convinced there is intelligent life out there, and was an advocate of searching for it, but he was also an outspoken critic of sending out any communicative signals on the grounds that this may actually be very dangerous (Cofield 2015). Any civilizaton that is able to pick up our signals and tell where they are coming from is likely to be more technologically advanced than we are—possibly even billions of years more advanced. And they might not be friendly. As Hawking put it, an advanced civilzation coming to Earth could be like Columbus coming to America—and that, of course, "didn't turn out so well."

WHERE IS EVERYBODY?

In 1950, Italian-born physicist Enrico Fermi, architect of the first nuclear reactor and a Nobel Prize winner in physics, sat down to lunch with his colleagues in the Fuller Lodge at the Los Alamos National Laboratory in New Mexico (image 25.2). It was a high-powered intellectual bunch, all of them eminent physicists of the day. The topic of discussion turned inquiringly to a spate of recent UFO sightings and then moved on to other things. Fermi went quiet for a time, mulling over a question that was forming in his mind. Finally, he asked the table that startling and now famous question: "Where is everybody?" (Gray 2015). Although the question seemed to come from out of nowhere, everyone seemed to know what Fermi meant. If the probability of extraterrestrial life is so great, then where is everybody? Why don't we have any hard evidence of intelligent extraterrestrial life yet? Intelligent life in remote solar systems should have had billions of years to evolve and develop the technology needed to overcome the time constraints of interstellar travel, enabling the beings to colonize our galaxy. So why haven't we heard from the aliens yet? This is the conundrum known as **Fermi's paradox**.

There are three broad answers to Fermi's troubling question. First, perhaps we live on a "Rare Earth"—a planet with truly unique properties that have enabled it alone to support complex, intelligent life. Second, there are a number of highly improbable evolutionary steps that provide barriers to the evolution of intelligent life. Life on other planets may not have been able to penetrate these barriers. And third—and most troublesome—perhaps once complex, intelligent life becomes technologically advanced, it also becomes self-destructive before it has time to colonize its own and other galaxies.

Economist and futurist Robin Hanson has famously called these three barriers to the evolution of complex intelligent life the **great filter** (Hanson 1998). Let's take a closer look at these obstacles, starting with the first one—the possibility that Earth, a hospitable, life-supporting planet, is a rare thing in our universe.

Image 25.2: Enrico Fermi. Italian-born physicist Enrico Fermi, architect of the first nuclear reactor, asked the famous question "Where is everybody?" This initiated the conundrum known as Fermi's paradox.

RARE EARTH

I stated at the beginning of the chapter that there was a growing consensus among scientists that our planet was not likely unique in the universe. But esteemed scientists Peter Ward and Donald Brownlee think it may be. In their thoroughly researched book *Rare Earth*, they advocate that we live on a planet with conditions so specific, and so unusual, that the Earth could, in fact, be unique in supporting complex life (Ward and Brownlee 2000).

We discussed some of these unique properties outlined by Ward and Brownlee in chapter 3. Our sun is just the right size. Any smaller and it would generate too little heat; any larger and its lifespan would be too short to allow complex life to evolve. Earth is located in the habitable zone and has a nearly circular orbit, which keeps temperatures moderate year-round. Earth is large and massive enough to have gravity strong enough to retain an atmosphere that shelters us all from the airless abyss of space. Ours is a rocky planet large enough to support geologic activity, enabling a carbon cycle essential for carbon-based life and climate control. It has a liquid outer core that contains molten iron, which generates a magnetic field that protects the planet from cosmic radiation. And we are fortunate to have a unique moon that is large enough to stabilize Earth's rotational axis, which stabilizes the climate.

This is only a partial list, but it provides a sense that perhaps our planet is indeed a spectacular rarity—and that maybe complex, intelligent life is also very rare in our universe (see sidebar "What Happened to Venus and Mars?") (Strobel 2013; Cain 2015; Catling 2014; Astronomy Notes 2016). But what about simpler life forms? Ward and Brownlee make an important distinction between complex, intelligent life and microscopic, single-celled life. As we have seen, the evolution of complex intelligent life requires a planet with a host of unique properties, navigation through a number of highly improbable biologic and geologic bottlenecks, and billions of years of trial and error. Microscopic life, on the other hand, has evolved here on Earth over a relatively short geological time period and has not had to overcome so many bottlenecks. Furthermore, microbes are hardy. They evolve quickly, and they can adapt to live in wildly inhospitable environments—at crazily hot or cold temperatures, in acid and salt, or at crushing ocean depths without sunshine. Some of them can steal energy from inorganic chemicals if need be. Microbes don't need the cushy, gentle environment that complex animal life—including us—depends on. They are much more adaptable and can live under a variety of conditions.

Such conditions may even exist in our own solar system. NASA's unmanned Cassini spacecraft, which explored Saturn for thirteen years, photographed sprays of salty water bursting through cracks on the surface of Saturn's diminutive moon Enceladus (Hansen et al. 2011) And the Hubble Space Telescope has seen plumes of water vapor billowing up from the surface of Jupiter's moon Europa (Roth et al. 2014). Beneath their icy crusts on the surface, these moons are thought to be hiding saltwater oceans heated by infernally hot cores. Scientists infer that these oceans may harbor hydrothermal vents, similar to the hydrothermal vents on Earth's ocean floors that teem with microbial life. It is not hard to imagine that extraterrestrial bacterial life could exist in hidden oceans on these moons, right here in our own solar system.

EVOLUTIONARY BARRIERS TO COMPLEX INTELLIGENT LIFE

Let's go back to Fermi's paradox. Why haven't we heard from the aliens? There is a second alterative explanation. Perhaps there is no intelligent life out there because life has not been able to navigate the evolutionary barriers required to produce complex intelligent life. We have visited some of these improbable evolutionary events in earlier chapters—the formation of the first replicating molecules, the emergence of oxygen-producing photosynthesis by cyanobacteria, the merger of a bacterium and an archaeon to form the first complex cell, and the evolution of intelligent life. These and other biologic barriers are part of the great filter, which may have blocked and may continue to block the evolution of intelligent life throughout the universe.

IS INTELLIGENT LIFE SELF-DESTRUCTIVE?

There is a third possible explanation for Fermi's paradox, and this may be our greatest concern. Let's suppose there are other planets with unique properties that support complex intelligent life. And let's suppose life elsewhere in the universe has been able to navigate the highly improbable evolutionary steps to complex life. Maybe the reason we have not heard from intelligent life out there is because once life reaches a stage of advanced intelligence and advanced technology, it

WHAT HAPPENED TO VENUS AND MARS?

Venus and Mars formed about the same time as Earth, about 4.5 billion years ago, and were composed of very similar material. They also orbit the sun at distances not too different from Earth's. And yet Earth evolved into a comfortable oasis, with balmy temperatures and abundant water, teeming with life—while Venus became a scalding pressure cooker, and Mars a frigid desert wasteland. What happened to Venus and Mars?

Venus is about the same size as Earth, but it has a thick atmosphere composed almost entirely of carbon dioxide. This produces an intense atmospheric greenhouse effect, heating Venus's surface to a scorching 477°C (Strobel 2013). The "air" is so thick that the atmospheric pressure is ninety-two times that of Earth's—enough to crush any creature that might live on the surface of the planet. Venus might seem Earthlike in some ways, but our neighbor's surface is hot, heavy, and toxic to life.

Venus wasn't always this way. Venus formed out of ingredients similar to Earth's and likely had oceans in its early days. But Venus orbits the sun more closely than Earth, and heat from the sun caused its surface water to evaporate, warming the planet as a result of the greenhouse effect. As the planet warmed, more water evaporated, adding to the greenhouse effect and further warming the planet. This vicious cycle continued in a **runaway greenhouse effect** until all the oceans had boiled away (Cain 2015).

Volcanic eruptions spewed CO_2 into the atmosphere, adding to Venus's greenhouse effect and further increasing temperatures. Venus, unlike Earth, has no plate tectonics and no water, and it is unable to remove CO2 from the atmosphere through the weathering process. It has been left with a dense atmosphere composed almost entirely of CO_2, resulting in furnace-like temperatures not suitable for life.

Mars is smaller than Earth, with a radius about half and a mass about one-ninth of Earth's. Its atmosphere, while composed of 95 percent CO_2, is so thin there is minimal greenhouse effect, and temperatures are frigid, averaging about -57°C (Catling 2009).

Mars, like Venus, had a promising start. Mars had flowing water on its surface in its early days and was almost certainly suitable to harbor life. But Mars's small size and weak gravitational field plus the lack of a magnetic field to protect its atmosphere against solar radiation made it difficult to hold on to a blanket of atmospheric gas, and much of its atmosphere drifted into space. Mars is farther from the sun than Venus, and as the meteorite bombardment in its early days abated, it began to cool. Water vapor condensed on its surface, reducing the greenhouse effect and further cooling the planet. Cooling begot more condensation of water vapor and more cooling in a freezing cycle known as a **runaway refrigerator effect**, leaving Mars the frigid wasteland it is today (Astronomy Notes 2016).

The tragic lives of our closest planetary neighbors help us to appreciate the precariousness of the conditions for life and help us to understand that Earth may indeed be a very rare planet.

becomes self-destructive. The late cosmologist Carl Sagan and others have speculated that technological civilizations may become self-destructive shortly after they develop space and radio technology (Sagan 1980, 315; Bostrom 2012) There are so many ways advanced alien civilizations could destroy themselves: nuclear holocaust, biological warfare, climate change, artificial intelligence gone awry, and self-destruction by other horrible weapons of mass destruction unknown to us. This is an alarming prospect.

If there is a great filter that blocks the survival of advanced technological civilizations, it must be absolute. If even 1 percent of civilizations squeezed through the filter and survived to become capable of colonizing remote galaxies, we should know about it. And yet all communications have been silent until now. This is a sobering thought. If all civilizations with advanced technology are self-destructive, then what does that say about our own future? Maybe we will be the exception to a universal rule. Maybe we will be wise enough to squeeze through the filter of self-destruction. Or maybe we won't, and our glorious civilization—like other potential civilizations across the vastness of space—will eventually meet an inevitable demise.

WOULD ALIENS LOOK LIKE US?

If the universe is teeming with Earthlike planets—which at this point, it appears to be—then it seems highly likely that intelligent life must exist out there somewhere. We may not have detected that life because of the vast distances between us, or perhaps because most intelligent life has self-destructed before we could hear from it—but someday in the far-off future we are likely to encounter aliens from remote planets. What will they look like? Will they look like us?

Sci-fi movies and TV shows would have us believe so. Witness "E.T." in *E.T. the Extra-Terrestrial*; the two-legged, two-armed aliens with oversize brains in *Independence Day*; or the Na'vi, the humanoid species indigenous to the Alpha Centauri star system in *Avatar*. A rare exception to these depictions of humanlike aliens are the squidlike six-armed Heptapods that came to visit Earth in *Arrival*.

British paleontologist Simon Conway Morris, whom we met in chapter 8, has championed the belief that evolution is deterministic—that certain outcomes are inevitable (Conway Morris 1998). This hypothesis is based on many examples of convergent evolution here on Earth—the evolution of similar traits by distantly related species along separate independent evolutionary pathways. We have already seen many examples: the evolution of flight by such diverse groups as birds, bats, and insects; and the independent evolution of camera-like eyes by cephalopods (squids and octopuses) and vertebrates. If natural selection produces similar solutions to common environmental challenges, we might predict that life on extraterrestrial Earthlike planets would produce life that looks a lot like it does here.

But there are also many examples of nonconvergence in our big, diverse family of life on Earth. There are myriad potential methods by which any creature can solve a given problem posed by its environment, and more often than not, different clades adapt in their own specific, unique ways, rather than converging on a common way of adaptation (Losos 2017). The woodpecker and the aye-aye (a rare nocturnal lemur indigenous to Madagascar), for example, both live off wood-eating grubs but have evolved entirely different methods for obtaining their meals (Losos 2017). The woodpecker has a tough beak to chisel into wood, a long tongue to pull out the grubs, and a thick skull to withstand jackhammering, while the aye-aye has evolved rodent-like teeth to gnaw holes in wood and elongated twig-like fingers to pull the grubs out.

The late astronomer Carl Sagan believed that extraterrestrial intelligent life will turn out to be very different from us: "They may live on the land or in the sea or air. They may have unimaginable chemistries, shapes, sizes, colors, appendages and opinions. We are not requiring that they follow the particular route that led to the evolution of humans. There may be many different evolutionary pathways, each unlikely, but the sum of the number of pathways to intelligence may nevertheless be quite substantial" (Sagan 1995).

The late evolutionary biologist Stephen Jay Gould, whom we met in chapter 11, emphasized how chance events—both highly improbable evolutionary events and geological events resulting in mass extinctions—have changed the natural history of life on our planet (Gould 1989). "Rewind the tape," as he so famously said, and we would certainly see a different outcome. There is no guarantee that intelligent life would evolve again on Earth, but if it did—given the immense diversity of life on Earth—it would evolve differently. Look to *Arrival*, not *Independence Day* or *Avatar*, to get an idea of what aliens might look like.

There is one more possibility. If and when we find life elsewhere in the universe, it may not be biologic at all. We, *Homo sapiens*, have been here for only about three hundred thousand years—a very short time in geologic terms. It is likely that many alien civilizations have been around much longer than us, by millions to billions of years. It appears there is a relatively brief time period in the life of civilized societies—perhaps a few centuries—between the technological achievement of producing radio waves and developing advanced artificial intelligence. Many experts believe that once artificial intelligence surpasses human intelligence, there is a strong possibility that artificial intelligence will take over. If this is true, there may be only a vanishingly brief window of opportunity to detect, contact, and communicate with highly developed extraterrestrial biological life. If we miss that brief window of opportunity, and we probably already have for most extraterrestrial civilizations, we may find that biologic life has been replaced by robotic artificial intelligence.

CHAPTER 26

CLIMATE CHANGE

> We probably could have saved ourselves, but we're too damned lazy to try very
> hard … and too damned cheap.

> —Kurt Vonnegut, satirical novelist

Global warming is not new. Earth's climate has oscillated wildly between hot and cold since its beginning 4.5 billion years ago. Temperatures fell so drastically during two ancient glacial epochs that glaciers reached all the way to the equator and the world's oceans froze. No part of the "Snowball Earth," as these time periods are now called, was spared from ice. At the other extreme, temperatures soared during the Eocene hothouse. Balmy tropical climates extended into the arctic, supporting palm trees and alligators across land that today is icy, frigid tundra, covered in snow.

Given all the publicity around global warming and the current melting of our ice cap glaciers, you might be surprised to learn that our planet is actually in the middle of a long ice age. For most of Earth's 4.5-billion-year history, the world has been warmer than it is today. For the most part, our planet has been ice free, even at the north and south poles. Those warmer periods have been the norm, but they have been interrupted by at least five major ice ages (technically known as **glacial epochs**), one of which persists today. Climate changes have been a major driving force in the evolution of life on our planet. Understanding the forces shaping climate change and the history of climate change will add perspective to our current climate crisis and help provide answers how to deal with it. Let's start with our current glacial epoch.

THE HISTORY OF CLIMATE CHANGE BURIED IN ICE

Thermometers have been available for only 150 years, but innovative scientists have found clues to Earth's ancient temperatures buried in ice. The history of climate change over the past eight hundred thousand years has been documented by international and Russian scientists from analyses of layers of ice built up in Greenland and Antarctica (EPICA Community Members 2004; University of Michigan Global Change Program 2017).

Researchers have been able to estimate Earth's temperatures by analyzing the distribution of oxygen isotopes in ancient ice core samples. Heavy isotopes of oxygen in water vapor ($H_2^{18}O$) precipitate at higher temperatures than lighter isotopes ($H_2^{16}O$), so a higher ratio of heavy to light isotopes ($H_2^{18}O/H_2^{16}O$) in ancient ice core samples indicate a warmer climate at that time. Fortunately, ice cores preserve annual layers, making it easy to date the ice being analyzed. Using such data, climate scientists have been able to construct temperature plots, which have documented the wide temperature swings of glacial and interglacial periods that cycle about every one hundred thousand years during the Quaternary glacial epoch—just as Milankovitch predicted.

Atmospheric carbon dioxide (CO_2) levels can also be measured from ice core samples by sampling air bubbles trapped within the cores. Plots of CO_2 levels during the Quaternary glacial epoch have shown a strikingly high correlation with cyclic temperature changes.

THE QUATERNARY ICE AGE

Earth's climate over the past eight thousand years has made life here a walk in the park. Over all those millennia, average temperatures have not varied more than about 1°C from temperatures measured in 1950 (Fagan 2004). In his insightful book *The Long Summer,* Brian Fagan describes how the climate of this temperate period has allowed agriculture and human civilization to flourish. Who knows how human civilization would have developed without such favorable climatic conditions. Fagan concludes, "Civilization arose during a remarkably long summer. We still have no idea when, or how, that summer will end" (Fagan 2004, 25).

Although it doesn't feel like it, this "remarkably long summer" has graced us in the middle of an ice age, known as the Quaternary ice age, or more properly the Quaternary glacial epoch, which began about 2.6 million years ago. It doesn't feel like an ice age now because we are in a moment of warm respite called an **interglacial period**. Within the greater context of the current ice age, warm interglacial periods alternate with cold **glacial periods** about every one hundred thousand years.

The last glacial period, often inaccurately referred to as the "last ice age," ended about 11,700 years ago, but it left marks that persist today. Glaciers scored the Earth, carving out gorges that are now the Great Lakes and a host of smaller lakes. Huge ice sheets covered much of Europe and North America. The massive Laurentide ice sheet, which covered Canada and the Northern part of the United States, was up to several kilometers thick, blanketing regions that now contain the great cities of Boston, New York, Chicago, and Saint Louis. (Can you imagine—Chicago and New York City under more than a mile of ice?) Legendary megafauna like the Irish elk, the European elephant, and the iconic woolly mammoth, roamed far and wide in Europe and North America, but as the glacial period came to an end and the glaciers began to retreat, these megafaunas vanished from the fossil record. There is still debate as to whether their extinction was primarily due to climate change or to overkill from hunting by *Homo sapiens*, who migrated to North America about twenty thousand years ago.

The history of glacial movement is written in the landscapes of North America and Europe, and clues about Earth's ancient temperatures are buried in the ice of Greenland and Antarctica. Analysis of oxygen isotopes in ancient ice samples allows scientists to estimate the temperature at the time the ice formed (Sidebar "The History of Climate Change Buried in Ice"). From these data, scientists have constructed extensive plots of Earth's temperatures over the past eight hundred thousand years, documenting global temperature swings through glacial and interglacial periods every one hundred thousand years or so (EPICA Community Members 2004; University of Michigan Global Change Program 2017; US EPA 2017).

MILANKOVITCH CYCLES

Serbian astronomer and mathematician Milutin Milankovitch dedicated his career at the University of Belgrade in the early twentieth century to developing a mathematical theory of climate change. His theories sought to predict climatic change based on the amount of solar radiation reaching Earth at any given time—an amount that varies, drastically, based on Earth's position in relation to the sun. Changes in the

---

MILANKOVITCH CYCLES

Serbian astronomer Milutin Milankovitch predicted in 1938 that variations in the amount of solar energy reaching Earth result in cyclic climate changes. He made these predictions based on mathematical models before there was any geologic information about the history of temperature and climate change on our planet. Here is the basis of his theories: Earth's orbit around the sun varies from nearly circular to slightly elliptical over about a one-hundred-thousand-year cycle. When the orbit is more elliptical, there is greater variation in solar energy reaching Earth because there is more variation in Earth's distance from the sun. Earth's tilt, or spin axis, in relation to the plane of its orbit varies from about 22 degrees to 24.5 degrees from the vertical over a cycle of about forty-one thousand years. When the tilt is smaller, there is less seasonal temperature variation at middle and high latitudes—and the summers are cooler. Cooler summers at the poles allow ice to persist there year after year, and massive ice sheets accumulate over time. Now here's the most interesting bit: as the polar ice sheets grow, more solar radiation is reflected back into space by the glistening ice, causing less heat absorption and further cooling of the Earth. Based on all of these factors, Milankovitch's mathematical models predicted that glacial periods would occur when Earth's phases of orbit, axial tilt, and wobble were all aligned to give the northern hemisphere the least exposure to sunlight. He used these data to estimate Earth's cyclic temperatures over the prior six hundred thousand years.

Milankovitch's theories were initially met with skepticism but have now been collaborated with evidence of Earth's global temperature history from ice core samples in Greenland and Antarctica. There is broad acceptance today that the cyclic changes in Earth's orbit, axis tilt, and axis wobble, now called Milankovitch cycles, affect the amount and distribution of solar radiation reaching Earth. These changes trigger cyclic alterations in temperature, which are amplified by secondary changes in atmospheric CO_2 concentrations and result in glacial and interglacial periods alternating about every one hundred thousand years.

---

trajectory of Earth's orbit around the sun, in the tilt of Earth's axis of rotation, and in the wobbliness of its axis of rotation (like the wobbliness of a spinning top) all effect the amount of the sun's heat we feel on the face of the planet (sidebar "Milankovitch Cycles") (Graham 2000).

Milankovitch's mathematical models of climate change, published in 1938, predicted wide swings in global temperatures and were initially met with great skepticism. Could our climate really have changed so much and still be changing? It was not until forty years later that his theories gained traction, when studies from ocean-floor sediments documented climate change over the past four hundred fifty thousand years consistent with his predictions (Hays, Imbrie, and Shackleton 1976). His theories were further corroborated with evidence of Earth's global temperature history from ice core samples in Greenland and Antarctica. The scientific world has now accepted Milankovitch's premise that the glacial and interglacial periods of the Quaternary Ice Age, which have temperature swings of roughly 10°C (18°F) about every one hundred thousand years, are due to changes in Earth's orbit and axis tilt, which affect the amount of solar radiation reaching our planet (Hays, Imbrie, and Shackleton 1976). These cyclic variations in solar radiation reaching our planet and the resulting periodic temperature swings are now known as **Milankovitch cycles**.

GREENHOUSE EFFECT

French mathematician and physicist Jean Baptiste Joseph Fourier, who is best known for developing the mathematical process known as Fourier analysis, is also credited with the discovery of the greenhouse effect—how **greenhouse gases** modulate temperatures on our planet (Fleming 1999). Fourier was orphaned at the age of nine after the death of both of his parents, and he entered a school run by monks, where he immediately showed great talent in mathematics. He began his career as a math teacher but was soon caught up in the politics of the French Revolution and joined the Revolutionary Committee, after which he was arrested and narrowly escaped the guillotine. He served in a number of prominent positions under Napoleon but finally returned to teaching and research later in his life, which was when he published his theories on climate change.

Fourier calculated that, based on Earth's size, its distance from the sun, and the sun's luminosity, it should actually be much colder than it is. Puzzled by the discrepancy, he began casting about for other factors that could be missing from his calculations. What about Earth's atmosphere? Could it be acting as some kind of insulator, like a warm blanket in winter? His publication in 1824 described how this might work (figure 26.1). As sunlight penetrates the atmosphere and reaches Earth's surface, it warms the planet. As Earth warms, it emits heat back into the atmosphere and toward outer space in the form of thermal infrared radiation. Infrared radiation has a longer wavelength than visible light and is selectively absorbed by several gases in the atmosphere: carbon dioxide (CO_2), methane, and water vapor. These are known as the greenhouse gases—gases that trap radiated infrared heat from Earth, keeping it in our atmosphere and warming the planet. Of these gases, fluctuations in CO_2 levels have usually been the primary driving force in greenhouse-induced climate change throughout our planet's history. CO_2 warms the Earth through its greenhouse effect, but by raising temperatures, it also causes ocean waters to evaporate, increasing the atmospheric concentration of water vapor and further enhancing the greenhouse effect.

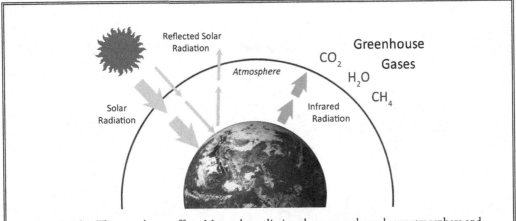

Figure 26.1: The greenhouse effect. Most solar radiation that passes through our atmosphere and reaches Earth's surface is absorbed and warms the planet, while some is reflected and passes back through our atmosphere and into space. Earth's warmed surface emits infrared radiation, most of which is absorbed by greenhouse gases, which act like a blanket to keep Earth warm. Without greenhouse gases, infrared radiation and the heat generated from it would escape into space.

The term "greenhouse effect" was coined because of the analogy with greenhouses, the buildings used to nurture plants. As sunlight passes through the glass walls of a greenhouse and strikes surfaces within, they are warmed. These surfaces in turn radiate infrared heat, but the heat is unable to escape the greenhouse because glass selectively blocks infrared radiation. The greenhouse heats up dramatically. Our planet, covered by a blanket of greenhouse gases, heats up in the same way.

There is general agreement that the major cyclic variations of temperature within our current ice age—the glacial and interglacial periods—are triggered by variations in solar radiation reaching Earth (the Milankovitch cycles). But these cycles are amplified by the effects of greenhouse gasses. Atmospheric CO_2 levels and temperatures cycle up and down together over periods of one hundred thousand years, with a very tight correlation between the two (figure 26.2). Climate scientists initially thought the CO_2 levels were responsible for the temperature swings, but they changed their minds when they learned that CO_2 levels do not rise until several hundred years after temperatures rise (Skeptical Science 2018). Scientists now believe that the cyclic rise and fall of atmospheric CO_2 levels are secondary to the rise and fall of temperatures due to Milankovitch cycles. As temperatures rise, CO_2 is released from the oceans into the atmosphere, and rising atmospheric CO_2 levels accentuate the rise in temperatures. Likewise, when temperatures fall as a result of Milankovitch cycles, CO_2 is absorbed by the oceans, reducing the greenhouse effect and accentuating the fall in temperatures.

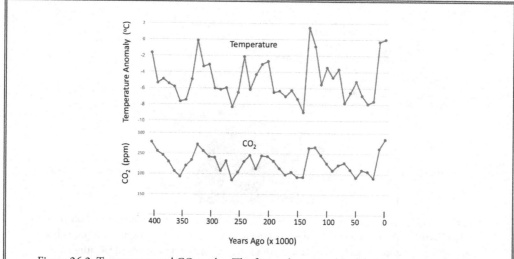

Figure 26.2: Temperature and CO_2 cycles. The figure shows Earth's global temperatures (relative to today's temperatures) and atmospheric CO_2 levels over the past four hundred thousand years from Vostok Antarctica ice core data (US EPA 2017). Temperatures cycle from warm interglacial periods to cold glacial periods about every one hundred thousand years owing to changes in solar radiation reaching Earth (Milankovitch's cycles). The temperature cycles trigger secondary cyclic swings in CO_2 levels, which amplify the temperature cycles. We are currently in a warm interglacial period (right of the graph) after recovering from our last glacial period, which ended about eleven thousand years ago. Due to human-induced CO_2 emissions, CO_2 levels reached a peak of 410 ppm in 2017, well above their usual cyclic highs of 250–300 ppm.

Apart from the amount of energy Earth receives from the sun, the greenhouse effect is the most important determinant of Earth's temperature and climate. Alterations in atmospheric greenhouse gases—especially CO_2, which is governed by the carbon cycle—are responsible for the extreme climate changes throughout Earth's history, from the five glacial epochs to the Eocene hothouse. Let's see how the carbon cycle works and how it has impacted Earth's climate over its 4.5-billion-year history.

THE CARBON CYCLE

Carbon atoms connected end to end make up the backbone from which all life is built. Carbon is in the oceans, atmosphere, rocks, and fossil fuels, and it is constantly on the move from one reservoir to another in an exchange called the carbon cycle. CO_2 cycles back and forth between Earth's crust and its atmosphere through geological processes driven by changes in plate tectonics. Tectonic plates, which make up Earth's rigid outer shell (the crust and upper mantle), eternally move, slide, and collide with one another. As they do, they buckle upward to form mountain ranges and often crumple downward into Earth's mantle, carrying calcium carbonate deposits with them. The inferno-hot environment of the mantle burns up calcium carbonate, releasing CO_2 into the atmosphere through cracks in the Earth's crust and through volcanic eruptions.

At the same time, CO_2 is removed from the atmosphere by a process known as **weathering**. When atmospheric CO_2 dissolves in rainwater, it becomes a weak acid (carbonic acid) and rains to the Earth. The weak carbonic acid reacts with rocks and minerals, and the products are carried by rivers and streams to the oceans, where they precipitate to the ocean floor as sediment, sequestering carbon for thousands of years.

When mountain ranges are created by the collision and buckling of tectonic plates, more rock is exposed to atmospheric CO_2, and the weathering process is accelerated. When atmospheric CO_2 levels increase, the rate of removal of CO_2 by weathering also increases, and this serves as a natural feedback mechanism that helps to control CO_2 levels and moderate Earth's temperatures. These interchanges of CO_2—removal of CO_2 from the atmosphere through weathering and addition of CO_2 to the atmosphere through volcanic eruptions—are called the slow carbon cycle, or **geological carbon cycle**, because they take place over millions of years. It is this slow carbon cycle, driven by plate tectonics, that is responsible for most of the epic climate swings that have created the great glacial epochs and hothouses in the history of our planet.

A faster exchange of carbon occurs through the **biological carbon cycle**. Plants absorb CO_2 in their daily work of photosynthesis, and animals and bacteria release CO_2 with their own breathing. When plants and animals die, their organic remains—those that are not buried and sequestered—are oxidized, releasing CO_2 back into the atmosphere. The biological carbon cycle is so rapid that it produces seasonal fluctuations in atmospheric CO_2 levels. In the Northern Hemisphere, during the summer, when photosynthesis is actively absorbing CO_2, CO_2 levels fall; and during the winter, when many plants have lost their leaves and are no longer absorbing CO_2, CO_2 levels rise. The Northern Hemisphere drives this process, because most of Earth's landmass, and therefore most of its trees and plants, sits in the Northern Hemisphere. The seasonal fluctuations in atmospheric CO_2 levels seem to show the Earth "breathing."

THE EOCENE HOTHOUSE

Fifty-five million years ago, as life was recovering from the aftermath of the asteroid collision that destroyed the dinosaurs, Earth entered its warmest period on record. This was the Eocene hothouse. Temperatures rose even higher than in the balmy days of the dinosaurs, reaching 10°C (18°F), higher than modern-day temperatures (Archer 2009, 58; Huber and Caballero 2011). Glaciers and polar ice had long since melted, and the moist, balmy environment allowed forests to cover Earth's landscape from pole to pole.

Temperatures in arctic regions averaged about 20°C (67°F) during the summer months and never fell below freezing in the winter. The arctic regions were rich in flowering plants, deciduous trees, and forests similar to those in southeastern regions of the United States (Harrington et al. 2011). A variety of reptiles and mammals made their homes in the arctic, including lizards, snakes, giant tortoises, tapirs, and primates. With such balmy temperatures in the arctic, one would expect scorching temperatures in the tropics. Surprisingly, this was not the case. The global climate was much more uniform than today, and the climate in the tropics was only moderately warmer than today's climate.

The Eocene epoch (from the Greek word *"eos,"* meaning "dawn") marked the birth of many of our diverse modern animal groups. The ancestors of modern ungulates—horses, tapirs, deer, cattle, and sheep—roamed between Europe and America over Greenland land bridges. They were the most prevalent mammals of the time. Some of the earliest members of our own order, the primates, appeared for the first time in the Eocene epoch.

How did Earth get so darn hot? Geoscientists at Penn State believe that it all began when tectonic plates in India and Asia collided shortly before the Eocene (Kerrick 2001). The plates crumpled together to form the Himalayan Mountains and the great Tibetan plateau, and at the same time, caused great quantities of limestone and dolomite rock to plummet from the plates down into the mantle, where they melted and released massive amounts of CO_2. In a process described as seismic pumping, CO_2 was released through faults and crevasses in the surface crust over millions of years, raising CO_2 levels to two to six times current levels and transforming Earth into the Eocene hothouse—the most intense greenhouse our planet has ever known (Pearson and Palmer 2000).

The greenhouse effect from carbon dioxide may have been magnified by the simultaneous release of methane, which is an even more potent greenhouse gas. Methane is produced by deep-sea microorganisms, and it typically forms stable complexes with water called methane hydrates. These hydrates are temperature sensitive and break down with increased heat. As temperatures rose because of the CO_2 greenhouse effect in the Eocene, these hydrates dissociated and released methane into the oceans and atmosphere, adding to the greenhouse effect and accelerating the temperature rise (Sloan et al. 1992; Gu et al. 2011).

The Eocene Hothouse has become a focal point of climate research today because of potential analogies with the global warming we are currently experiencing. Although the sources of the carbon flowing into the oceans and atmosphere are different today, the potential for secondary methane release and the impacts of ocean acidification hold lessons that may help us to understand what the future may hold for us.

SNOWBALL EARTH

While most of Earth's history has been glacier free with balmy weather, there have been five great glacial epochs with temperatures so low that much of life was severely threatened. We have already discussed our current Quaternary Glacial Epoch, in which we are currently in a mild interglacial period. The most severe glacial epoch was the Huronian glaciation that occurred about 2.4 billion years ago, when life on Earth was still entirely microscopic. The Huronian glacial epoch was so severe that Earth was covered by glacial ice from pole to pole. Average global temperatures plummeted to an estimated -20° to -50°C (-4° to -58°F) (Ward and Brownlee 2000, 115). Microbial life was at the brink, and likely only a few pockets of microbes living in deep hydrothermal vents were able to survive. This glacial epoch earned the term **Snowball Earth.**

We know about Snowball Earth from telltale geologic evidence—grooves, scratches and characteristic rocky deposits called tillites left by the migration of ancient, now long-dead glaciers. The glacial remnants from Snowball Earth are unique because they cover the *entire* planet, including equatorial regions; ice was everywhere.

The Huronian glaciation appeared to result from a perfect storm—the convergence of a transformational biologic phenomenon and a prolonged geological event (Tang and Chen 2013). Cyanobacteria developed the capacity for photosynthesis not long after life emerged on our planet, and oxygen began to accumulate in the atmosphere about 2.3 billion years ago. (Remember the great oxygen catastrophe from chapters 6 and 18, when newly released oxygen killed almost every living thing on Earth?) Back in those days, methane was the predominant gas in the atmosphere, and it had a major role in warming the planet through its greenhouse effect. The newly released oxygen reacted with and removed methane gas from the atmosphere, triggering a dramatic cooling of the planet.

At the same time, major shifts in plate tectonics broke up the supercontinent Kenorland and exposed massive amounts of rock to the atmosphere. This resulted in extensive weathering, with removal of large quantities of CO_2 from the atmosphere. With the removal of both methane and carbon dioxide, the greenhouse effect all but vanished, and the planet rapidly cooled. And as the planet cooled and glaciers expanded, Earth's reflectivity increased, and much of the sun's light and warmth were reflected back into space. The confluence of these events at a time early in Earth's history, when the sun's energy output was 30 percent less than it is today, led to the longest and most severe great ice age in the history of our planet.

With Earth completely covered with ice, a thawing of this historic icehouse was a tricky proposition. Because of the frozen tundra and oceans, there was little evaporation and condensation of water—and therefore no weather cycle. With the landscape covered in ice, most solar radiation was reflected, preventing absorption of life-saving warmth that the sun might bring. The planet was locked in solid ice for two hundred million years. Fortunately, volcanoes finally came to our rescue. New volcanic eruptions spewed CO_2 into the atmosphere, jump-starting global warming again and beginning the process of melting Snowball Earth.

Following the thaw, more than a billion years later, Earth experienced a second Snowball Earth, the Cryogenian glaciations, consisting of two glaciations lasting together almost one million years. Once again life was at the edge; pockets of microorganisms were huddled around hydrothermal vents, and patches of algae somehow survived with very little water. But with the eventual thaw came a spectacular revival. Peter Ward and Donald Brownlee speculate in *Rare Earth* that as disastrous as these glaciations were, they may have been triggers for the evolution of complex life (Ward and Brownlee 2000). The harsh conditions of the Cryogenian Glaciation separated microbial and algal groups into separate pockets, facilitating their evolution into separate species. As the ice melted and exposed the mineral-rich oceans to sunlight, blooms of cyanobacteria flourished and produced huge quantities of oxygen. Life adapted rapidly from cold to warm, and from low to high oxygen levels. With new environments and vacant habitats, there was an adaptive radiation, with rapid diversification of new species. The Cambrian explosion soon followed.

THE OTHER GREAT GLACIAL EPOCHS

Apart from the two Snowball Earths, our planet has experienced three other great ice ages or glacial epochs: the Andean-Saharan glaciation, the Karoo glaciation, and our current Quaternary glaciation (figure 26.3). We'll talk briefly about the Karoo glaciation because of its relationship to our current climate crisis.

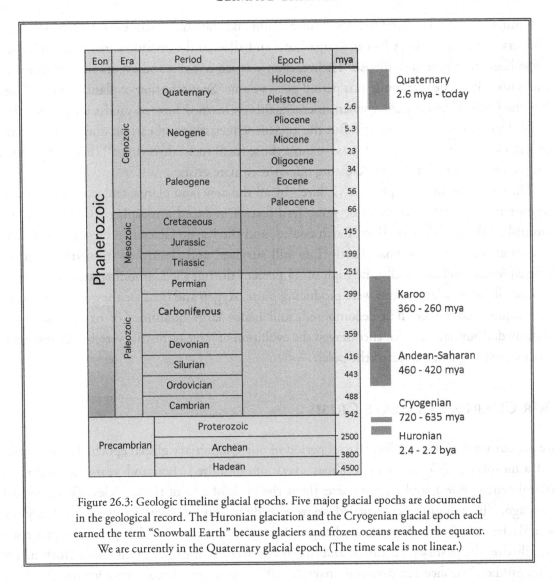

Eon	Era	Period	Epoch	mya
Phanerozoic	Cenozoic	Quaternary	Holocene	
			Pleistocene	2.6
		Neogene	Pliocene	5.3
			Miocene	23
		Paleogene	Oligocene	34
			Eocene	56
			Paleocene	66
	Mesozoic	Cretaceous		145
		Jurassic		199
		Triassic		251
	Paleozoic	Permian		299
		Carboniferous		359
		Devonian		416
		Silurian		443
		Ordovician		488
		Cambrian		542
Precambrian		Proterozoic		2500
		Archean		3800
		Hadean		4500

Quaternary
2.6 mya - today

Karoo
360 - 260 mya

Andean-Saharan
460 - 420 mya

Cryogenian
720 - 635 mya

Huronian
2.4 - 2.2 bya

Figure 26.3: Geologic timeline glacial epochs. Five major glacial epochs are documented in the geological record. The Huronian glaciation and the Cryogenian glacial epoch each earned the term "Snowball Earth" because glaciers and frozen oceans reached the equator. We are currently in the Quaternary glacial epoch. (The time scale is not linear.)

The Karoo glaciation, which was named after glacial tills found in the Karoo regions of South Africa, began about 360 million years ago at the beginning of the Carboniferous period. Not long before the Karoo glaciation, land plants evolved and colonized the continents, producing oxygen and consuming and removing vast quantities of CO_2 from the atmosphere. At the same time, carbon-containing organic remains of the great fern forests, which would usually be oxidized, were sequestered beneath Earth's surface. Usually when plants die, fungi and bacteria chew the dead wood into small pieces and oxidize their organic remains, releasing their carbon as CO_2 back into the atmosphere. But the great fern forests that amassed during the

Carboniferous period evolved the wood fiber lignin and the waxy bark-sealing suberin—plant polymers that resisted decay by microorganisms and allowed the organic remains enough time to fossilize rather than decompose. Their carbon was sequestered and eventually converted to coal. This is how the Carboniferous period got its name. Sequestration of plant carbon abated after the Carboniferous period when microorganisms learned how to produce enzymes that would degrade lignin and suberin. But the ancient sequestration of carbon during this period has left a legacy that endures today as we burn those fossil fuels, throwing all that carbon back into the atmosphere at once and creating our own climate crisis.

The reduction in atmospheric CO_2 levels by all the new land plants, and the underground sequestration of carbon from those great fern forests, reduced atmospheric CO_2 levels and diminished the greenhouse effect. Earth cooled, and the Karoo Glaciation began. The Karoo Glaciation was not a Snowball Earth. Life still survived and flourished in warmer regions. Oxygen levels reached an all-time high of 35 percent during the Carboniferous period, both because all the new land plants were producing more oxygen and because the great fern forests were sequestered rather than decomposed, and hence large quantities of oxygen were not consumed. Abundant oxygen encouraged the evolution of large terrestrial vertebrates and giant insects, even as glaciers covered the poles.

OUR CURRENT CLIMATE CRISIS

We are currently in a warm interglacial period of our Quaternary glacial epoch. Temperatures and atmospheric CO_2 levels cycle about every one hundred thousand years (according to Milankovitch cycles), and we are currently at the peak of one of those cycles. Two hundred years ago, climatologists were worried we were ready to enter the next ice age (technically the next glacial period), but things have changed dramatically since then. Temperatures have risen steadily over the past two centuries, and that rise is accelerating (NASA Goddard Institute for Space Studies, "Surface Temperature Analysis," 2017). Atmospheric concentrations of CO_2 have increased dramatically from preindustrial times, reaching levels of 410 ppm in 2017—levels we have not seen in over eight hundred thousand years (NASA, "Relentless Rise," 2018; Global Greenhouse Gas Reference Network 2018). Drastic climate change is upon us.

Humans are responsible for the current rise in CO_2 levels. The burning of fossil fuels—oil, coal, and gas—has added billions of tons of carbon to the atmosphere every year. (US EPA 2017). While this is only a small part of total CO_2 emissions from natural sources, the excess emissions exceed the capacity of natural CO_2 sinks to fully compensate, and over time this has resulted in steep increases in atmospheric CO_2 levels.

We know the increased atmospheric concentrations of CO_2 are due to human emissions because the carbon emitted from burning fossil fuels carries a distinctive isotopic marker that indicates its anthropogenic source. Specifically, fossil fuel–related carbon dioxide does not contain the carbon isotope ^{14}C, as other sources do, because the small proportion of ^{14}C that was sequestered in coal three hundred million years ago has long since decayed. When CO_2 enters the atmosphere from burning fossil fuels, the proportion of ^{14}C in the atmosphere (which is less than 1 percent) goes down, and analysis of the changing carbon isotope ratios can provide an estimate of CO_2 contributions from human industrial sources in particular (NOAA 2018). Experts also believe rising CO_2 levels are human induced, because there is no other explanation. There have been no major changes in geologic activity (volcanic eruptions) or biologic contributions (organic decay). Taking all this into consideration, there is an overwhelming consensus among climate scientists that human-induced carbon emissions are responsible for rising atmospheric CO_2 levels and our current climate change (Cook et al. 2013; NASA, "Scientific Consensus," 2018).

THE CONSEQUENCES OF OUR CURRENT CLIMATE CHANGE

In many ways, our current climate change doesn't sound too bad—summery temperatures for hundreds and thousands of years? Why not? It certainly sounds better than the next dreaded ice age, from which we might expect another layer of ice two miles thick to cover Chicago and New York. A temperature increase of two to three degrees Celsius by the end of the century, as is currently predicted, does not seem extraordinary to most of us. Seasonal and daily temperatures variations are far more than this. And put in the larger perspective of our entire geologic climate history—as I have tried to do in this chapter—this warming is certainly not an extraordinary change. For most of history, Earth has been much warmer than it is today.

Why, then, all the hysteria about global warming? Why not just put away our sweaters and wear more bathing suits? Make no mistake; this time the climate change is different. It is different in three especially dangerous ways:

1. The temperature rise is occurring over decades and centuries rather than over thousands or millions of years, and there will be less time for us and other species to adapt.
2. The current climate change is occurring for the first time in an age of advanced human civilization. Advanced societies, with cities built on ocean shores and infrastructures with many networks of interdependence, are highly vulnerable to the consequences of climate change. We do not yet know how fragile our structures might be.

3. Creatures with high levels of consciousness and great awareness suffer most. We humans, with the highest level of consciousness, will certainly suffer more than the microorganisms that succumbed to the mass extinctions of the great oxygen catastrophe or the Huronian glaciation.

The effects of climate change will be amplified and felt for millennia because carbon dioxide released into the atmosphere will take a long time to be reabsorbed. For most of us, about one hundred years is the farthest into the future that we can easily comprehend. We can envision impacting the lives of our children and our grandchildren, and maybe even our great-grandchildren, but not beyond that—not really. That must be somebody else's responsibility, right? They'll figure it out then. Perhaps because of this, most of our analyses of climate change look at consequences that will be felt by the end of the century.

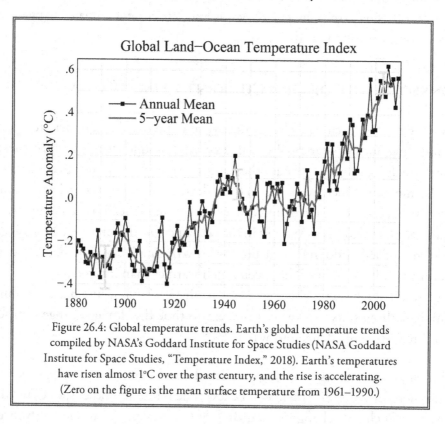

Figure 26.4: Global temperature trends. Earth's global temperature trends compiled by NASA's Goddard Institute for Space Studies (NASA Goddard Institute for Space Studies, "Temperature Index," 2018). Earth's temperatures have risen almost 1°C over the past century, and the rise is accelerating. (Zero on the figure is the mean surface temperature from 1961–1990.)

Average temperatures have increased about 1.0°C (1.8°F) since preindustrial times (1850–1900), and experts have forecast that average global temperatures will likely increase 2.0° to 4.9°C (3.6° to 8.8°F) relative to preindustrial times by the end of this century, even accounting for current mitigating policies (figure 26.4) (NASA Goddard Institute for Space Studies, "Temperature Index," 2018; Raftery et al. 2017). It seems unlikely that the target set by the 2015 Paris Climate Accord—holding temperature increases to less than 2.0°C by the end of the century—can be reached. The consequences of this failure are likely to be profound.

William Nordhaus, in his informative book *The Climate Casino*, emphasizes the four major consequences of our current climate change: sea level rise, ocean acidification, accelerating hurricanes, and species and ecosystem loss (Nordhaus 2013). Of these, sea level elevations may pose the greatest short-term consequences. The seas will rise for two reasons. First, as temperatures rise, ocean waters expand as a result of physical interaction between their molecules; and second, melting ice from glaciers and ice caps—especially from the Greenland and Antarctic ice sheets—add freshwater to the sea (image 26.1). Forecasting elevations of sea levels from melting ice can be very difficult, because when huge icebergs break off from an ice sheet and drift toward the equator, vast amounts of ice can melt very quickly. Unpredictable events such as this can cause rapid changes in sea levels. Global sea levels have risen more than seven inches since 1870 and are projected to rise by another one to four feet by 2100 (US EPA 2017).

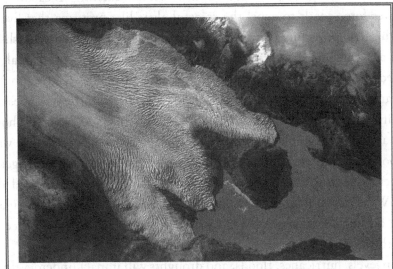

Image 26.1: Gray Glacier, Southern Patagonia. This image of the Gray Glacier in the Southern Patagonian Ice Field of Chile was photographed from the International Space Station. The glacier has retreated over the past twenty-two years, with acceleration in recent years.

With vast civilizations built on ocean fronts, rising oceans may be disastrous. Parts of Manhattan will be underwater, and some coastal cities may disappear entirely. Some cities might be temporarily rescued by dikes and water gates but will still be vulnerable to storms, as

Manhattan was in 1992 when a storm surge flooded the subways and paralyzed the city. Low-lying, heavily populated coastal regions, such as Bangladesh, the Mekong Delta of Vietnam, and the Nile Delta in Egypt will likely be swamped by the rising tides. Bangladesh may be the most vulnerable of these. The monsoon climate brings heavy rains and annual flooding in late summer and fall, and since most of the country is less than one meter above sea level, there is no place for the flood waters to drain. Bangladesh, with its dense population and few resources to spend on flood prevention and recovery, may be among the most vulnerable regions on Earth to climate change. Millions of refugees seeking higher ground could create conflict between nations, resulting in great human suffering.

Ocean acidification and rising temperatures present major threats to marine life. As atmospheric CO_2 levels rise, CO_2 levels in the oceans also rise. CO_2 dissociates to form weak carbonic acid, and in an increasingly acidic ocean environment, many biochemical reactions essential for life do not work well. The building of skeletons and shells in marine animals, such as corals, mussels, and oysters, is inhibited by an acidic environment, and if there is enough acidity, the acid can dissolve existing shells. Rising oceanic temperatures have also induced widespread deaths of coral reefs through a process known as coral bleaching. Bleaching occurs when high temperatures force corals to expel the colorful algae-like protozoa that live in symbiotic relationship with them, providing them with food and giving them their pinkish-orange color.

Hurricanes are spawned by warm ocean surface temperatures, and global warming has already increased their activity and severity. Monstrous Hurricanes Katrina in New Orleans, Sandy in the Caribbean and New Jersey, Harvey in Texas, Irma in the Caribbean and Florida, Maria in the Caribbean, Florence in North and South Carolina, and Super Typhoon Haiyan in the Philippines are recent examples—all with devastating consequences. Since warm air holds more water vapor than cold air, rainfall will increase as temperatures soar, begetting stronger storms and severe flooding. Paradoxically, the warm climate will also breed droughts and megadroughts. What happens is that a warming climate alters atmospheric circulation patterns such that there are more extremes of wet and dry atmospheric regions, hence an increased incidence of both floods and droughts

Rising ocean levels, hurricanes, floods, and droughts will impact underdeveloped countries the most because these countries do not have the resources to adapt quickly. On a planet already struggling with limited resources, conflict and social unrest are sure to follow. While the current climate change does not appear to be an existential threat to our species, it is and will be a major source of economic loss and human suffering and is a major threat to the well-being of mankind. And make no mistake—while it won't wipe out us humans, the current global warming is an existential threat to many species on our planet. We'll explore the effects of climate change on Earth's ecosystems and diverse species when we discuss the sixth extinction in the next chapter.

WHAT CAN WE DO?

Theoretical physicist Michio Kaku, in his prophetic book *Physics of the Future*, has outlined several potential technical solutions for dealing with rising CO_2 levels (Kaku 2012). They sound a little nuts, but new technology often does. One proposal is to launch rockets that will scatter sulfate aerosols into the stratosphere. The sulfates can reflect sunlight back into space, cooling the planet. Another proposal is to create massive algae blooms that would consume huge quantities of CO_2. This could be done by dumping iron chemicals into the ocean to stimulate algae growth. Entrepreneur J. Craig Venter has proposed using genetically engineered microorganisms capable of consuming and removing CO_2 emissions from coal-burning power plants. And Princeton physicists have advocated genetically engineering fast-growing trees that will consume large quantities of CO_2. While none of these new technologies is yet proven to work, they illustrate a host of possibilities that could help decrease atmospheric CO_2 levels until we have figured out how to transition our main energy production away from fossil fuels.

As Kurt Vonnegut projected at the beginning of the chapter, we can probably save ourselves—if we are not too damned lazy or too damned cheap to try. The problem of controlling our greenhouse gas emissions is more economic and political than it is technical. The critical period is now. Hopefully, over the next few decades we will transition our energy production from fossil fuels to renewables, or possibly to nuclear energy. How quickly we make the transition will largely determine the course of climate change. The 2015 Paris Climate Accord, signed by 195 nations, represented a historic global effort to combat human-induced climate change. Unfortunately, the accord suffered a huge setback in 2017 when the US government, under the Trump administration, pulled out of the agreement, making the United States currently the only nation not agreeing to the accord. If nothing else, this dark moment served to highlight the economic and political obstacles we face in dealing with climate change. At the time of this writing, the political will to solve this huge problem is not favorable. Hopefully, this will soon change, but our time is short.

What can we as individuals do? Perhaps the biggest impact we can have on climate change is to become informed so we can inform others and make wise choices when we choose our leaders, who will make the big decisions affecting climate change. The Environmental Protection Agency has put forth a number of recommendations to help individuals reduce energy use. Individual efforts have little impact, but if enough people are involved, we may reach a tipping point where we can make a difference and give our leaders the support and courage they need to steer us in the right direction.

CHAPTER 27

THE SIXTH EXTINCTION

The love for all living creatures is the most noble attribute of man.
—Charles Darwin, in *The Descent of Man*

On July 18, 1989, Kenya's head of state, Daniel arap Moi, put a torch to 2,500 elephant tusks soaked in gasoline. The destruction of $3 million worth of ivory was witnessed by over 850 million people around the world. Anthropologist Richard Leakey, who served as director of the Kenya Wildlife Service and made the decision to burn the ivory, was interviewed on ABC's *Good Morning, America*, and his message to the world was clear: "Don't buy ivory, or the elephant will soon be extinct" (Leakey and Lewin 1995, 201).

Leakey, the son of the famous fossil hunters Louis and Mary Leakey, whom we met in chapter 19, has devoted his life to the preservation of Kenya's wildlife and sanctuaries. After he was appointed director of the Kenya Wildlife Service in 1989, he committed his energies to cracking down on ivory poachers, attacking corruption, and restoring Kenya's national parks. He eventually came into conflict with the corrupt governing political party and was forced to resign his post. He later accepted a position as professor of anthropology at Stony Brook University in New York, and he has continued to support conservation in Kenya and throughout the world. In his gripping book *The Sixth Extinction: Patterns of Life and the Future of Humankind*, he provides a lurid description of the tragic victims of elephant poaching:

> I had seen countless corpses of poachers' victims: great, gray bodies lying bloated under the hot sun, the white excreta of vultures spelling the story of death on their backs. Soon the distended torsos would explode as putrefaction inexorably churned the once-living guts into foul-smelling slime and gas. Worst of all was their faces, transformed from majestic visages into bloody pulp within a few

seconds as poachers ripped off their tusks with axes or chain saws. It was a truly sickening and emotion-roiling sight. Add to this the common occurrence of an emaciated, perplexed infant pathetically trying to rouse its murdered mother, and the rational urge to put an end to the slaughter became an emotional obsession. (Leakey and Lewin 1995, 202)

Image 27.1: African elephant. The African elephant population has declined dramatically over the past century due to habitat destruction and ivory poaching. The species' survival is now threatened.

Elephants are very much like us. They are social animals; females and offspring live in herds helping and protecting one another. They are among the most intelligent animals, exhibiting sophisticated learning ability, compassion, and cooperation. Like us, they live to about seventy years. Throughout history, elephants have played important roles in world economies and religions. Because of their size and strength, we have used them to transport materials, to perform heavy jobs, and even as formidable instruments of war. In many cultures, elephants represent strength, power, and wisdom. And yet we are killing them in droves (image 27.1).

The African elephant population has declined dramatically over the last century, from around four million animals in the 1930s to about four hundred thousand today (IUCN 2016). Their decline is due to the destruction of essential habitats and, even more importantly, to illegal poaching for the ivory trade (image 27.2). The creation of protected reserves, prohibitions against the trading of ivory, and enforcements against poachers have helped to stem the decline, but the African elephant is still vulnerable and in danger of extinction.

The African elephant is just one of many species that may now be on their way to oblivion. Our planet is losing ecosystems, populations, and entire species at an alarming rate. We face the sixth great extinction event in Earth's

Image 27.2: Confiscated ivory. Ivory from elephant tusks assembled for exportation.

history. The cause is unsettling and unprecedented. It is us. Seven billion human beings have spread to all corners of the Earth, breeding, consuming, polluting, preying, warring, and destroying habitats. We, *Homo sapiens*, are waging war against nature.

EXTINCTION IS PART OF LIFE

Extinction is an inevitable part of the natural history of life. If species fail to adapt to changing environments, they will be left behind. An estimated 99.9 percent of all species that have ever existed on Earth have gone extinct, and the remaining species living today will one day be extinct. The fossil record shows that the average life span of any one species ranges from one to ten million years.

These "natural" extinctions due to failure to adapt are what we call background extinctions. As we have seen, species can also become extinct through catastrophic events resulting in mass extinctions. Over the past 540 million years there have been five mass extinctions documented in the fossil record, each due to a major geological or cosmic event that resulted in sudden, drastic environmental changes, giving species little chance to adapt. Each of these mass extinctions wiped out 75 percent or more of Earth's existing species. In contrast to background extinctions, which result in the elimination of individual species over a prolonged time period, mass extinctions impact a broad spectrum of species—usually entire higher taxa—over a relatively short geological time period.

Most experts agree we are currently in the midst of a mass extinction—what has been called the sixth extinction. This mass extinction is different from the rest. It is not due to geological forces but is a consequence of human behavior. *Homo sapiens* is wreaking havoc on our environment and destroying the creatures that share our planet. While the sixth extinction is not due to geological forces, it does share common features with other mass extinctions. It is occurring rapidly, and it is having a broad impact on entire taxa of life.

THE SCOPE OF THE SIXTH MASS EXTINCTION

The diversity of species on our planet today is richer than ever, but this plethora of species is being depleted at an astounding rate. Hundreds of species are driven to extinction every year. Historically, extinction rates—including both mass extinctions and baseline extinctions—have been assessed by observing the disappearance of species in the fossil record. Past baseline extinction rates have been estimated to be about 0.1 to 1.0 extinctions per million species per year (0.1–1.0E/MSY) (Ceballos et al. 2015). Modern-day extinctions are assessed by simple observation. A species is considered extinct when we can no longer find individual members of that species, despite diligent and sustained effort (Barnosky et al. 2011; IUCN 2001). The golden toad, for example, recognized by its brilliant golden orange color, was observed in 1987 in well-known breeding sites in tropical forests surrounding Montverde, Costa Rica. Its breeding pools dried up, decimating all but a few of the toads, and by 1989 only a single male toad could be found. This was the last contact with the now extinct species. Both methods of evaluating extinction rates—the fossil record and simple observation—are thought to underestimate actual rates of extinction, and comparisons are tricky because of the different methodologies, but these comparisons provide the best information we have on current trends.

It has been estimated that over eight million species inhabit our planet, but only a measly 1.8 million of these species have been identified and classified (Mora et al. 2011). In an effort to understand current rates of extinction, investigators formally assess subgroups of animals or plants to determine whether their members have become extinct or are in imminent danger of extinction. Based on such methods, extinction rates over the past century are estimated at one thousand times baseline rates, or about one hundred to one thousand extinctions per million species per year (100–1000E/MSY)—and this rate is believed to be accelerating (Pimm et al. 2014) At this rate, 75 percent of our species will become extinct in the next millennium or next several millennia. While this may seem like a long time to us, it is just a heartbeat on a geologic time scale. And this is a very conservative estimate; the extinction of three-quarters of our species could come much sooner. There is little doubt we are headed toward the sixth mass extinction.

The focus on extinction rate might give the impression that the problem of the sixth extinction is not an imminent threat—that we have time to deal with this and turn it around. Losing a small fraction of a percent of our eight million species per year has not always generated great public concern, especially when many of the species are obscure or live in remote areas. Most of us will not miss the pipistrelle bat (extinct in 2009) or the Catarina pupfish (extinct in 2014) (Ceballos, Ehrlich, and Dirzo 2017).

But researchers at Stanford University and the National Autonomous University of Mexico have looked at the problem in a different way and provide a much more alarming perspective (Ceballos, Ehrlich, and Dirzo 2017). The investigators divided the planet into twenty-two thousand regions and found stark evidence of declining populations and decreasing habitat ranges over the period from 1900 to 2015. Terrestrial vertebrates showed widespread population decline, and many vertebrate populations became extinct in regions they had previously populated, decreasing their range across the globe. The lion, for example, historically inhabited most of Africa, southern Europe, the Middle East, and even India. It is now confined to sub-Saharan Africa, with a small remnant population in India. In its homeland of Africa, the lion historically occupied two thousand of the study's regions but now occupies only six hundred of them. The same narrowing of range and population decline is happening to most species across the globe.

Perhaps the most striking statistic from this study is the investigators' estimate that we have decimated as much as 50 percent of the animal population that once shared the planet with us. This alarming statistic is supported by a more recent study by the World Wildlife Fund, which found that since 1970 we have lost 60 percent of all animals with backbones—fish, amphibians, reptiles, birds, and mammals (Karlekar 2018). The current mass extinction is accelerating at an alarming rate and, in the investigators' words, has become a "biological annihilation" (Ceballos, Ehrlich, and Dirzo 2017).

HISTORY OF HUMAN-INDUCED EXTINCTION

We think of the sixth extinction as a modern phenomenon, and in many respects, this is true, because the extinction has accelerated greatly over the past century. But human-induced extinction began long ago. Fossil evidence shows that as *Homo sapiens* migrated out of Africa and into the Middle East and Europe one hundred thousand years ago, extinction followed in its wake. Neanderthals, who had formerly populated all of Europe, abruptly disappeared a short time after the arrival of *Homo sapiens*. Australian megafauna disappeared from the fossil record shortly after the arrival of humans forty thousand years ago. The woolly mammoth and other American megafauna disappeared not long after the arrival of humans to North America twenty thousand years ago, although their demise is thought to be partly related to climate change. Everywhere humans came, fauna—especially large fauna—disappeared. These early extinctions were caused mostly by big game hunters armed with tools, but the disruption of ecosystems and introduction of alien species contributed to the killing.

When humans turned to agriculture about twelve thousand years ago, our impact on the environment and ecosystems escalated. Massive habitat destruction occurred as humans cleared land for crops and grazing. All plants not considered crops were called weeds and destroyed, and all but a few animals were regarded as pests and were similarly eliminated. In this process, entire ecosystems were destroyed. The introduction of agriculture increased food supply, brought populations closer together, facilitated rapid reproduction, and triggered population growth, resulting in further encroachment on wildlife habitats. In the eighteenth century, the Industrial Revolution and advances in technology further stimulated population growth and accelerated humans' impact on Earth's ecosystems, hastening the rate of extinction of our planet's species.

CAUSES OF THE SIXTH EXTINCTION

Humans have found many ways to destroy Earth's ecosystems and decimate our planet's species. First among these is habitat destruction—the biggest anthropogenic cause of extinction (Wilcove et al. 1998; Rewilding Institute 2017). We degrade and encroach on essential habitats in lots of ways. We clear land for farming and pastures; we pave land for cities, highways, and parking lots; we log; we mine; we dam; and we burn. Forests, which provide habitat for most of Earth's terrestrial species, currently cover about 30 percent of Earth's land surface and have declined 1 percent over the past twenty-five years (World Bank 2018). Tropical rainforests, which are home to a staggering 50 percent of Earth's plant and animal species, are being destroyed at an alarming rate (Nature Conservancy 2017; Lindsey 2007). In 1950, 15 percent of Earth's land surface was covered by tropical rainforests, but by the turn of the century over half of the world's rainforests had fallen victim to chainsaws, fires, and development. At the current rate of destruction, rainforests are projected to disappear by the end of this century (Think Global Green 2017).

The introduction of alien species—including plants, animals, and microorganisms—by humans as they travel to new lands is the second-largest current cause of extinction (Wilcove 1998). Foreign species often propagate unchecked because of a lack of natural enemies in new places, and they may prey on or compete with native species. Travel and trade introduce pathogenic microorganisms to new locations where the native species lack historic immunity and may succumb to epidemics. Hawaii's tropical weather and central location as a hub for pacific trade have made it a paradise for invasive alien species (Kahn 2018). The introduction and proliferation of mongooses on the Big Island have decimated many local bird populations. *Miconia*, a genus of flowering plant introduced from Central America and known as the "purple plague" or "green cancer," grows like a weed, smothering most vegetation in its path, and has driven a number of native plants to extinction. These are just two examples.

Human-induced pollution degrades ecosystems and habitats across the globe, both on land and in water. The runoff of sediments and fertilizers from farms into streams, lakes, and oceans blocks sunlight and thereby inhibits the growth of photosynthesizing organisms like plants and algae. At the same time, nutrients in fertilizers can cause massive destabilizing blooms of harmful algae, which kill fish and their predators. A staggering eight million metric tons of plastic—plastic bottles, candy wrappers, straws, syringes, and so much more—enter the ocean globally each year. That's equivalent to putting five bags of plastic trash in the ocean for every foot of coastline in the world (Joyce 2015). Scientists have identified seven hundred different marine species that are threatened by this plastic pollution, usually by entanglement and ingestion (Good 2017; Gall and Thompson 2015). A video gone viral of a sea turtle bleeding from its nostril as researchers work to extract a four-inch plastic straw has made the sea turtle the poster child against plastic pollution (Cuda and Glazner 2015).

Overexploitation of species by unregulated hunting, fishing, and trapping has decimated populations and often led to extinctions. As we have seen, ivory poachers have drastically reduced elephant populations in Africa. The demand for tiger pelts and body parts on the black market, some of which are used for medicinal purposes with undocumented benefits, has fostered illegal poaching and endangered tiger populations in southeast Asia, India, and western China. Unregulated cod fishing off the coasts of Newfoundland and Nova Scotia reduced the cod population to 1 percent of its historic levels and brought a collapse of the cod-fishing industry there in 1992. The list goes on.

Finally, as we saw in the last chapter, human-induced climate change is taking its toll on Earth's species. Elevated CO_2 levels are acidifying the oceans and reducing the ability of corals, mussels, oysters, and other shell-building marine animals to build shells from calcium. Warming ocean temperatures lead to coral bleaching. If temperatures become warm enough, the oceanic circulation of oxygen may become impaired and lead to massive deaths of marine animals, such as what was seen in the end-Permian extinction. Climate change has triggered intense drought in southern Africa, compromising animal populations. Rising temperatures have led to proliferations of insects, upsetting the ecological balance and jeopardizing other species. Moose populations in North America, for example, are declining because of the profusion of winter ticks, and amphibian populations worldwide are being jeopardized because of severe fungal infections brought on by rising temperatures.

As the world heats up, the survival of many species will depend on their ability to migrate from their old home habitats to new areas that meet their biological needs. For most, this may prove impossible. Many species are isolated because of the destruction of adjacent habitats, and they will not be able to move. For other species, suitable, hospitable habitats may simply vanish from the Earth. The loss of species diversity due to climate change will accelerate as temperatures keep rising, and this will interact with and exacerbate the other causes of species extinction (image 27.3).

Image 27.3: Threatened polar bears. The polar bear (*Ursus maritimus*) is a poster child representing the threat climate change poses to many species. Polar bear populations are in decline as a result of climate change and historic poaching, and the species is vulnerable to extinction.

WHY DOES IT MATTER?

Extinction is a natural process. Even mass extinctions, which have happened five times in the past five hundred million years, have had a positive role in evolution. Following each of the five great mass extinctions, there has been an expansive rebirth—a burgeoning of new species that lit up the seas and landscapes with never-before-seen life, bringing new diversity to our planet. The last of these mass extinctions, the end-Cretaceous extinction that ended the reign of the dinosaurs, opened new space and habitats for the emergence of mammals—and this, of course, eventually led to the evolution of us. As Stephen Jay Gould told us in chapter 18, "Mass extinctions are not unswervingly destructive in the history of life. They are a source of creation as well" (Gould 1985).

Is the sixth extinction any different? Some think not. Biology professor R. Alexander Pyron argues that the movement to save our species from extinction has taken on an unnecessary urgency (Pyron 2017). He contends that extinction is the engine of evolution—pruning the weak and selecting the strong. All species eventually go extinct. We are all endangered species. He and some other experts believe we can use technology to overcome our interdependence on other species and complex ecosystems to survive the sixth extinction.

Conservationists and most experts disagree with this perspective. All species evolved as part of complex ecosystems with many interdependent parts. We and each of the plants, animals, and microorganisms with which we share this planet are heavily dependent on one another and our ecosystems for our survival. Without plants, we would have no food source. Without diazotrophs, the special bacteria that make nitrogen available to plants, we would have no source of nitrogen to build proteins and DNA. Without birds and bees to pollinate, we would have no flowering plants. All life is so interconnected and interdependent. The destruction of wildlife and ecosystems threatens our survival. As Richard Leakey warned, "*Homo sapiens* might not only be the agent of the sixth extinction, but also risk being one of the victims" (Kolbert 2014, 268).

The Apollo astronauts have given us a new perspective of our beloved planet Earth. On Christmas Eve 1968, the *Apollo* spacecraft circled the far side of the moon and the astronauts watched as the Earth disappeared from view. They were suspended in the darkness of space with no radio contact, isolated from the control center at home. As they emerged from behind the moon, the crew looked up and saw a spectacular sight. Earth was rising above the moon's horizon, emerging as a peaceful, beautiful blue marble laced with swirling veils of white against the dark abyss of space (image 27.4). The contrast of this magnificent living planet with the stark, barren landscape of the moon that framed the foreground was striking. The image created a deep impression of just how precious our planet is and how alone and fragile it is. Some years later, Mercury 7 astronaut Scott Carpenter put it well: "This planet is not terra firma. It is a delicate flower and it must be cared for. It's lonely. It's small. It's isolated, and there is no resupply. And we are mistreating it. Clearly, the highest loyalty we should have is not to our own country or our own religion or our hometown or even to ourselves. It should be to, number two, the family of man, and number one, the planet at large. This is our home, and this is all we've got" (Spacequotations.com 2018).

Earth is our cherished home, and we need to do all we can to protect it, its ecosystems, and its wildlife. We don't want a planet with nothing but pavement, parking lots, and farmland. As E. O. Wilson argued, we don't want to live in a world of crows and rats. Our world is so much richer with a diversity of wildlife and habitats.

There is another compelling reason to preserve Earth and its ecosystems. We have a moral obligation. We and all the other species on our planet evolved from common ancestry, and we are all one family. Each of our lineages has survived five mass extinctions and many other catastrophes, and we are lucky to be here. Humans may have evolved a superior intellect, and they may be in a

Image 27.4: Earthrise. On Christmas Eve 1968, the *Apollo* spacecraft emerged from behind the moon, and the crew looked up and saw this spectacular sight. This iconic image of our planet Earth, floating in the dark abyss of space with the barren lunar landscape in the foreground, created an enhanced appreciation of the uniqueness and fragility of our planet and our need to take care of it.

dominant position because of it, but we truly have no more right to this planet than other species. Protecting *Homo sapiens* at the expense of all other creatures is anthropocentricity at its worst. We know that we are responsible for the carnage of the sixth extinction—and only we have the power to stop it. The extinction of each species ends a long genetic line dating back billions of years to our common ancestors and snuffs out a unique part of life on Earth that will never return. We owe it to our children and to our children's children not to destroy the wondrous diversity of life of which we are only a part. We have a moral obligation to turn back this Anthropocene extinction.

WHAT CAN BE DONE?

The destruction of our environment and ecosystems is driven by a growing human population and growing consumption. Earth's population has grown from 4.0 billion in 1980 to 7.6 billion in 2017 (United Nations Department of Economic and Social Affairs 2017). More people means more encroachment on nature's habitats and more consumption of Earth's precious resources. Yet billions of citizens in developing countries have come out of poverty through capitalistic economic growth, and billions more poor people in developing countries cannot and should not be denied the opportunity to share in the world's growing prosperity. The challenge is to bring a level of prosperity and a decent quality of life to all citizens of the world, especially those in developing countries, while at the same time preserving the natural environment and the diversity of species on our planet. At times there appears to be a standoff between environmentalists and people-first advocates. But eminent biologist and conservationist E. O. Wilson believes neither side wants to win entirely: "Down deep, I believe, no one wants a total victory. The people-firster likes parks, and the environmentalist rides petroleum-powered vehicles to get there" (Wilson 2002, 152).

Despite all our dire predictions, the sixth mass extinction is not a done deal. It is true that we have lost about half of our wildlife over the past half-century, and about one-third of our species are threatened, but in the grander perspective over the last century, we have lost about 1 percent of our species to extinction (Pimm et al. 2014). We are in trouble, but the situation is not hopeless.

Our rapidly increasing population has been the major driver of the sixth extinction. Concern about overpopulation came to widespread public attention in the late 1960s when global population growth rates peaked, severe famines ravaged much of Africa, and environmentalists became increasingly aware that overpopulation was threatening our ecosystems. Biologist Paul Ehrlich published his bestselling book *The Population Bomb* and ignited the zero population growth (ZPG) movement. The movement sought to raise public awareness of the link between population growth and destruction of ecosystems, and to encourage people to have smaller families. ZPG's mantra: "Stop at two."

Over the past few decades, the zero population growth movement has waned. China's one-child policy, adopted in 1979, entailed a number of human rights violations, and this created a stigma around nationally enforced population policy. Economists argued that population growth was needed to fuel economic growth, and social scientists cited the need for a younger population of workers to support social security and other safety net programs. Meanwhile, fertility rates in developed countries fell, suggesting that further efforts to control population growth were not needed (Wise 2013). The United Nations, in its 2017 report, has predicted that Earth's population will increase from its current 7.6 billion to 11.2 billion in 2100, and that it will crest early next century (figure 27.1) (United Nations Department of Economic and Social Affairs 2017). This provides some hope that we will eventually reach the goal of zero population growth, but our growing population will be with us for some time to come and will remain perhaps the major force driving the sixth extinction.

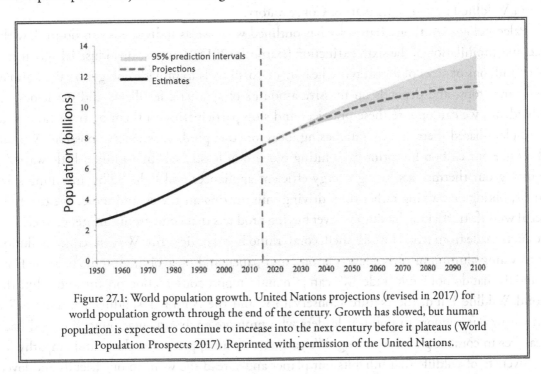

Figure 27.1: World population growth. United Nations projections (revised in 2017) for world population growth through the end of the century. Growth has slowed, but human population is expected to continue to increase into the next century before it plateaus (World Population Prospects 2017). Reprinted with permission of the United Nations.

In 1973, in response to growing concerns about the effects of aggressive hunting and development on American wildlife and ecologies, Congress passed (with bipartisan support) and President Nixon signed the Endangered Species Act (ESA). The law established rigorous protection measures for threatened species and required government agencies to take measures to save species at risk of extinction. But it soon became clear that trying to save every endangered creature was unrealistic and inefficient. And many of these "saved" creatures who were able to survive in captivity could not survive when reintroduced into their previous habitats. This fostered a movement to save endangered species based on their value (their ecological importance, utility, place in our heritage, and beauty) and the economic cost of saving them, but these are very difficult judgments to make (Kahn 2018).

We should applaud efforts to save our endangered species, but these last-ditch efforts are expensive and too little too late to have a major impact. Our most effective efforts at slowing and halting the sixth extinction will be to attack the root causes. We should focus on saving and restoring habitats, controlling alien species, reducing pollution, regulating overexploitation, and moderating climate change. This can be done by supporting international conservation organizations, such as the Convention on Biological Diversity (CBD), the International Union for Conservation of Nature (IUCN), and other nongovernmental organizations, including the World Wildlife Fund and the Nature Conservatory.

Paleontologist Anthony Barnosky has outlined what we as individuals can do to slow the creeping annihilation of the sixth extinction (Barnosky 2014). Large-scale industrial agriculture clears millions of acres of forest, relies heavily on fossil fuels to provide energy to raise animals and grow crops, and depends on massive amounts of synthetic fertilizers and pesticides. As individuals, we can oppose these practices and stop participating in them by transitioning to more plant-based diets and by purchasing local, organic produce whenever possible. We can all reduce our carbon footprint by limiting our use of fossil fuels in a million little ways: by adjusting our thermostats, using energy-efficient appliances and light bulbs, installing solar panels, biking or walking rather than driving, minimizing air travel, and more. We can fight illegal wildlife trafficking by simply never buying products that come from threatened species—products made from ivory, tortoiseshell, coral, rhino horn, or tiger fur. We can reduce pollution by recycling aluminum cans, glass, plastics, and paper, and by avoiding products made from materials that do not biodegrade. We can join and support conservation organizations like the World Wildlife Fund, the Nature Conservatory, and the National Wildlife Federation. We can throw our support to political leaders who promote environmental protection and back measures to combat global warming. We can nurture an appreciation, love, and empathy for the diversity of wildlife that inhabits our planet and spread the word to our friends and loved ones that the extinction crisis is real—and that we can do something about it. As the number of concerned citizens grows, we may reach a tipping point where our collective efforts make a measurable difference.

Finally: don't give up. The fight to save the world from ourselves will be a long-term uphill battle. Slowing the sixth extinction will take decades of work and continued efforts to maintain. We humans are a massive destructive force. But we don't have to be.

CHAPTER 28

TECHNOLOGICAL EVOLUTION

> By the time of the Singularity, there won't be a distinction between humans and technology. This is not because humans will have become what we think of as machines today, but rather machines will have progressed to be like humans and beyond.
>
> —Ray Kurzweil, in *The Singularity Is Near*

Evolution is a slow grind of random mutations and natural selections that has shaped our genome over billions of years. Now we have entered a new phase of accelerated cultural and technological evolution that is outpacing biological evolution by orders of magnitude. It is nigh on impossible to predict the future. Who would have known just a few decades ago that we would soon be able to talk to someone on the other side of the planet through a little glass-fronted handheld device? And who would have thought that we'd be able to talk to this little device and get answers to almost any question imaginable? No one could have predicted this—at least, not exactly. British cognitive psychologist Geoffrey Hinton scoffed at those who confidently predict the future: "Seeing into the future is like looking through fog. When you're in fog, you can see short distances quite clearly. When you look a bit further, it's fuzzier. But then if you want to see twice as far as that, you can't see anything at all. That's because fog is exponential" (Dormehl 2017, 221).

We may not be able to predict the future, but we can be certain that our future will be driven by advances in technology, and that the pace of change will be much more rapid than we might expect. Genetic engineering, nanotechnology, and artificial intelligence—often abbreviated with the acronym GNR (R for "robotics")—will likely dominate technological and societal change in the foreseeable future. Let's take a look at how our technological destiny might unfold.

GENETIC ENGINEERING

Humans first began to meddle with evolution about ten thousand years ago, at the onset of the Agricultural Revolution, when we began selectively breeding plants and domesticating animals. Our ancestors mated the stoutest bulls with the finest cows, generation after generation, until they obtained breeds with plenteous meat for the dinner table. They selected wild grasses with especially large and tasty kernels, and selectively bred them into what we now know as corn. Selective breeding is now a mainstream practice in agriculture and domestication, producing hybrids with special characteristics to meet our needs. This process, which steers variation and natural selection to meet our own needs, was perhaps humanity's first biotech innovation.

Genetic engineering took things a step further, enabling investigators to directly manipulate species' genomes (image 28.1). Ingenious scientists at the pharmaceutical giant Eli Lilly were among the first to do this, when they engineered a way to produce human insulin (Humulin, as they called it) outside of the body. They isolated human genes responsible for insulin production and attached them to small pieces of DNA called plasmids that they had removed from *E. coli* bacteria. They inserted this hybrid DNA into new host *E. coli* cells, and as the bacteria multiplied, the plasmids, with their attached insulin genes, also multiplied. The engineered *E. coli* produced huge quantities of what we call recombinant human insulin. Humulin was approved for use in humans by the FDA in 1982 and has since replaced porcine and bovine insulin for the treatment of diabetes.

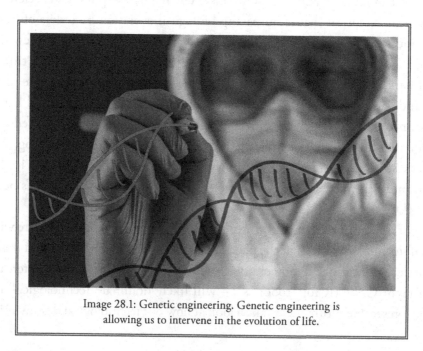

Image 28.1: Genetic engineering. Genetic engineering is allowing us to intervene in the evolution of life.

Next came **cloning**. In 1996 Ian Wilmut and his colleagues at the University of Edinburgh made headlines around the globe when they announced that they had produced the first fully cloned mammal, a sheep named Dolly (Campbell et al. 1996). The scientists obtained a donor cell from the mammary gland of an adult donor sheep and removed the cell's nucleus with its load of DNA safely tucked inside. They then retrieved an unfertilized egg from a female sheep, removed the egg's nucleus, and replaced it with the nucleus from the donor cell. The egg was given an electric shock that induced the cell to begin dividing. This produced a cloned embryo that was placed in the uterus of a surrogate mother sheep. Dolly, a baby clone of the donor sheep, was born months later. With her birth, we had learned how to successfully copy life.

Next scientists simulated the creation of life itself—almost. In May 2010, maverick biologist and billionaire entrepreneur Craig Venter and his associates announced that they had produced the first synthetic cell. To accomplish this feat, they copied the DNA code of a simple bacterium, *Mycoplasma mycoides*, and used this code, with a few modifications, to synthesize a new genome from basic nucleotide building blocks (Gibson et al. 2010). They then inserted the new genome into a bacterium whose DNA had been removed, and the new cell, under the direction of its new genome, produced many daughter cells through replication. Among the modifications to the new genome were cheeky additions. Using a code devised from the letters corresponding to the nucleotides (A, T, C, G), they embedded DNA sequences into the synthetic genome coding for Venter's website, along with a quotation from the legendary physicist Richard Feynman: "What I cannot build, I cannot understand" (Ward 2010).

Venter and his team called their new life-form Synthia. Its creation held great promise for the future: the potential to engineer customized organisms that could produce necessaries like fuel, food, specialized chemicals, and more. Large-scale genome synthesis using advanced editing technologies could create a massive biotech industry, which could totally transform our future in the same way that digital technology has shaped our world and our global economy today (Enriquez and Gullans 2015, 240)

Genetic modification took a great leap forward when scientists introduced a new gene-editing technology, **CRISPR-Cas9**, which was able to splice DNA sequences out of any genome and replace them with new codes. This new technology was stolen from bacteria. Bacteria are able to cut invading viruses out of their own genomes and replace them with harmless codes—similar to what antivirus programs do to protect our computers. Scientists repurposed this bacterial process to create CRISPR-Cas9, which is far more efficient at editing genes than recombinant DNA technology (the process used to produce Humulin) and which has greatly enhanced gene editing.

CRISPR-Cas9 and other technologies have been used to genetically modify a growing list of animals (Reardon 2016). Pig embryos have been implanted with genes that convert unhealthy fatty acids to healthy omega-3 fatty acids, creating transgenic pigs with healthier omega-3 fats for our consumption (Lai 2006). Mice embryos have been implanted with genes that enhance signal detection by neuroreceptors in the brain, creating transgenic mice with enhanced memory and learning abilities (Tang et al. 1999; Yang, Wang, and Jaenisch 2014). As we learned in chapter 22, modification of longevity genes in yeast, worms, and fruit flies have increased their lifespans by 50 to 100 percent (Guarente and Kenyon 2000; Tissenbaum and Guarente 2001). And salmon have been genetically modified to produce increased levels of growth hormone, enabling them to grow to full size in eighteen months rather than the usual three years. These salmon (patented as AquAdvantage salmon by AquaBounty Technologies) became the first genetically modified animals approved by the FDA for human consumption in the United States in 2015 (Ledford 2013; Pollack 2015). This discovery provided a glimpse at how new technology might help us to meet the food demands of our growing populations, but it also triggered immediate protests from environmental advocacy groups.

It was only a matter of time before genetic engineering would be applied to humans. In 2016, the United Kingdom approved a controversial treatment, **mitochondrial replacement therapy** (MRT), to prevent the transmission of mitochondrial diseases from mothers to their children. As you may remember from earlier chapters, mitochondria—the energy factories in the cytoplasm of cells—harbor their own genes, which are transferred from the mother's egg to the zygote at the time of fertilization. (Mitochondrial DNA in sperm is degraded at conception and is not passed on to offspring.) In rare cases, mitochondrial DNA contains unhealthy genes that can lead to devastating diseases of childhood, often causing stunted development, neurologic disorders, and heart disease. Women identified as having diseased mitochondrial genes, either because they inherited the disease or had a previous child with one of these disorders, may be candidates for MRT to prevent transmission of the disease to intended offspring. In MRT, the mother's egg is harvested, and its nucleus—which contains her nuclear DNA—is removed and transferred to an egg from a donor whose nucleus has also been removed, but whose mitochondrial genes are healthy. The new egg, which has healthy nuclear genes from the mother and healthy mitochondrial genes from the donor, is then fertilized with the chosen father's sperm through **in vitro fertilization** (IVF), and the resulting embryo is implanted into the mother's uterus. The first of these so-called "three-parent babies" was delivered in the summer of 2016, ushering in a new era of genetically altered babies.

Today IVF is practically a household term. The technology is commonly used to enable infertile couples to have children, producing what are known as "test-tube babies." Eggs harvested from the mother are fertilized by the father's sperm outside the body. The resulting embryos are implanted into the mother's uterus. IVF has also been used to screen embryos for defective genes to avoid passing on heritable diseases; only embryos with healthy genes are implanted into the uterus. In the future, IVF may be used not just to screen and eliminate unhealthy genes but also to select genes that code for desirable traits and to edit genes to produce advantageous traits.

NANOTECHNOLOGY

In 1959, theoretical physicist and Nobel laureate Richard Feynman gave a talk at Caltech entitled "There's Plenty of Room at the Bottom." In it he famously prophesied engineering at the level of atoms and molecules, and thereby forecast the birth of nanotechnology. It was not until two decades later that a major breakthrough jump-started the field. The invention of the scanning tunneling microscope in 1981 allowed unprecedented visualization of atoms and molecules—something we thought we would never be able to see. Scientists now had the tools to start working "at the bottom," as Feynman had predicted. It is hard for us to imagine the smallness of this scale. Atoms are about 0.1 to 0.5 nanometers in size, and one nanometer is one-billionth of a meter, or 1×10^{-9} meters. Comparing an atom to a meter is like comparing a marble to the Earth.

In 1986, American engineer K. Eric Drexler published a prophetic book entitled *Engines of Creation: The Coming Era of Nanotechnology*. In it he forecast a science fiction–like future of atom-scale technologies. In this world, the entire Library of Congress could fit on a chip the size of a fingernail. He projected "molecular assemblers" using nanosized robots to assemble products atom by atom—making everything from clothes to computers. He created hope for a utopian society—a world in which we could make anything out of anything else and therefore have unlimited resources. But at the same time, he realized the potential dark side of this technology. He envisioned a wild "gray goo" end-of-the-world scenario, in which out-of-control self-replicating nanoscale robots would consume all matter in their paths and use the atoms to build nanoscale machines and copies of themselves. Needless to say, there were and are many skeptics of this dark scenario, and of the molecular assembler in general.

Because of nanotech's immense potential for medical, industrial and commercial use, the US federal government has invested heavily in its research, providing over $24 billion in funding to the National Nanotechnology Initiative between 2001 and 2017. Many new applications are emerging. Nanotechnology may soon revolutionize digital processing by replacing transistors, which are etched onto silicon computer chips, with nanoscale molecular transistors that can change configuration into an on or off mode (Kaku 2011, 221). This will bring ultimate miniaturization to electronic devices. The ultimate hypothetical product of nanotechnology is that hypothetical "molecular assembler." Envision trillions of nanobots converging to take apart raw materials, molecule by molecule, and quickly reassemble them into something entirely new. If this seems like an impossible feat, take a look at your own arm. Your muscular contractions, driven by the actin and myosin filaments in your muscle fibers sliding over one another, are powered by tiny molecular machines analogous to this fictional assembler. And in the ribosomes of each of your cells, proteins are being pieced together from amino acid building blocks by molecular assemblers. These are just two examples of molecular machines that have been designed and produced by nature. Someday we will be able to produce our own molecular machines.

The molecular assembler and other innovations of nanotechnology could provide all our material needs and radically alter society. People, rid forever of production and assembly jobs, would be free to pursue loftier pursuits—art, music, literature, and the whims of their fondest dreams. Yet again, technology offers us the promise of a new utopian world. Of course, things rarely turn out so optimistically.

ARTIFICIAL INTELLIGENCE

On February 10, 1996, amid countless headlines and great anticipation, IBM's chess-playing computer Deep Blue faced off against world chess champion Garry Kasparov in the ultimate test of man versus machine (History.com editors 2009). In a historic upset, Deep Blue triumphed in the first game—to the frustration of Kasparov. But the tenacious and brilliant Russian changed his strategy and rebounded to win the six-game match four to two.

IBM computer scientists went back to the drawing board. In designing Deep Blue, they had taken advantage of the computer's immense processing speed and memory space and created algorithms that would sift through every possible game move to find the best one (Aung 2017). The evaluation of each move required an analysis of an almost infinite number of possibilities for subsequent future moves and their consequences. They had also tailored Deep Blue's algorithms to Kasparov's particular style of play. But Kasparov, shrewd as he was, simply had changed his style of play after game four, issuing new moves that Deep Blue was not familiar with. This turned the tide of the match. Deep Blue—as is typical of early generation **artificial intelligence**—had complex algorithms capable of sorting through myriad possible responses and choosing the best one but was incapable of forming memories and learning from past experiences. It could not respond to Kasparov's changes in strategy. Over the next year, IBM scientists updated Deep Blue so that programmers could adjust strategies between each game. They also increased the computer's capacity to evaluate different chess positions from one hundred million to two hundred million positions per second.

Image 28.2: Deep Blue versus Kasparov. IBM's chess-playing computer Deep Blue faced off against chess champion Garry Kasparov in the ultimate test of man versus machine.

Deep Blue was ready for a rematch. The heavily publicized second six-game match between the computer and its human nemesis began on May 3, 1997. Kasparov won the first game, and Deep Blue the second. Then followed three tie games in a row. On May 11, Deep Blue won the sixth and final game, and the match. It was a triumph of machine over man (image 28.2). While Kasparov complained that Deep Blue had been given human assistance—and therefore an unfair advantage—IBM rightfully celebrated their triumph. The future of humanity would be inextricably linked to artificial intelligence (AI), and now the public was very much aware.

Deep Blue represents first-generation purely reactive artificial intelligence (AI), designed for a specific purpose. Deep Blue sees all the pieces on the chessboard, determines all the possible moves for itself and its opponent, and selects the best one based on preprogrammed algorithms. But it does not have any memory of the past and cannot learn from experience if one of its moves turns out not to work out well.

MACHINE LEARNING

In almost every way, computer hardware outcompetes the biological hardware of the human brain. Computer chips can transfer information ten million times faster than neurons. Computers have unlimited long-term memory and hard drive storage, compared with the brain's restricted memory space. Computer transistors are reliable and can run 24-7. Brains need sleep. Finally, computers can connect and sync with other computers online, providing unlimited access to data and knowledge. It is not surprising that computers have already surpassed humans in performing many specific tasks. But still, the human brain is the most complex structure on Earth, and creating human-level intelligence—known as artificial general intelligence, or AGI—has seemed an almost impossible task.

One essential aspect of human intelligence is the ability to learn. Traditionally, machine learning has required a human programmer to guide the machine through any learning process. For example, let's say a programmer wants to "teach" a computer to recognize images. The programmer gives the computer a large number of images, tells the computer what each image is, and creates algorithms to help the computer identify each image. Through trial and error and adjustments in the algorithms, the computer will soon be able to identify all the images in the set. But if a new image with new features appears, the human programmer must intervene and supply new algorithms, thus "teaching" the machine about the new image. This is **supervised machine learning**. Deep Blue required supervised learning. Unable to innovate or learn on its own, it was unable to cope with Kasparov's new chess strategy without supervised learning.

In the early 2000s, British computer scientist Geoff Hinton, working at the University of Toronto, pioneered the development of **artificial neural networks**—computer systems modeled on the human brain—and introduced the new concept of **deep learning**. In this new **unsupervised machine learning**, the machine is thrown large volumes of data but is given no explanation as to what it is. Given a large enough dataset, the machine will gradually recognize patterns on its own. The machine looks for a preprogrammed variety of features in the data input and places those features in separate, or "deep," layers, hence the name deep learning. Hinton's pioneering work in deep learning and his revolutionary visual image recognition software have earned him the title "Godfather of AI."

Deep machine learning has been widely employed in speech recognition (Geitgey 2016). Sound waves from voice input are analyzed and transformed into a digital format that can be processed by the computer. Algorithms are created to identify vowels or consonants correlating with the digitized voice input, and the computer creates a text output based on these algorithms. Initially the computer will make many mistakes. But with time, as the voice recognition system is corrected for its mistakes, it will learn, and accuracy will improve. The computer also learns which letters commonly go together. If the first three letters are "HEL", the computer expects the next two letters to be "LO." As the computer accumulates more data, it learns and remembers what letters and what words usually go together, and this greatly enhances its ability to convert voice to text more accurately. The more exposure it gets, the better it learns, and the better it performs. Anyone who has used a deep learning voice recognition system cannot help but be impressed by how well the system learns to get it right.

These new neural networks with deep learning have overtaken traditional supervised machine learning systems in many critical data-processing tasks. They provide the basis for much of today's AI and hold the greatest promise for the future.

MACHINES WITH EMOTIONS

In his fascinating and surreal book *Heart of the Machine: Our Future in a World of Artificial Emotional Intelligence,* Richard Yonck describes a possible future in which intelligent machines will understand and even manipulate our emotions (Yonck 2017). Already developers have begun including emotional detection and expression as an integral part of the design of intelligent machines. Smart watches can measure heart rate, blood pressure, and temperature, all of which can be used to help define an individual's emotional state. Home devices may soon detect emotional states using voice analyses and facial expression recognition. Once this is done, the machines will be able to respond to the user's emotional cues. Artificial intelligence is not yet able to exhibit and express an emotional state of its own, but that will come before long. As these machines learn to perceive and express emotional states, they will become more human-like, and our interactions with them may begin to mimic human interactions. Yonck cites a humorous story that illustrates what our future could look like:

> Archie slams his hand down on the dining room table. "No! You are not marrying Michael and that's final!"
> "But Daddy!" Gloria shouts, on the edge of tears. "No one's ever made me feel like this! Michael treats me like I'm the most wonderful person in the world."
> Her mother Edith chimes in. "Archie, you really haven't given Michael a chance."
> "No, Daddy," Gloria says, standing her ground. "You most certainly have not. Your bigotry toward Michael has been obvious from day one."
> "I am not a bigot!" Archie snarls through clenched teeth, seething at the accusation. "No daughter of mine is marrying a robot and that's final!"
> Gloria cringes and jumps up from the table. "Michael is a cybernetic person with the same rights you and I have! We're getting married and there's nothing you can do to change that!" In tears, she storms out of the room. (Yonck 2017 194–95).

We are, of course, not there yet. But AI researchers are working toward such a future. AI expert Arend Hintze summarized the challenge: "If AI systems are indeed ever to walk among us, they'll have to be able to understand that each of us has thoughts and feelings and expectations for how we'll be treated. And they'll have to adjust their behavior accordingly" (Hintze 2016).

IMPACT OF ARTIFICIAL INTELLIGENCE ON SOCIETY

Our lives are already irreversibly intertwined with artificial intelligence, although we usually don't think of our technology that way. Once we integrate AI into our daily routines, we often no longer call it that. Our cell phones are loaded with apps that guide our daily activities. Map apps guide us to unknown places, and Pandora and Spotify select music tailored to our tastes. Amazon provides us with "recommendations" and pop-up ads when we search for products online. Computers control airplane flight patterns and landings, and electronic medical record systems provide data and communicate orders in hospitals and doctors' offices. Self-driving cars are beginning to populate our roads and highways and will soon make the roads safer for all and provide mobility for those that are incapable of driving. As Elon Musk argued, we will reach a point where it will be unethical for humans to drive vehicles: "It's too dangerous. You can't have a person driving a two-ton death machine." (Dormehl 2017, 139). AI has provided conveniences that have been integrated into our daily lives and now have become necessities. Our emerging reliance on these intelligent computers has begun to compromise our independence, but so far this has been relatively harmless.

Image 28.3: Artificial intelligence. Artificial intelligence has already been integrated into our daily lives and is projected to replace nearly half of the jobs in the United States over the next two decades (Frey 2013).

AI has a downside. We have already seen many jobs replaced by automation, and advanced AI could replace so many more—not just blue-collar manufacturing and clerical jobs, but the entire spectrum (image 28.3). A 2013 study from Oxford projected that 47 percent of jobs in the United States could be replaced by automation in the next two decades (Frey and Osborne 2013). AI will replace bank tellers, customer service representatives, paralegals, and accountants. With the advent of self-driving vehicles, AI will replace cabdrivers, delivery workers, truck drivers, and more. Part of the solution will be to retrain people for jobs requiring creativity and nuanced human interaction—skills AI is not very good at. We will still need people to work as social workers, psychologists, and trial attorneys. We need to encourage people to enter jobs that give them a sense of purpose—mentoring, volunteering, and teaching. But in the end, there will be far fewer jobs and a shorter workweek. The Agricultural and Industrial Revolutions replaced jobs, but they created many new ones. Many experts believe AI will replace jobs with few replacements. What are we to do when machines are doing our work? For many, this will provide welcome leisure time and create new opportunities to pursue creative activities. For others, work is their passion, without which they may be lost. As the world grows more complex, and as machine intelligence surpasses ours, we will have to adapt.

Widespread automation through AI could result in concentration of wealth in the hands of the few who control the new technology and could push displaced workers to the bottom of the socioeconomic hierarchy. This could create an enormous wealth disparity, which could foster a climate conducive to social unrest and conflict. The new AI technology will create not only huge economic gaps between individuals within societies but also huge gaps between rich and poor nations. Poor nations unable to develop AI of their own will become economically dependent on those with AI. The two current leaders in AI development, China and the United States, may be in the dominant role (Lee 2017).

THE SINGULARITY

Futurist Ray Kurzweil has been described by Bill Gates as "the best person I know at predicting the future of artificial intelligence" (Kurzweil 2006). In his provocative book *The Singularity: When Humans Transcend Biology*, Kurzweil describes technological **singularity** as the point at which technological change becomes so rapid and profound that machine intelligence exceeds human intelligence. When we reach this tipping point, it will be impossible to predict how the future might unfold. Intelligent machines will then be able to perform any intellectual task as well as a human can—able to make a cup of coffee or graduate from college, perform at high levels in economically important jobs, and converse with humans with a facility that will make them indistinguishable from our human friends. This seems like science fiction to most of us and appears to be a long way off, but technological progress is increasing at exponential rates, and Kurzweil and some AI experts believe that the singularity will arrive much sooner than we think—within decades, not centuries.

Kurzweil is widely respected as a futurist, but his surreal vision of the singularity occurring in the next few decades is not without challenge. Greek physicist and futurist Theodore Modis charges Kurzweil and the growing techno-cult of singularitists are indulging in parascience lacking in scientific rigor. Modis believes that despite impressive technological progress, Kurzweil's predictions of exponential, runaway technological growth are misguided. He believes there is no convincing evidence that Kurzweil's singularity will ever take place (Modis 2006). The late Microsoft cofounder Paul Allen has also been skeptical. He argued that building the complex hardware and software that would create a machine with human-level intelligence requires a nearly perfect understanding of how the brain works. This means knowing not just the physical structure and physiology of the brain but also how billions of neuron interactions result in consciousness and cognition. Allen believed the exponential growth of technology is not uniform across all fields, and that progress toward superhuman artificial intelligence will be slowed by the complexity of the problem. He believed that by the end of the century we will still be wondering if the singularity is near (Allen 2011).

Just as we can't see very far through the fog, we can't see very far into the future. We don't know when or if the singularity will occur, or how it might unfold. What we do know is that our future will be driven by technology, and in the near term this will be dominated by the big three—genetic engineering, nanotechnology, and artificial intelligence. Let's hope we have the wisdom and fortitude to channel these technologies for the common good.

CHAPTER 29

EXISTENTIAL THREATS

> The twenty-first century may be a defining moment. It is the first in our planet's history where one species—ours—has Earth's future in its hands and could jeopardize not only itself but also life's immense potential.
>
> —Martin J. Rees, astrophysicist, in the foreword to *Global Catastrophic Risks*

The accelerated pace of technology has brought us to a critical phase in the evolution of our species. Genetic engineering, nuclear power, nanotechnology, and artificial intelligence have provided us with unprecedented opportunities in providing for our needs and improving the well-being of human-kind, but each of these innovations has brought us grave new risks. The future of our species and our planet will depend on how we can work together to meet the challenges of these new technologies (Kurzweil 2006; Bostrom 2002).

Remember the Fermi paradox from chapter 25? If the probability that complex extraterrestrial life exists elsewhere in the universe is very high, as many experts think it is, why is it that we have not seen or heard from the aliens? Many authorities believe that we have not seen any evidence of extraterrestrial life because once intelligent life achieves advanced technology, it becomes self-destructive. This is a disturbing thought as we contemplate the future of our species.

Humans are currently facing, and will continue to face, a number of existential threats that can push our civilization back to the stone age or lead to our extinction. Some of these threats, such as nuclear holocaust and climate change, have been recognized for some time, while others, such as nanotechnology and artificial intelligence, are just beginning to be appreciated as potential threats. Let's take a look.

NUCLEAR HOLOCAUST

In October 1962, the United States and Soviet Union faced off in the Cuban missile crisis. The Soviets had secretly installed nuclear missiles in Cuba, just ninety miles from US shores. When President Kennedy became aware of the threat through our intelligence services, an unprecedented nuclear crisis emerged and threatened world peace. President Kennedy addressed the American public on television to inform them of the imminent threat, ordered a naval blockade around Cuba, and warned the Russians that he was prepared to use military force to remove the threat. The world was on the brink of nuclear war. Disaster was averted when Nikita Khrushchev backed down and agreed to remove the missiles, in exchange for assurances that the United States would not invade Cuba.

Ever since the conclusion of World War II, nuclear war has been one of the most feared and imminent existential risks to humanity (image 29.1). Few other threats have the ability to annihilate so many so quickly. The United States and Russia still have large stockpiles of nuclear weapons despite efforts to reduce them, and England, France, Israel, India, Pakistan, China, and North Korea all have nuclear capabilities.

An all-out nuclear war would devastate our planet. Nuclear explosions would wipe out millions in war zones and release tons of thick black smoke that could block sunlight for years, plunging our planet into a nuclear winter. Radiation fallout could persist for years, leading to widespread long-term bodily damage through cancer and genetic defects. While widespread nuclear war may not completely decimate the human population, it could send our civilization back to the Stone Age and destroy our potential for future development.

Over the past decade, many experts and international panels have advocated for sharp reductions in nuclear arsenals, with the goal of eventual abolition. The consensus is that these weapons are simply too dangerous in any hands. President Obama, following his election in 2008, prioritized the cause of nuclear disarmament, and in 2010 the United States and Russia signed the Strategic Arms Reduction Treaty, which reduced the size of both US and Russian nuclear stockpiles. In 2015, the six great nuclear powers reached an agreement with Iran to limit Iran's potential to develop nuclear weapons. Despite this progress, the United States and other nuclear states, while acknowledging that the world would be better off without nuclear weapons, are not yet ready to move to a policy of complete abolition. Nuclear deterrence has made all-out nuclear war and worldwide nuclear holocaust unlikely in the near future, but regional nuclear conflict initiated by terrorist groups or rogue states remains a major threat and could lead to horrific casualties and suffering.

At the time of this writing, North Korea remains an active nuclear threat, Russia and China are increasing their stockpiles, and the United States is threatening to do the same, and the

Trump administration has withdrawn from the Iran agreement. While there has been limited public dialogue on these trends, we are on the brink of a new nuclear arms race (image 29.1) (Schlosser 2018).

BIOENGINEERED PANDEMIC

Biological warfare is nearly as old as war itself. Mongol warriors hurled plague-infected cadavers over city walls in the siege of Caffa in Crimea in 1346, seeding the city with the great bubonic plague (Wheelis 2002). British colonists deliberately triggered a smallpox epidemic in Native American tribes in 1763 by distributing contaminated linens during the siege of Fort Pitt (Gill 2004). And the Japanese attacked multiple Chinese cities in World War II with ceramic bombs containing plague-infected fleas, resulting in over two hundred thousand deaths (World Future Fund 2017).

Image 29.1: Hydrogen bomb. The first hydrogen bomb test, code-named Ivy Mike, was conducted in 1952 by the United States in the Marshall Islands. After a period of reduction in nuclear arsenals, we are now entering a new nuclear arms race (Schlosser 2018).

After World War II, as scientific knowledge about biological agents grew, nation-states became increasingly interested in germ warfare. Concerned about the risks, the United Nations established the Biological Weapons Convention, which was signed by 178 countries in 1972 and prohibited the development, production, and stockpiling of biological weapons. Unfortunately, the treaty has no verification process to monitor compliance and has no enforcement provisions, so it has had limited effectiveness. Recent advances in biotechnology and bioengineering have created new threats, increasing the risks from biological agents.

Genetic engineering is capable of producing "superbugs" that are more proliferative, more contagious, more lethal, and more resistant to environmental influences and antibiotics than natural pathogens. These agents can be produced in large numbers without them losing their virulence, and they can be delivered effectively to the enemy with safe guards for one's own peoples. Such superbugs could be engineered with long latency periods, allowing them to infect entire populations and remain undetected before individuals become symptomatic. These genetically engineered superbugs may be the ultimate hidden killer.

While bioweapons certainly pose great risks, until now they have been used infrequently and isolated to particular exposed populations. Only a few countries are suspected of having biological weapons today, but rapid advances in biotechnology provide most countries that have pharmaceutical and medical industries with the knowledge and capability to develop biological weapons. As the technology spreads, the risk from the surreptitious development and deployment of superbugs as weapons of mass destruction remains a growing threat that could become an existential threat.

ASTEROID IMPACT

Sixty-six million years ago, an asteroid three times the width of Manhattan smashed into the Yucatan Peninsula, vaporizing all life in the region and filling the skies with dust and debris that obliterated the sun and caused temperatures to plummet. This event caused the death of the dinosaurs and destroyed an estimated 75 percent of species on our planet.

Because of the historic example of the end-Cretaceous extinction, we know that potential future asteroid impacts still represent a threat to life on our planet. Although small asteroid impacts occur on a regular basis, large impacts are quite rare. It is estimated that large asteroids (over one kilometer in diameter) collide with Earth only once in five hundred thousand years, so based on probability alone, we don't expect one to hit us anytime soon (Bostrom 2002). The asteroid that killed the dinosaurs was ten to fifteen kilometers in diameter, and we haven't seen anything close to that size in the intervening sixty-five million years. By the time an asteroid does threaten our planet, we will hopefully have the technology to divert it from a collision course with Earth. This strategy was used in the science fiction film *Meteor* (1979), in which Russia and the United States joined forces to destroy an asteroid on a collision course with Earth. As an existential threat, an asteroid collision appears very unlikely.

CLIMATE CHANGE AND MALTHUSIAN CATASTROPHES

Of all the potential species-wide catastrophes that could face us in the near future, human-induced climate change may be the one with the greatest public awareness. As was discussed in chapter 26, global warming will likely lead to widespread famine, mass displacement, and great loss of life and property, but it is unlikely to result in our global extinction. It is an existential threat to many species, but not to *Homo sapiens*. Mankind will have time to adapt to climate changes, and new technologies should help us gain control over our climate. Unless climate change causes such international strife that it triggers massively destructive global war, it is unlikely to culminate in our total demise.

In 1779, English economist Thomas Malthus warned that uncontrolled population growth could result in widespread famine. Agriculture simply would not be able to keep up with demand, and people would starve. Economists and ecologists have given warnings about this exact scenario for almost two centuries, calling it a **Malthusian catastrophe**. But after World War II, new technology enabled dramatic increases in agricultural production, which expanded the world's food supply and lowered food prices. And by the twenty-first century, most developed countries had experienced a fall in fertility rates, and population growth slowed. The world population increased by 400 percent in the twentieth century but is expected to increase by only 50 percent in the twenty-first century, from 7.5 billion in 2017 to about 11 billion by 2100 (Roser and Ortiz-Ospina 2017). While, sadly, there remain pockets of famine and starvation across the globe, fears about a Malthusian catastrophe have mostly abated, and it does not appear to be an existential threat. Still, the negative impacts of population growth, economic growth, and climate change on the environment are dramatic and potentially devastating, as we so pointedly saw in chapters 26 and 27.

NANOTECHNOLOGY

Eric Drexler's famous end-of-the-world gray goo scenario, in which out-of-control nanosized robots consume all matter in their path, has received great publicity over the years but is no longer considered a credible threat. While molecular machines exist in nature, we are nowhere close to producing our own molecular machines, and out-of-control nanobots will remain science fiction, probably for a very long time.

The big societal threat from nanotechnology is not the gray goo scenario but rather the potential for the creation of weapons of mass destruction (Treder and Phoenix 2006; Sandberg 2014). Molecular manufacturing could produce cheaper and more destructive conventional weapons in such quantities that they would eventually pose a species-wide existential threat. Future nanotechnology engineers might produce microscopic, self-replicating nanobots capable of injecting toxins into enemy personnel—a man-made imitation of viruses and bacteria. The weapons only need to be small enough, deadly enough, and cheap enough to make this possible. The only defense against such nanoweapons would come from rival nanotechnology.

Nuclear weapons have a paradoxical deterrent effect. In times of international conflict, enemies are hesitant to use nuclear weapons because of their immense, indiscriminate, destructive power—and because of the threat of mutually-assured destruction. In contrast, nanotech weapons can be developed in secret, could single out specific targets, and could be activated after some delay. Because nanotech weapons will be more focused, they will have less deterrent effect than nuclear weapons and will be more likely to be used. And defending against such a precise, subtle attack could prove very difficult.

ARTIFICIAL INTELLIGENCE

The year is 2050. Just as robots have replaced assembly line workers in factories, computers with highly intelligent algorithms have replaced most of our human assistants—including travel agents, bank clerks, investment advisers, accountants, and even many doctors. In this imagined future, most jobs have been replaced by AI and robots—and yet the people are happy. They have an abundance of material goods, perhaps a guaranteed income, and a lot more time to pursue their creative interests.

In this future, your personal assistant—part of a new generation of software in the wake of Apple's Siri, Microsoft's Cortana, Amazon's Alexa, and Google's assistant—accesses your files, reminds you of appointments, provides suggestions for gifts and orders their delivery, organizes meetings, and provides a wide variety of secretarial services. As your confidence in this ideal assistant grows, he takes on increasing responsibilities. (The program uses a soothing male voice and, in your mind, is unmistakably a *he*.) He manages your investments, searches for jobs uniquely suited to your skill set, and even helps you to find a mate. This smart technology becomes so knowledgeable and so intelligent, and knows you so well, that you no longer really question his advice or actions. As he learns how to assess your emotional state, he begins to respond accordingly. By this point, your AI has become something more than just an assistant. He has become a friend and coworker, a colleague and agent. The lines begin to blur as to who is in charge. Will AI take over?

In April 2000, Bill Joy, cofounder of Sun Microsystems, captured the sense of this potential scenario in a widely quoted article in *Wired* magazine with the fear-filling title "Why the future doesn't need us," in which he warned,

> As society and the problems that face it become more and more complex and machines become more and more intelligent, people will let machines make more of their decisions for them, simply because machine-made decisions will bring better results than man-made ones. Eventually a stage may be reached at which the decisions necessary to keep the system running will be so complex that human beings will be incapable of making them intelligently. At that stage, the machines will be in effective control. People won't be able to just turn the machines off, because they will be so dependent on them that turning them off would amount to suicide. (Joy 2000)

This takeover scenario, while hypothetical, has been a topic of discussion for decades. The best minds contemplating the future of our species have recently placed AI near the top of the list of existential threats to the human race. The late Stephen Hawking warned, "Success in creating AI would be the biggest event in human history. Unfortunately, it might also be the last, unless we learn how to avoid the risks … humans, limited by slow biological evolution, couldn't compete and would be superseded by AI" (Sainato 2015). Elon Musk, CEO of Tesla and SpaceX, explicitly called AI "our greatest existential threat" (Sainato 2015). And even the optimistic futurist Ray Kurzweil concedes the risks of AI: "There is no purely technical strategy that is workable in this area, because greater intelligence will always find a way to circumvent measures that are the product of a lesser intelligence" (image 29.2) (Barrat 2013, 242).

Nick Bostrom, director of the Future of Humanity Institute at the University of Oxford, has argued that the future impact of AI is the greatest challenge humanity will face. He opens his book *Superintelligence: Paths, Dangers, Strategies* with a parable about a group of sparrows and an owl (Bostrom 2014). A group of sparrows, while laboring to build their nests, tweet with each other about how easy life would be if they just had an owl to help them build their nests, defend them from other predators, and look after their old and their young. With such a powerful assistant, they could live out their sparrow lives in leisure and prosperity. Inspired by this revolutionary idea, the sparrows can hardly contain their excitement. But then a single dissenter speaks up: a one-eyed sparrow with a fretful temperament named Scronkfinkle. The irritable Scronkfinkle cautions his companions, saying, "This will surely be our undoing. Should we not give some thought to the art of owl domestication and owl taming first, before we bring such a creature into our midst?" But his protests fall on deaf ears. The sparrows agree that owl taming will be very complicated but that they should go get their owl first and work out these details later. With this decision made, they fly off in search of their savior. Bostrom leaves the rest of the parable to the imagination of the reader—but his book, which includes a discussion of the darker side of artificial intelligence, is dedicated to Scronkfinkle.

Image 29.2: Artificial intelligence / robots. Intelligent machines and humans: cooperation or dominance? Artificial intelligence may be our greatest existential threat.

Bostrom believes that the progress of advanced AI may be Darwinian (Adams 2016; Bolstrom 2014). Once we create an intelligence that is superior to our own, with the ability to autonomously grow and learn through access to the internet, we may expect it to self-evolve and even become dominant over us—just as highly intelligent humans have dominated the biological world. An inferior intelligence, he argues, will always depend upon the benevolence of beings with superior intelligence for survival. The great apes now depend upon humans for their survival. Someday we may depend upon artificial intelligence for our survival.

How soon might this takeover happen? As we discussed in the last chapter, some experts think the technological singularity could happen within decades, but there are many skeptics who do not see this in the foreseeable future, if at all. Intelligent machines designed for a specific purpose that are able to learn without supervision can certainly do many things better than we can, but get these machines out of their comfort zones and they fail miserably. Ask an image-recognizing machine to drive a car or make a cup of coffee, and it won't know where to start. Could these intelligent machines be endowed with consciousness and free will to obtain some level of autonomy? We have no clue. We have so little understanding of our own consciousness and free will (if indeed we have free will at all!). How could we presume to embody AI with such traits? Any superintelligent AI will need to develop countless accurate theories about the world and how it works (Marcus 2017). This is a huge challenge; it took us billions of years of evolution to design our own intelligence, after all. AI may represent one of the major existential threats to our species, but the threat does not appear to be imminent.

UNFORESEEN RISKS

Concerns about diverse global threats seem to come and go like fads. When I was a kid in the United States in the 1950s, the threat of the day was nuclear Armageddon. We were taught to "duck and cover," hiding under our desks at school. Then, in the late 1960s, Paul Ehrlich's groundbreaking book *The Population Bomb* made Americans suddenly aware of the looming threat of overpopulation, leading to the Zero Population Growth movement. Now we are primarily concerned with climate change and terrorism. Much of the world is oblivious to what we have described here as the major potential threats of the twenty-first century—genetic engineering, nanotechnology, and artificial intelligence (GNR, R for "robotics"). But that is rapidly changing.

Knowing this history gives us cause for humility. It would be foolish to be overconfident in our ability to predict the future. The things we suspect to be problematic now may prove to be irrelevant—and we may soon be threatened by new existential risks that we have not imagined. A mere one hundred years ago, we had no computers, no genetic engineering, no nanotechnology, and no nuclear weapons. One hundred years from now, we will be living in a different world, dealing with very different threats.

STRATEGIES FOR DEALING WITH EXISTENTIAL THREATS

It appears that GNR, the three giants of technology that hold the greatest promise for a utopian future, ironically may also pose the greatest existential risks. Nuclear winter and climate change are certainly grave threats, but they are not likely to wipe out mankind, and an asteroid collision is unlikely to happen anytime soon.

The public is used to seeing these new technologies "go bad" in sci-fi movies and books, but for the most part these risks are not taken seriously. We humans tend to give a lot of attention to events that we have experienced before, while discounting unprecedented or imagined risks as unrealistic. We worry about things that have already happened—not things that *will* happen. If we want to protect ourselves, we need to change. We first must create public awareness about the potential risks of these new technologies so that we can allocate appropriate resources and priorities to defensive strategies.

There has been much attention given to prioritizing safety in the early stages of technological development, especially with AI. Once AI surpasses human intelligence, it will be out of our control. We won't be able to turn a switch to shut it off, just as a chimpanzee can't turn a switch to shut us off. Philosopher Nick Bostrom, renowned expert in the risks of AI, advocates that, for our future safety, we must create AI with goals that explicitly embrace human values (Bostrom 2015). But programming human values in AI is no easy task. The list of values is almost endless. And who decides what our values are? Each value is situationally dependent, and situations change. Bostrom believes the only way to create AI with human values is to use AI's superhuman intelligence to learn and understand what we value and modify these values as situations change. He believes this can be done. There is concern that some AI investigators, in a rush to create advanced AI, will not provide the necessary safeguards. Bostrom has advocated the crucial importance of AI investigators prioritizing safety in parallel with working toward advanced AI. A few centuries from now, our descendants may thank us for getting this perhaps most important of all accomplishments right.

ESCAPE FROM EARTH

The late cosmologist Stephen Hawking believed our species will need to populate a new planet within the next one hundred years if we are to escape the numerous existential threats we face here on planet Earth. "Life on Earth," he said, "is at the ever-increasing risk of being wiped out by a disaster such as sudden global warming, nuclear war, a genetically engineered virus or other dangers ... I think the human race has no future if it doesn't go into space" (Kazan 2009).

But leaving Earth will be no easy task. The staggering obstacles of time and distance during interstellar travel are currently prohibitive; even obstacles to travel within our own solar system are immense. Providing oxygen, nutrition, a temperate environment, and protection against solar radiation will be huge challenges. Ultimately, robots may prove to be our best pioneers.

Despite the many obstacles, interplanetary colonization may become a reality sooner than we imagine. If the history of life on Earth were compressed into one year, humans have been here for only thirty-nine minutes, and modern civilization for only one minute (beginning with the Agricultural Revolution). The colonization of new planets, and even new solar systems, could begin mere seconds from now.

CAN WE GET ALONG?

David Livingstone Smith's disturbing book *The Most Dangerous Animal*, which was mentioned in chapter 22, chronicles the gruesome horror of man's inhumanity to his fellow man and makes it clear that we are our own worst enemy. We are our greatest existential threat. If we are to survive as a species, we must overcome our tribal instincts and finally learn to get along.

Observing the repeated episodes of gun violence and mass shootings in the United States, we can't help but believe our world is becoming more violent. Most of us would be surprised to learn that the opposite is true. Psychologist Steven Pinker, in his enlightening book *Better Angels of Our Nature: Why Violence Has Declined*, promotes the thesis that our world has become less violent, less cruel, and more humane (Pinker 2011). Archaeological studies of gravesites have found that the proportion of people dying violent deaths from murder or tribal conflict was much greater in hunter-gatherer times than in agricultural and industrial societies. These trends have continued, and the frequency of death from violent causes since World War II has reached an all-time low. Pinker attributes these trends to the influence of commerce, the power of the state, and the rule of law. He also believes that the "humanitarian revolution," which has condemned our age-old habits of slavery, torture, and despotism, has moved us toward more tolerance and empathy as a species.

The number of democracies is increasing. In 1950, 14 percent of the world's population was governed by democracies; in 2002, this had grown to 65 percent (Christian, Brown, and Benjamin 2014, 299). Democracies promote egalitarian societies be emphasizing respect for the individual, contrasted with dictatorships and global corporations that are most often governed by self-interest.

At the same time, the world is getting smaller. Globalization has been stimulated by technological advances in transportation and telecommunications, and spans economic, cultural, political, and ideological dimensions. Facebook, Twitter and Instagram connect us through social media. Political globalization has been represented by supranational institutions such as the United Nations, the European Union, the International Criminal Court in the Hague, and the World Trade Organization, and may ultimately reduce the importance of nation-states. And nongovernmental organizations (NGOs), such as the Carter Center, Doctors without Borders, the Clinton Foundation, and Amnesty International, cross borders to provide assistance and help bring us all together.

Nations, at times, have shown the ability to cooperate in dealing with global threats. The Montreal Protocol, signed in 1987, helped save the ozone layer by phasing out the production and release of chlorofluorocarbons and other substances responsible for ozone depletion. It has been hailed as an example of exceptional international cooperation because of its widespread adoption and implementation. In 2015, 195 nations came together and committed to the Paris Climate Accord, agreeing to new standards for reducing carbon dioxide emissions. Unfortunately, the Trump administration withdrew the United States from the accord in 2017.

These trends give us some hope that we may be able to get along, but we have a long way to go. In an era of advanced technology and weapons of mass destruction, we are our own greatest enemy. We will need to overcome our innate tribal instincts and xenophobia and learn to cooperate and work together on an international scale if we are to avoid mutual destruction. Changes in innate destructive behavior could be mediated through cultural influences or possibly through genetic engineering. Neuroscientists have recently identified genes in mice and humans associated with an aggressive nature, raising the possibility that such behavior could be mitigated through genetic engineering (Betuel 2018). Our ability to "evolve ourselves" may provide opportunities to change our behavior so that we may be able to come together to solve our many existential problems.

WHEN THE SUN DIES

In this chapter, we've discussed a number of existential threats that could imperil life on our planet by the end of the century. As difficult as it is to make these predictions, once we look beyond the end of this century, given the pace of technological change, predictions become mere speculative science fiction.

However, if we consider the very remote future—long after our species has evolved into something entirely new and different—our knowledge of the cosmos allows us to predict a bit of what the future holds. Someday our sun will die. The sun produces energy and light through nuclear reactions, fusing hydrogen to form helium. This process is gradually speeding up and has caused the sun's energy output to increase by 30 percent since its origin 4.5 billion years ago. In the next billion years, the sun will become another 10 percent hotter, and in 3.5 billion years it will be 40 percent hotter. Earth's oceans will boil, and all of the water vapor will drift away into space. Earth will become a hot, dry world not suitable for life. Finally, in five billion years, as the sun finally dies, it will expand into a red giant, ultimately engulfing Mercury, Venus, and Earth (Williams 2016). We can only hope that our descendants will have long since left for a new home.

Meanwhile, the universe continues to expand. The universe has been expanding since the big bang, and cosmologists have long believed that the forces of gravity would eventually reverse this expansion, causing the universe to collapse into a new singularity—perhaps to start a new universe with another big bang. But new observations have found that the expansion of the universe is not slowing down but in fact is *accelerating* (Moskowitz 2012). Cosmologists now believe, as Einstein once predicted, that there is a force permeating the universe that acts in opposition to gravity and pushes objects apart. This means that over the ensuing hundreds of billions of years, our universe will continue to expand and will eventually spread out into nothingness. Star formation will cease, and ultimately there will be no stars left and no planets in orbit. Where will our descendants live then?

Don't get depressed, dear reader. Let's try to look on the bright side. Our universe is still in its infancy. We are alive because this is an age in which our universe holds trillions of galaxies, each with billions of stars, and who knows how many planets with an abundance of habitats suitable for complex life. As our universe gets older and continues to expand, life may run its course and no longer be possible. Let's be thankful, then, that we have the good fortune to live in the springtime of our universe. And perhaps when our own universe is no longer habitable, other universes will be in their springtime.

CHAPTER 30

A NEW SPECIES OF MAN

We have directed the evolution of so many animal and plant species. Why not direct our own? Why wait for natural selection to do the job when we can do it faster and in ways beneficial to ourselves?

—Peter Ward, paleontologist

All **species** evolve, and all species eventually become extinct. Are we any exception? What might a new species of man be like? We don't know the answers to these questions, but it seems clear that the future of our species will be driven by technology. Our evolution will no longer be controlled by the slow, random process of biological evolution. Genetic engineering and artificial intelligence will shape the future of our species, and we will potentially be in control of these technologies. As Juan Enriquez and Steve Gullans have so famously said in their prescient book by the same title, we will be evolving ourselves (Enriquez and Gullans 2015).

EVOLUTION OF NEW SPECIES BY NATURAL SELECTION

All species, including us, *Homo sapiens*, can potentially evolve into new species by splitting into two or more groups that are separated geographically so they can no longer interbreed. The groups then evolve separately and independently, and at some point, they become genetically different enough that they can no longer interbreed even if they should be allowed to come back together. At this point the populations are considered separate species. Being able to breed with one another, but not with individuals in other groups is a major defining characteristic of a species. This process of splitting and evolving into two or more new species is called **cladogenesis**.

A species can also become a new species—without splitting into two separate species—by gradually evolving new traits as it adapts to a changing environment. It becomes so different that we call it a new species. In this process, called **anagenesis**, the species remains together so that all members continue to interbreed, keeping the genome relatively homogeneous even as the entire population changes together. This is what has happened to *Homo sapiens*. Although we have scattered to the four corners of the globe, we remain connected and continue to interbreed with one another. This is why we have not evolved into separate species and appear to be unlikely to do so. Remember from chapter 19 that at one time, as recently as fifty thousand years ago, there were four human (*Homo*) species across Europe and western Asia: the Neanderthals, the cave-dwelling Denisovans in Siberia, *Homo floresiensis* (the midget "Hobbits" of Indonesia), and we, *Homo sapiens*. This speciation occurred by cladogenesis and was facilitated by geographical separation—but for humans in the interconnected modern world, evolution in this way no longer seems feasible. Now, of course, we are the only *Homo* species left. A new human species is unlikely to evolve by cladogenesis, but evolution by anagenesis, with our entire species shifting to become something new, is already happening today.

There is a widespread misconception that for humans, biological evolution has slowed to a crawl and is no longer relevant. In fact, the opposite is true. Human evolution has actually accelerated. By examining data from an international genomics project, anthropologists have estimated that over the past five thousand years, humans have evolved new genes with beneficial adaptations at a rate about one hundred times faster than at any other period in human evolution (Mattmiller 2007; Hawks et al. 2007). This has been driven by rapid changes in our environment. We have had to adjust accordingly. Many of our new genes have evolved as defenses against microorganisms responsible for epidemic diseases and as adjustments to dietary changes. More than two dozen new genes have been identified related to malaria resistance; and, as was discussed in chapter 21, new genes have evolved to allow adults to produce the enzyme lactase, enabling the digestion of cow's milk in response to domestication of animals and the introduction of milk into the adult diet.

As human evolution has accelerated, we might expect *Homo sapiens* to evolve into a greatly changed and enhanced new species by natural selection through the process of anagenetic change. But there are two forces working against this. Through advances in public health, medicine, and technology, we can now keep more people alive longer than ever before. Those who might otherwise die in famine, by virulent disease, or from their own crippling genetic mutations are now kept alive long enough to bear their own children. Their genes remain in the gene pool. Life is no longer survival of the fittest. But there is an even more important reason why natural selection is unlikely to be the driving force in the future evolution of our species. We are dealing with new paradigms. Biological evolution through natural selection has brought us to where we are today, but despite its accelerating course, it is being transcended and outstripped by fast-moving cultural and technological evolution. The evolution of our species going forward will almost certainly be driven by genetic engineering and artificial intelligence.

DESIGNING OUR SUCCESSORS

Genetic engineering has advanced to the point where it is widely used to guide the evolution of plants and animals to meet our needs. It is now poised to intervene in the process of human evolution (image 29.1). We are on the threshold of evolving ourselves. We have seen how in vitro fertilization (IVF) is used to prevent the transmission of inherited genetic diseases by selecting healthy embryos for implantation while discarding those with harmful mutations. This preimplantation genetic screening has already been used to select the sex of offspring, and it won't be long before embryos can be chosen for other desirable traits, such as intelligence, longevity, and athleticism. We need only to learn the genetic correlates of these traits to begin making such selections.

The next logical step is the modification of embryos with genetic editing, using new editing tools such as CRISPR-Cas9, to remove harmful diseased genes and replace them with healthy ones (image 30.1). Once we've begun actively editing the human genome in this way, it may not be long before we begin replacing normal genes with supergenes that enhance specialized favorable traits. Let's give a hypothetical example. Scientists may be able to identify and synthesize genes responsible for producing cones that sense ultraviolet light in the extraordinary eyes of the mantis shrimp. They could then insert these genes into the human embryo with new gene-editing tools and enable the developing embryo to see ultraviolet light. The possibilities are endless.

Scientists currently have much of the technology needed to modify the genome of human germ cells and embryos, but the complexity of our genome creates a number of obstacles. Genes impact multiple traits, and likewise traits are usually coded by multiple genes. Tinkering with one gene can have the unintended consequence of altering unspecified traits, and any of these changes would be transmitted to future generations, becoming a permanent part of the human genome. For obvious reasons, this is currently considered unethical, and in most countries it is illegal to perform gene editing on human embryos.

Despite these technical and ethical constraints, it won't be long before we transition from altering our genome for the prevention of Alzheimer's disease to altering our genome to make us smarter. We will program our progeny to be stronger and faster, to live longer, and to be healthier. We will transcend the course of biological evolution from the blind, slow process of natural selection to human-directed, genetically engineered rapid change. We, *Homo sapiens*, will soon be in control of our destiny. Momentum for these technologies will continue to build. While there will be ethical debates and challenges and great potential for abuse and misuse, any society that tries to delay these uses of genetic engineering will ultimately fall behind in health care, longevity, intelligence, and human capability. It is a recipe for societal obsolescence. The potential benefits of human genetic engineering are too great to be ignored.

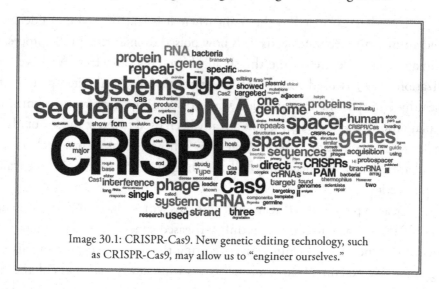

Image 30.1: CRISPR-Cas9. New genetic editing technology, such as CRISPR-Cas9, may allow us to "engineer ourselves."

ARTIFICIAL INTELLIGENCE AND CYBORGS

Many futurists believe that the destiny of humanity lies not with our genes but with artificial intelligence (AI). If and when artificial intelligence reaches human-level intelligence, it will be able to learn independently and will rapidly acquire intelligence superior to that of its creator. It could very well become dominant. Whether this future superbeing will be our friend or foe is a subject of fierce debate and will be dependent upon how we design the technology.

Would this new dominant artificial superintelligence be considered a new species? By our current definitions, certainly not; technological beings are not currently considered alive. But we will be dealing with entirely new paradigms, and in the future, life may well be nonorganic. We may create an entirely new definition and classification of life, labelling AI as a new nonorganic domain. A new technological species would certainly not be in the genus *Homo*, as it would not be human—and yet it would still be considered a descendant of our lineage, since we will be the ones who created it. This all sounds like science fiction, but there are many experts who believe our future will indeed be dominated and controlled by intelligent machines and robots (Bostrom 2014; Kurzweil 2006; Rajendra 2017).

If superintelligent AI does not emerge to dominate the human race, we will certainly learn to augment ourselves with AI in entirely new ways. Many experts, including futurist Ray Kurzweil and philosopher Nick Bostrom, believe that our future lies in the merging of man with machine—that we will become **cyborgs**. This change has already begun. Anyone who has had a medical device implanted into his or her body—artificial joints, cochlear implants, lens implants, artificial hearts, cardiac pacemakers, and brain stimulators—is already a cyborg. It won't be long before devices are implanted not just to treat medical problems but also to enhance physical and mental capabilities.

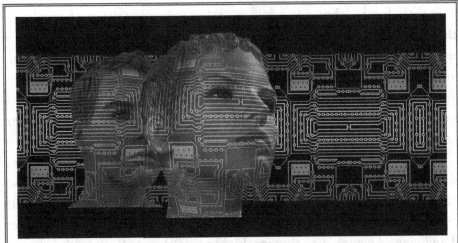

Image 30.2: Brain–computer interface (BCI). New technologies may facilitate the merger of man with machine, transforming the next species of man into a cyborg.

The culmination of the cyborg age will be the interfacing of human brains with computers. We humans have already greatly expanded our mental capacities with cell phones and the internet, giving ourselves nonstop immediate access to almost limitless data. Scientists are now working to connect our brains directly to computers using brain–computer interface (BCI) technologies (image 30.2). Researchers have implanted electrodes directly into the brains of patients with paralyzed limbs, allowing the conversion of their brain waves into electronic signals. These patients are then able to "think" a command, and their brain wave signals can be transmitted wirelessly to control a robotic arm or pick out letters on a keyboard (Rao and Wu 2017). Elon Musk's company Neuralink has developed a computer–brain interface consisting of microelectrodes implanted in the motor cortex of the brain, allowing a patient with amyotrophic lateral sclerosis (ALS) to type with her mind (Levy 2017). Ultimately, the goal of BCI is for the brain to "talk" to a computer directly and have the computer talk back to the brain. This direct communication is not an easy technology to develop, but with enough time, great progress is sure to happen.

Kurzweil envisions an ultimate brain–computer interface using microscopic robots (nanobots), which will be injected into the bloodstream and circulate through the brain, where they will be able to communicate with external AI wirelessly. The merger with AI will provide the brain with unlimited access to information and the capability to operate orders of magnitude faster than the human brain acting alone. This partnership with artificial intelligence at the very center of our consciousness will change what it means to be human in ways we cannot yet imagine.

Many see these future cyborgs as posthuman, but remember that we have already merged with machines using our many medical devices. We are still human. And even when our bloodstreams are filled with nanobots—initially for medical purposes, and later to interface directly with neurons, enhance our senses, and assist our memories—we will still be human. We will still share the same genome, and we will remain the same *Homo sapiens* species, though greatly enhanced. We may need to overcome biases against our nonorganic parts. Nick Bostrom, director of the Future of Humanity Institute, put it well: "Substrate is morally irrelevant. Whether somebody is implemented on silicon or biological tissue, if it does not affect functionality or consciousness, it is of no moral significance." He went on to say, boldly, that "Carbon-chauvinism is objectionable on the same grounds as racism" (Bostrom 2005).

Investigators developing BCI technology assume that human brains will remain the command and control center, directing machines to do their will. But this is by no means a given. The late Stephen Hawking warned, "[We] urgently need to develop direct connections to the brain, so that computers can add to human intelligence rather than be in opposition." (Kurzweil 2006, 309). The great concern is that AI will morph into a superintelligence that will not join with us but rather replace us. It is to be hoped that as our brains merge with AI, we can integrate our values and priorities into our new hybrid species. If the new cyborgs share our values, and if these values offer respect for all life on our planet and beyond, we have the opportunity to create a new future together.

COULD HUMANS DIVERGE INTO SEPARATE SPECIES?

Conventional wisdom has held that humans are not likely to evolve into two separate species because international travel and communication have brought us close together, facilitating widespread interbreeding. Genetic engineering and artificial intelligence could change this. As genetic engineering becomes more sophisticated and allows us to design ourselves, and as we begin to merge with machines, not everyone will participate equally. Only peoples in technologically advanced nations may have access to gene editing and computer brain interfaces, and in those nations, access will probably be limited by social and financial status. Some will opt out for religious or moral reasons, and some cultural groups will reject the technology altogether. It is very possible that we may develop communities in which genetically engineered humans with computer–brain interfaces are nearly universal, and other communities in which these technologies are not available or are rejected almost entirely. As these genetically engineered kids grow up with their uniquely good looks, high IQs, prolonged lifespans, and computer enhanced brains, they may choose to mate preferentially with others of their kind and over time create social groups with enhanced physical and mental qualities, and enhanced wealth and influence (augmented, no doubt, by the wealth that they will inherit from the parents who had enough money to engineer their offspring in the first place). If the barriers for mating outside the group become great enough, this new genetically engineered, computer-enhanced community could evolve into a new and separate superelite species.

The implications are profound. The rich and connected could access greater intelligence, longevity, and physical prowess, while the poor would be left behind—broadening the huge economic and social divide that exists today into a more fundamental biological divide. As people become more alienated from one another, eugenics-like movements could dominate the planet, shaking the foundations of democracy and fostering totalitarian states in which dutiful soldiers and obedient workers serve the engineered superelite. When we look at our long human history of ethnic strife and genocide, it is not hard to imagine the scale of potential conflict between two diverging human species. After all, the divide between species is much greater than the divide between ethnic, racial, and religious groups. How would a new genetically engineered, brain-enhanced *Homo* species treat an inferior *Homo sapiens* species? Remember what happened to the Neanderthals.

If we are to preserve human rights and the ethical and moral fabric that we value in our current societies, we must avoid the creation of separate species—one of which would surely be dominant, the other subordinate and possibly enslaved. To ensure this future, we must guarantee a broad access to genetic engineering and other technologies that will bring new and enhanced traits to our species, regardless of economic and social status. We must strive to maintain an egalitarian society, even as we evolve into something new. This will not be an easy task, and our success is by no means assured. But the responsibility is ours. We will be evolving ourselves. As we evolve into a new species of man, we must remain inclusive. We must evolve into a single species. Our humanity depends upon it.

EPILOGUE

We are floating in a purposeless cosmos, confronting the inevitability of death, wondering what any of it means. But we're only adrift if we choose to be.

—Sean Carroll, in *The Big Picture*

On Valentine's Day 1990, the *Voyager 1* spacecraft had just finished its journey across our solar system and was ready to enter interstellar space. Although it was not in the plans, cosmologist Carl Sagan, a member of the Voyager team, suggested pointing *Voyager*'s camera back toward Earth for a final look (Landau 2015). The photograph it captured was immediately iconic: An image of a pale blue dot—a barely visible speck that we call home. This image has continued to inspire awe and wonderment ever since it was taken (image E1). Sagan expressed that wonderment in his book *Pale Blue Dot: A Vision of the Human Future in Space*:

> That's here. That's home. That's us. On it everyone you love, everyone you know, everyone you ever heard of, every human being who ever was, lived out their lives ... There is perhaps no better demonstration of the folly of human conceits than this distant image of our tiny world. (Sagan 1994)

The majesty of the photo captures Earth as a mere "mote of dust, suspended in a sunbeam," hanging adrift in our vast solar system (Sagan 1994). It highlights the incredible smallness and vulnerability of our fragile planet, and the precariousness of humanity and all other life on Earth. Ever since we humans began to think and reason and reflect upon ourselves, we have asked, "Why are we here?" There is a sense that we must have a purpose, but at the same time we are overwhelmed by our feeling of insignificance. We are but a tiny speck on one planet circling one star, and we are here for only a few heartbeats. Can any one of us really matter?

But we do matter. Each of us is a thinking, feeling being searching for meaning in his or her own life. The search for meaning is elusive—so elusive that the pursuit of answers is often a subject of parody. In Douglas Adams's novel *The Hitchhiker's Guide to the Galaxy*, a giant computer named Deep Thought is asked to find the "Answer to the Ultimate Question of Life, the Universe, and Everything." After 7.5 million years of calculation, Deep Thought spits out the obvious answer: 42! (Adams 1980). The British comedy film *Monty Python's the Meaning of Life* gave us another answer. The host moderator is handed a note in the final scene, from which she announces the meaning of life: "Try to be nice to people, avoid eating fat, read a good book every now and then, get some walking in, and try to live in harmony with people of all creeds and nations." (IMDb 2017). The meaning of life remains elusive, but a little humor sometimes helps.

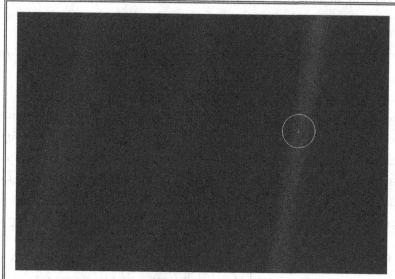

Image E1: *Pale Blue Dot*. This iconic image of Earth, taken from *Voyager 1* in 1990 as it looked back from more than four billion miles away, shows our planet as a tiny speck in the darkness of space. The image captures the vulnerability of our fragile planet and the precariousness of life on Earth as we enter the next stage of our evolution.

The French philosopher Albert Camus famously rejected scientific and religious answers to the question of the purpose of life and concluded there simply is no answer. He called the condition of human existence absurd, as we have a human need to search for the meaning of life and yet determining that meaning is an impossible task because the finite human brain is incapable of fully comprehending the infinite universe. His essay *The Myth of Sisyphus* chronicles the Greek legend in which Sisyphus is cursed by Zeus to spend eternity pushing a huge bolder up a mountain only to have it roll back down again. The metaphor for life without purpose is clear. But Camus gives us some hope at the end of his essay: "But Sisyphus teaches the higher fidelity that negates the gods and raises rocks. He too concludes that all is well. This universe henceforth without a master seems to him neither sterile nor futile. Each atom of that stone, each mineral flake of that night-filled mountain, in itself forms a world. The struggle itself toward the heights is enough to fill a man's heart. One must imagine Sisyphus happy" (Camus 1995, 124).

While Camus believes that we are incapable of understanding the purpose of life, he does not believe that life is meaningless or not worth living. We can live in the moment and find pleasure, happiness, and meaning in our own personal world.

THE MEANING OF OUR LIVES

So how can we find this meaning? Questions about the meaning of our lives have been pursued by theologians and philosophers for millennia. Theologians seek to understand life's purpose through God and their religion, but this is of little help to nonbelievers. Philosophers have had difficulty in dealing with life's purpose because they have not had an advanced understanding of psychology and human nature. Contemporary psychologists have helped us to understand how evolution has shaped human nature, and this understanding has given us a framework with which to address the existential questions concerning the meaning and purpose of our individual lives.

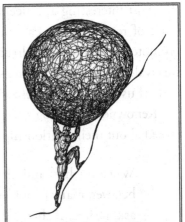

Image E2: Sisyphus. Albert Camus's essay *The Myth of Sisyphus* describes a Greek legend in which Sisyphus is cursed by Zeus to spend eternity pushing a rock up a mountain only to have it fall back again. Sisyphus symbolizes man's futile search for meaning.

Human nature has been shaped by natural selection for millions of years, dating back to the hunter-gatherer days of our hominin ancestors on the plains of Africa. Natural selection has given us selfish traits that help us in the struggle for resources, status, prestige, and mates, and has also fostered traits that have made us industrious and creative. As we joined together in large societies, we evolved prosocial traits that promoted cooperation with other members of our group, enhanced social attachments, and gave us the desire to seek causes greater than ourselves. The need to connect with something larger than ourselves gives us a sense of grand purpose and promotes self-sacrifice for the betterment of the group. In contrast, as we have seen, tribal instincts, while promoting cooperation within the group, foster xenophobic and hostile behavior toward those outside the group. Both history and current events are riddled with examples of ethnic, racial, and religious conflict and genocide. We all have angels and demons within us.

Eric Fromm, in his classic bestseller *The Art of Loving*, wrote that "love is the only sane and satisfactory answer to the problem of human existence." (Fromm 1956, 123). Erotic love, romantic love, maternal love, and brotherly love have been bred into us through our evolutionary past. Erotic love is a trick of nature designed to help us propagate our genes, but it is short-lived. Romantic love is the love between two people whose lives are deeply intertwined and who care for and trust one another. Romantic love likely evolved to facilitate child rearing, as was discussed in chapter 11. Maternal love is unconditional love between a mother and child, which facilitates survival of offspring and propagation of parental genes. And perhaps the most interesting and deeply human kind of love is brotherly love, which carries with it a sense of responsibility, care, and respect for other human beings. Brotherly love evolved to promote cooperation and altruistic behavior between members of a group to promote the group's survival. We are social animals and have inherited essential needs for strong relationships, and each of us must satisfy these social needs to find fulfillment and meaning in life.

Renowned cosmologist Carl Sagan died in 1996. Some years later, his wife, Ann Druyan, talked about their relationship and the meaning it brought to their lives:

> When my husband died, because he was so famous and known for not being a believer, many people would come up to me—it still sometimes happens—and ask me if Carl changed at the end and converted to a belief in an afterlife. They also frequently ask me if I think I will see him again. Carl faced his death with unflagging courage and never sought refuge in illusions. The tragedy was that we knew we would never see each other again. I don't ever expect to be reunited with Carl. But, the great thing is that when we were together, for nearly twenty years we lived with a vivid appreciation of how brief and precious life is. We never trivialized the meaning of death pretending it was anything other than a final parting. Every single moment that we were alive, and we were together was miraculous—not miraculous in the sense of inexplicable or supernatural. We knew we were beneficiaries of chance … That pure chance could be so generous and so kind … That we could find each other, as Carl wrote so beautifully in *Cosmos*, you know, in the vastness of space and the immensity of time … That we could be together for twenty years. That is something which sustains me and it's much more meaningful … The way he treated me and the way I treated him, the way we took care of each other and our family, while he lived. That is so much more important than the idea I will see him someday. I don't think

I'll ever see Carl again. But I saw him. We saw each other. We found each other in the cosmos, and that was wonderful. (Druyan 2003)

Ann Druyan conveys the miraculous beauty of their loving relationship, which was made ever more precious by its fleeting nature. We as a species and we as individuals have arrived on this Earth by chance and are so lucky to be here. We are so lucky to live on a miraculously beautiful, welcoming planet, and we are so lucky to have inherited social behavior as part of our human nature, allowing us to foster meaningful human relationships. From these relationships, we derive much of our meaning in life.

We also have inherited as part of our human nature a basic drive to create and make things happen (White 1959). The Hungarian-born psychologist Mihaly Csikszentmihalyi described the ultimate experience of creating—a highly focused mental state in which there is such complete absorption in an activity that nothing else enters awareness. He calls this "flow." It is the feeling athletes describe as "being in the zone." Csikszentmihalyi has described a process common among successful, happy people—whether they be poets, dancers, novelists, physicists, biologists, or psychologists—called "vital engagement," characterized by experiences of creativity, flow, and dedication to a cause greater than oneself (Haidt 2006 223–24). Many find this vital engagement in their work: work that is more than a job, more than a career—work that serves a larger cause and is a calling. Others find vital engagement outside of work through their churches, through volunteer work, and through hobbies. These activities foster relationships, provide experiences of flow, and give us a sense of purpose larger than ourselves—an individual purpose that is unique to each of us.

Meaning in our lives is tied to personal happiness. The traditions of Buddha and the Stoic philosophers of ancient Greece and Rome teach us that peace and happiness come from within and cannot be found by trying to make the world conform to our desires. We should focus on what we can truly control: our own thoughts and reactions. Shakespeare put it best: "There is nothing either good or bad but thinking makes it so."

So how do we find meaning in our own lives? It comes from within and without. We must find peace from within. We must learn to control what we can, accept what we can't, and learn to know the difference. And we must cultivate creativity and meaningful relationships from without. Psychologist Jonathan Haidt summarized this in his best-selling book *The Happiness Hypothesis*: "Just as plants need sun, water and good soil to thrive, people need love, work, and a connection to something larger. It is worth striving to get the right relationships between yourself and others, between yourself and your work, and between yourself and something larger than yourself. If you get these relationships right, a sense of purpose and meaning will emerge." (Haidt 2006, 239).

It is up to us, as individuals, to create our own meaning and purpose in life in the brief time that we are here on Earth. No one else can do this for us. As Sean Carroll said in the epigraph at the beginning of the chapter, "we're only adrift if we choose to be."

Is There a Larger Purpose?

The search for meaning in our own lives is an individual journey. Each of us has to find his or her own way. But is there a larger purpose? Why do the Earth, life on Earth, our solar system, and all the galaxies exist? Does the universe have a purpose? Theologians and philosophers have searched for answers to these questions through the ages. As theism has given way to naturalism, our confidence that there is a purpose to the universe has begun to wane. I think Camus had it right. Our search for a purpose in the universe is absurd. We are finite beings with finite brains, and it is impossible for us to understand a potentially infinite universe, much less understand its purpose.

But maybe we need to ask a more limited and more specific question—what is the purpose of *human* existence? After all, the question of purpose is very subjective and dependent upon who is asking the question. Meaning, morality, and purpose are not inherent in the universe—this is one of humanity's oldest misconceptions—but rather are human-based concepts. Science cannot answer these questions for us. The answers to the questions of meaning and purpose are not found in the outside world but rather within us.

Darwin has provided an explanation for how we came to be here through the process of random variation and natural selection. While Darwin's evolution creates an illusion of purpose with the evolution of complex creatures highly adapted to their environments, the evolutionary process is random—driven by random variation in the traits of offspring and altered by chance mass extinctions that change the course of evolution. There is no orderly march to perfection and no guarantees of a perfect outcome. Darwin's natural selection entails a bloody struggle for survival with immeasurable suffering and casualties left in its wake. Only the fittest survive to reproduce and pass their genes to the next generation. Throughout the long evolutionary journey, the sole purpose of living organisms has been to survive and reproduce. But with us, this is now changing.

As has been made clear throughout the final chapters of this book, we human beings have entered an age of rapid cultural and technological evolution that has surpassed the slow pace of biological evolution and has completely transformed our societies. We are embarking on a new paradigm in which, through our advanced technologies, we will be directing the evolution of life on our planet, including human life. Genetic engineering and artificial intelligence have the potential to promote health and longevity, enhance mental and physical abilities, and advance human cooperation and fulfillment. At the same time, our technology has become so powerful that it has the potential to destroy life on our planet. We now have an awesome responsibility. We can leave behind the ruthlessness, the callousness, and the suffering of natural selection and survival of the fittest, and act with purpose to shape the future course of life on our planet. Life, which has had no goal beyond survival and reproduction throughout all its long history, may have finally found a purpose in us. As Enriquez and Gullans conclude in *Evolving Ourselves*,

> We embark upon the greatest of all human adventures: the creation of our own successors. For better and worse, we are increasingly in charge. We are the primary drivers of change. We will directly and indirectly determine what lives, what dies, where, and when. We are in a different phase of evolution; the future of life is now in our hands. (Enriquez 2015, 261)

We are indeed in the next phase of our evolution. Our success in meeting the challenges we face will depend not just on our vision and our efforts but also on the ability of independent societies and independent nations to work together in synergistic cooperation. Peter Corning, in his insightful book *Synergistic Selection: How Cooperation Has Shaped Evolution and the Rise of Humankind*, describes how the evolution of life on our planet has been characterized by greater and greater levels of synergistic cooperation and growing complexity (Corning 2018). At the microscopic level, bacteria cooperated in large numbers with quorum sensing to simultaneously release toxins to overwhelm a host. Individual cells joined together in multicellular organisms and differentiated to perform specialized tasks with enormous adaptive advantages for the new, larger complex organism. *Homo sapiens* joined together in synergistic cooperation and division of labor to form larger and more complex societies, creating advanced civilizations with great adaptive advantages. We are now on the precipice of the next phase of our evolution—the potential transition to a global society created by cooperation between nations and directed by global governance. Only through global cooperation will we be able to manage and control our rapidly advancing technology and address the global existential problems of climate change, nuclear holocaust, and more.

The outcome of this transition is far from certain. Many advanced civilizations in faraway galaxies may not have been successful in navigating this change. But there is hope that we can come together. Despite the recent wave of populist, isolationist, and xenophobic trends in national and international politics, trends over the last few centuries, and over the last millennia, have generally moved in the direction of globalization and international cooperation. But we have a long way to go. We will need to overcome our tribal instincts and xenophobic behavior.

Potentially this could be facilitated through genetic engineering, which could enhance our cooperative traits and suppress our aggressive traits. Artificial intelligence with brain–computer interfaces could provide opportunities for direct brain-to-brain communication and lead to unprecedented opportunities for cooperation. Strong leadership must emerge to bring all this together. Let's hope we can reach a "tipping point" at which we can all come together in a global society with global governance to solve our existential problems and steer us toward a better future. We are in control and have inherited the responsibility to ensure the future for ourselves and for all life on our planet. This is our challenge. This is our purpose. This is why we are here.

GLOSSARY

access consciousness: A subjective awareness that results from accessing internal information, using it in thought and reasoning, communicating it to others, and using it to control behavior. This concept was popularized by philosopher Ned Block.

actin: A protein that forms (together with myosin) the contractile filaments of muscle cells and is also part of the cytoskeleton of eukaryotic cells.

adaptive radiation: A process by which organisms diversify rapidly into a multitude of new forms as a result of major changes in the environment.

aerobic capacity (VO_2 max): An animal's capacity or ability to consume oxygen during maximal exercise.

aerobic metabolism: A form of cellular respiration that uses oxygen in the production of ATP.

algae: Single or multicellular photosynthetic organisms inhabiting freshwater or saltwater and lacking true stems, roots, and leaves characteristic of most plants. Some are classified as marine plants, and some are classified as protists.

algorithm: A step-by-step procedure or set of rules for solving a problem or accomplishing some end. This is often done by a computer, but the analogy has been extended to the brain.

allele: One member of a pair of genes on homologous chromosomes that have the same location and code for the same trait but usually code for a different variant of that trait.

alternation of generations: The life cycle of algae and plants that encompasses two generations, a sporophyte (diploid) generation and a gametophyte (haploid) generation.

amniotes: A group of tetrapod vertebrates that house their developing embryo inside an amniotic sac. This includes birds, reptiles, and mammals.

amphiphilic molecule: A bipolar molecule with a polar water-soluble group attached to a nonpolar water-insoluble hydrocarbon chain

anaerobic metabolism: a form of cellular respiration that generates ATP without using oxygen.

anagenesis: Evolution of a species without branching the line of descent. This is distinguished from cladogenesis, in which evolution occurs as a result of branching into separate species.

angiosperms: Vascular plants that have flowers and produce seeds. Angiosperms include herbaceous plants, shrubs, grasses, and most trees.

anisogamous sex: Sexual reproduction in which the male and female gametes are different in size and appearance.

Anomalocaris: A genus of large shrimplike creatures, now long extinct, that were the dominant predators in the Cambrian period.

anticodon: A three-letter code on transfer RNA that is complementary to a three-letter code (codon) on messenger RNA that specifies a specific amino acid.

apoptosis: Programmed cell death, which, in contrast to cell death from necrosis, is a controlled process that does not release toxic material into the environment. Apoptosis provides health maintenance by eliminating old cells, cancerous cells, unhealthy cells, and cells infected with viruses.

archaea: one of three domains of life consisting of single-celled microorganisms. The other two domains are Bacteria and Eukaryota.

archosaurs: A group of reptiles that survived the end-Permian extinction and gave rise to the dinosaurs, pterosaurs (flying reptiles), and crocodiles.

arthropod: A phylum of invertebrate animals characterized by an external skeleton (exoskeleton), a segmented body, and jointed appendages.

artificial intelligence: Computer systems able to perform tasks that normally require human intelligence.

artificial neural network: A type of artificial intelligence that uses hardware and software designed to mimic the function of the human brain.

asexual reproduction: Reproduction in which offspring arise from a single parent organism and are genetically identical to the parent organism. The offspring are called clones of the parent organism.

ATP (adenosine triphosphate): A high-energy nucleotide molecule that serves as the universal energy currency for the cell.

ATP synthase: A protein enzyme embedded in membranes of mitochondria and chloroplasts, and in cell membranes of Bacteria and archaea, that catalyzes the conversion of electrochemical energy from a proton gradient into ATP. This enzyme is universal to all life.

Bacteria: one of three domains of life consisting of single-celled microorganisms. The other two domains are Archaea and Eukaryota.

bacterial conjugation: The transfer of genetic material from one bacterial cell to another by direct cell-to-cell contact.

banded iron formation: Distinctive ancient sedimentary rocks containing layers of iron oxides formed when oxygen, released by cyanobacteria, reacted with iron and precipitated as iron oxides to the ocean floor.

big bang theory: The theory that the universe originated billions of years ago in an explosion from a singularity of nearly infinite energy density.

binocular rivalry: A phenomenon in which a subject is exposed to separate images for each eye and responds by having alternating subjective images of each.

biological carbon cycle: A process by which CO_2 is removed from the atmosphere through photosynthesis and added to the atmosphere through animal and bacterial respiration and organic decomposition of dead animals and plants.

bisexuality: Sexual attraction to both males and females.

brain plasticity: The ability of the brain to alter its structure and function in response to external stimuli.

broadcast spawning: An early form of sexual reproduction in which marine animals release massive amounts of sperm and eggs into water, where fertilization occurs.

bryophytes: Nonvascular land plants, which includes mosses, liverworts, and hornworts.

Burgess Shale: A remarkable historic fossil field in British Columbia with numerous well-preserved fossils of marine life from the Cambrian Period.

Calvin cycle: A metabolic pathway that is part of photosynthesis, in which carbohydrates are produced from carbon dioxide and water. This pathway takes place in chloroplasts of algae and plants.

Cambrian explosion: The sudden appearance in the fossil record of diverse macroscopic marine life dated to the Cambrian period.

carbon fixation: The process by which living organisms incorporate inorganic carbon dioxide in the synthesis of organic molecules.

Cartesian dualism: The doctrine that the brain and mind are separate entities acting in parallel and that their interaction is not well understood.

cellular respiration: A series of metabolic processes occurring in cells that converts chemical energy from the environment (from organic or inorganic molecules or sunlight) into usable energy in the form of ATP (adenosine triphosphate).

central dogma of molecular biology: The process by which the genetic code of DNA is used to produce proteins. This is universal to all cells.

chemical evolution: The formation of complex organic molecules from simpler molecules through chemical reactions in prebiotic oceans. This was the first step in the origin and evolution of life.

chemiosmosis: A process in cellular respiration in which electrochemical energy from a proton gradient across a membrane is used to produce ATP.

chlorophyll pigments: Green pigments in plants, algae, and cyanobacteria that transform the energy from sunlight into activated electrons as part of the process of photosynthesis.

chloroplast organelles: Organelles in plant and algae cells in which photosynthesis is performed.

choanoflagellate: A unicellular eukaryote organism characterized by a single flagellum surrounded by a collar of microvilli. Choanoflagellates are the closest living unicellular relative of animals and are descendants of the common unicellular ancestor of all animals.

chordate: An animal of the phylum Chordata whose members have a hollow dorsal nerve chord or notochord and bilateral symmetry. This phylum includes all vertebrates and two nonvertebrate subphyla.

chromosomes: Threadlike structures in the nucleus of cells consisting of DNA wrapped in a protein coat, which form when the cell is preparing to divide.

clade: A group of organisms that share a most recent common ancestor—a monophyletic group. The organisms of a clade are more closely related to each other than they are to any other group.

cladogenesis: Evolution of new species in which an ancestral species diverges or branches into two or more separate species.

cloning: Asexual reproduction in which the offspring are genetically identical to the parent.

closed circulatory system: A circulatory system in which blood is contained within the vascular system at all times.

codon: A three-letter code on DNA formed by three adjacent nucleotides that specifies an amino acid to be used in protein synthesis.

cones: Photosensitive cells in the vertebrate retina that sense light and generate signals that are transmitted to neurons. Cones can distinguish color but are not as sensitive as rods.

contingent evolution: Evolution that is influenced or constrained by improbable biological or geological events that are often random. The theory of contingent evolution was popularized by paleontologist Stephen Jay Gould.

convergent evolution: An evolutionary process by which organisms, not closely related, independently evolve similar traits as a result of adapting to similar environments.

CRISPR-Cas9: A gene-editing technology derived from bacteria that is used to edit, remove, and replace genes within cells and embryos.

culture: The collective sum of beliefs, behaviors, customs, laws, and other characteristics shared by members of a society that form a people's way of life.

cuticle: A hydrophobic, waxy, waterproof layer that covers plants and provides protection against water loss.

cyborg: An organic being enhanced with cybernetic or mechanical devices.

cynodonts: Carnivorous mammal-like reptiles of the order Therapsida who lived in the late Permian and Triassic periods and had well-developed teeth ("dog teeth"). They gave rise to all mammals.

cytosis: A transport mechanism for moving molecules and particles into and out of cells. Movement of molecules and particles into cells occurs by engulfing them with the cell membrane.

deep learning: A type of unsupervised machine learning that imitates the workings of the human brain. It has been incorporated into artificial neural networks.

dehumanization: The process of depriving a person or group of positive human qualities. This is a form of self-deception used against the enemy during time of war.

diazotrophs: Microorganisms capable of fixing nitrogen.

diploid cell: A cell that contains paired chromosomes, as in contrast to a haploid cell, which contains only one set of chromosomes.

DNA (deoxyribonucleic acid): a double-stranded self-replicating macromolecule that contains the genetic information that controls protein synthesis and operation of the cell and is passed to progeny in reproduction.

dominance hierarchy: A system of ranking in social animal groups (sometimes called a "pecking order") in which members are dominant to those ranking lower and submissive to those ranking higher than themselves.

dominant gene: A gene that expresses a variant of a trait and masks the effects of a recessive gene (allele) that codes for a different variant of that trait. This is contrasted to a recessive gene, which expresses the variant of a trait only when both alleles code for that variant.

ectotherms: Cold-blooded animals that are dependent on external sources for body heat.

egg: A female gamete cell that unites with a sperm cell to form a zygote. By definition, the egg is larger than the sperm, and it often contains nourishment for the developing zygote.

electron transport chain: A series of membrane-bound protein complexes that transfer electrons from one complex to the next, releasing energy that is used to pump protons across a membrane, creating a electrochemical gradient that is used to produce ATP. This process occurs in chloroplast membranes in plants, mitochondrial membranes in animals, and cell membranes in bacteria. It is a component of cellular respiration.

endoskeleton: An internal skeleton in vertebrates comprising bones that provide structure and support.

endosymbiosis: A symbiotic relationship in which one organism lives inside another and the two function as one organism. It is believed to be the mechanism by which mitochondria, chloroplasts, and other organelles arose within eukaryotic cells.

endotherms: Warm-blooded animals that generate heat from their metabolism and maintain a constant body temperature independent of the environmental temperature.

epigenetics: The study of how environmental factors can alter gene expression and how these changes in gene expression can be transmitted to future generations, providing new traits without changing DNA sequences.

Eukarya: one of three domains of life consisting of single-celled and multicellular organisms whose cells contain a nucleus and organelles.

eutheria: Mammals that have a placenta. This includes all mammals except monotremes and marsupials.

exaptation: An evolutionary process in which a protein or trait evolves for one particular function but subsequently is co-opted to serve another function.

exoskeleton: A rigid external covering of the body of some invertebrate animals that provides support and protection. Exoskeletons are most prevalent in mollusks and arthropods.

external fertilization: Fertilization of ova by sperm outside the body of the female.

extremophile: An organism capable of surviving in extreme physical or geochemical environments.

fermentation: A series of biochemical reactions that use the energy from organic molecules to produce ATP in the absence of oxygen.

Fermi's paradox: The apparent contradiction between the high probability of the existence of extraterrestrial civilizations and the lack of evidence for such civilizations.

fertilization: The union of a haploid sperm and a haploid egg to form a diploid zygote.

founder effect: The reduced genetic diversity that results when a population is descended from a small number of colonizing ancestors.

FOXP2 gene: A gene that codes for brain function in animals and is responsible for the development of speech and language in humans.

gamete: Haploid germ cells resulting from meiotic cell division that are able to fuse with gametes of the opposite sex (or type) to form a zygote.

gametophyte: The haploid generation in the life cycle of algae and plants.

gastrula: An early stage of embryonic development in which cells form a hollow cup-shaped structure and differentiate into three germ layers.

genetic bottleneck: A sharp reduction in the size of a population due to environmental events, which results in reduced genetic diversity.

genetic diversity: The total number of genetic characteristics (number of different alleles and their frequencies) in the genetic makeup of a given species.

genetic engineering: Modification or editing of genomes in organisms to produce new organisms or altered organisms with special traits.

genetic variation: The average difference in the genome of two individuals within a given species. It is usually measured as a percentage of the DNA code that is different between two genomes.

geological carbon cycle: A process by which CO_2 is removed from the atmosphere through the weathering process and added to the atmosphere through volcanic eruptions. These processes are driven by plate tectonics.

germ layers: Three layers of cell types in developing embryos—endoderm, ectoderm, and mesoderm.

giant impact theory: The theory that the moon formed from the impact of giant asteroid the size of Mars with Earth, resulting in rocky material being thrown outward from Earth and accreting to form the moon.

glacial epoch: A geologic time period marked by cold temperatures and glacier formation. It is sometimes called an ice age.

glacial period: That portion of a glacial epoch when temperatures are cold and glaciers are advancing.

Goldilocks zone: The zone around a star in which temperatures on the surface of a planet in that zone are moderate enough to support liquid water.

great filter: A term popularized by the economist Robin Hanson to describe the barriers to the origin, evolution, and survival of intelligent civilizations with advanced technology.

great oxygen catastrophe: An event occurring about 2.3 billion years ago when a rise in atmospheric oxygen content, caused by the release of oxygen by cyanobacteria, resulted in a mass extinction of microorganisms.

greenhouse effect: Trapping of heat from the sun in Earth's lower atmosphere, caused by the presence of greenhouse gases. Greenhouse gases, which include CO_2, water vapor, and methane, allow incoming radiation from the sun to pass through to Earth but absorb infrared radiation from Earth's surface.

greenhouse gas: An atmospheric gas that allows incoming sunlight to pass through but traps radiant heat from the Earth, resulting in warming of the Earth's atmosphere. Greenhouse gases include carbon dioxide, methane, and water vapor.

group selection: A proposed mechanism of evolution in which natural selection acts at the level of the group instead of at the more conventional level of the individual.

gymnosperms: Vascular plants that produce seeds but do not have flowers or fruit. They include conifers, cycads, and ginkgo.

haploid cell: A cell that contains unpaired chromosomes, in contrast to a diploid cell, which contains paired chromosomes.

Hayflick limit: The number of times a normal human cell population grown in culture will divide until cell division stops. The phenomenon is named after its discoverer, Leonard Hayflick.

hemocyanin: A copper-based oxygen-binding protein used by primitive circulatory systems for transporting oxygen.

hemoglobin: An iron-based oxygen-binding protein found in the red blood cells of vertebrates, used for transporting oxygen.

hemolymph: A fluid used in open circulatory systems to transport oxygen and nutrients, analogous to blood in vertebrates.

hermaphrodite: An organism that contains both male and female sexual organs and produces both sperm and eggs.

hominin: The lineage that includes all species on the human branch of the evolutionary tree after our split with chimpanzees.

Homo erectus ("the man who walks upright"): An extinct human species that lived from about 1.9 million years ago until 145 thousand years ago.

Homo habilis: The earliest human species known from the fossil record.

Homo heidelbergensis: A human species that descended from *Homo erectus* and is believed to have given rise to both Neanderthals and *Homo sapiens*.

homologous chromosomes: Paired chromosomes, one from each parent, that carry the same sequence of genes but with different alleles.

homologous: Showing similarities that indicate evolution from a common heritage. In biology this is usually applied when comparing genes, proteins, or structures of one organism with another.

homosexuality: sexual attraction to members of one's own sex.

Hox genes: Regulatory genes that control the body plan of animals in the anterior–posterior axis.

in vitro fertilization (IVF): A medical procedure in which an egg is fertilized by sperm outside the body in a test tube. The resulting embryo is implanted into the mother's uterus.

indirect reciprocity: A process in which one individual performs an action to improve the welfare of another individual without a necessary expectation of reciprocity but with the expectation that other unknown individuals will behave altruistically in return.

individual selection: Natural selection that acts at the level of the individual, as opposed to group selection, which is theorized to act at the level of the group.

intelligent design: The belief that the universe and living things were designed and created by the purposeful action of an intelligent agent, usually assumed to be God.

interglacial period: That portion of a glacial epoch when temperatures are warm and glaciers are receding.

internal fertilization: Fertilization of ova by sperm occurring inside the body of the female.

isogamous sex: Sexual reproduction in which the gametes are the same size and appearance.

kin selection: An evolutionary process in which individuals sacrifice their own welfare in an effort to improve the welfare of family members or kin. Since kin share common genes, altruistic genes that are likely shared by kin are selected and passed to the next generation.

Lamarckian inheritance: A theory of biological evolution holding that an organism can pass on characteristics that it has acquired during its lifetime to its offspring.

last universal common ancestor (LUCA): The last common ancestor of all life on Earth today.

legumes: A family of plants bearing nodules on their roots that contain nitrogen-fixing bacteria.

lichen: A hybrid of fungi and green algae or cyanobacteria that live together in a symbiotic relationship.

lignin: An organic polymer deposited in the vascular tissue of plants (xylem) that provides strength and structural support.

Lucy: An extinct hominin (*Australopithecus afarensis*) that lived between 3.9 and 2.9 million years ago. She is regarded as a crucial link between chimpanzees and humans and is often called the mother of all humanity.

Malthusian catastrophe: A catastrophe predicted by economist Thomas Malthus in the eighteenth century, in which population growth outpaces food supply, resulting in widespread famine.

marsupials: Mammals that deliver poorly developed embryos that complete their development in the mother's pouch.

meiosis: A special type of cell division that is part of sexual reproduction, in which a diploid precursor cell duplicates its chromosomes and then undergoes two cell divisions, producing four haploid gametes, each containing a random distribution of chromosomes and genes.

meme: A transmittable unit of cultural information proposed by Richard Dawkins in *The Selfish Gene.*

Milankovitch cycles: Cyclic changes in Earth's orbit, axis tilt, and axis wobble that affect the amount and distribution of solar radiation reaching Earth. These trigger cyclic changes in Earth's temperature about every hundred thousand years. The cycles are named after the Serbian astronomer Milutin Milankovitch.

mitochondria: Organelles present in the cytoplasm of eukaryotic cells in which ATP is produced through aerobic respiration. These are the "powerhouses" of the cells.

Mitochondrial Eve: A hypothetical ancestral mother of all humanity who lived in Africa two hundred thousand years ago. Her existence is based on analysis of mitochondrial DNA, hence the name Mitochondrial Eve.

mitochondrial replacement therapy: A special type of in vitro fertilization in which the future baby's mitochondrial DNA comes from third party. The procedure is used to prevent transmission of mitochondrial genetic diseases from mother to offspring.

mitosis: The process of cell division that results in two daughter cells each having the same DNA code as the parent cell. Ordinary tissues grow through mitotic cell division, and some organisms reproduce asexually using mitosis.

monotremes: Egg-laying mammals, which include the platypus and spiny anteaters (echidnas)

moral absolutism: An ethical view that specific actions are intrinsically right or wrong regardless of the circumstances.

moral relativism: The philosophy that recognizes the great diversity of morality across cultures, believes there are no objective moral values, and advocates nonjudgmental tolerance of this diversity.

morality: The belief that some behavior is right and acceptable, and that other behavior is wrong.

Muller's ratchet: The process of irreversible accumulation of deleterious mutations over multiple generations of asexual reproduction that eventually lead to the demise of the organism. The process gets its name from Hermann Muller, who proposed the process, and from a ratchet, a mechanical device that can only go in one direction.

multiregional evolution hypothesis: The hypothesis that *Homo sapiens* evolved from *Homo erectus* in various regions in the Old World and that speciation was prevented by extensive interbreeding.

mycorrhizal fungal colonies: Fungal colonies that are attached to plant roots in a symbiotic relationship. The fungi help with absorption of water and nutrients for the plant, and the plant provides organic products for the fungi.

myosin: A protein that forms (together with actin) the contractile filaments of muscle cells and is part of the cytoskeleton of eukaryotic cells.

natural selection: The process by which organisms better adapted to their environment tend to survive (are selected), reproduce, and pass their genes and traits to the next generation. This is part of Darwin's evolutionary theory of variation and natural selection.

natural variation: Variation in traits of offspring that result in differences in fitness and propensity to survive. This is a prerequisite for natural selection and an essential part of Darwin's theory of variation and natural selection.

neanderthals: A human species that resided in Europe and western Asia from about two hundred fifty thousand to forty-four thousand years ago. Neanderthals are our closest relative.

neo-Darwinism: A modern synthesis of Darwin's theory of evolution by variation and natural selection with theories of modern genetic inheritance. The synthesis rejects Lamarckian inheritance.

Neolithic Revolution: The wide-scale human transition from hunting and gathering to agriculture, domestication of animals, and settlement in towns and villages that began about 10,000 BCE. This is also called the Agricultural Revolution.

nerve net: A diffuse network of nerve cells that conducts impulses in all directions from the point of stimulus. Nerve nets are present in the most primitive animals—the cnidarians and flatworms.

neural correlates of consciousness (NCC): The minimal set of neuronal events (active areas of the brain) that correlate with a specific conscious experience.

Newton's law of universal gravitation: Every particle attracts every other particle in the universe with a force proportional to the product of their masses and inversely proportional to the square of the distance between their centers.

nitrogen fixation: A process by which nitrogen in the atmosphere (N_2) is converted to bioavailable nitrogen oxides and ammonia.

nitrogenase: The enzyme used by specialized bacteria (diazotrophs) to fix nitrogen.

nucleotide: a compound consisting of a nitrogenous base, a sugar, and a phosphate group that is a building block for DNA and RNA.

Opabinia: A bizarre-looking marine animal of the Cambrian period with five eyes and a proboscis.

open circulatory system: A circulatory system in which a primitive heart pumps hemolymph into open sinuses, where it bathes organs. Hemolymph is not confined to a vascular system, there is no true heart, and there are no capillaries.

out-of-Africa theory: The theory that Homo sapiens originated in Africa and migrated north to the Middle East, Asia, Europe, and finally to the rest of the world.

oxygen-evolving complex: A manganese-containing enzyme complex that catalyzes the splitting of water by chlorophyll ions in the process of photosynthesis, releasing oxygen as a by-product.

oxygenic photosynthesis: The process by which plants, algae, and cyanobacteria use energy from sunlight to produce organic molecules from carbon dioxide and water, releasing oxygen as a by-product.

panspermia: The theory that life on Earth originated from microorganisms seeded to Earth from extraterrestrial sources.

parthenogenesis: A form of asexual reproduction in which the female produces an egg that germinates and develops into a new individual without fertilization.

PAX6 gene: The regulatory gene that has been called the master regulator of development of the eye.

phagocytosis: The process of one cell ingesting a smaller cell or cell fragment.

phenomenal consciousness: The subjective component of sensory perceptions. Also known as qualia.

philosophical zombie: A hypothetical person physically identical to a normal human being but lacking consciousness.

phloem: A vascular tissue in vascular plants used to transport sugar, water, and minerals throughout the plant.

photopsin: A photosensitive pigment coupled with an opsin protein that senses light and generates a signal. Photopsins are embedded in cones in the vertebrate retina and are capable of distinguishing color.

photosynthesis: The process by which plants, algae, and some bacteria convert energy from sunlight into chemical energy in the form of ATP.

phylogenic constraint: The inability of animals to change their basic body plan through evolution because changes in the basic body plan are not compatible with survival.

phylum: A taxonomic grouping below kingdom and above class that includes organisms with similar body plans.

Pikaia: A primitive fishlike creature documented in the fossil record of the Cambrian period that resembled a worm as much as a fish. *Pikaia* is the first chordate documented in the fossil record and has been nominated as a possible ancestor of all current-day chordates, including humans.

placenta: An organ located between the uterus and the embryo in pregnant placental mammals that nourishes the embryo using the mother's circulation.

plate tectonics: The theory of geology that Earth's outer shell is divided into plates that glide over the mantle, the rocky layer beneath the outer shell. The collision and buckling of these plates causes geologic upheaval and creates mountains and valleys. This is the modern version of continental drift.

primary sexual characteristics: Body structures that are directly related to sexual reproduction, such as testes, ovaries, and external genitalia.

punctuated equilibria: A pattern of evolution in which long periods of very little change in diversity of species are interrupted by short periods of rapid change.

qualia: The subjective conscious component of sensory perceptions.

quorum sensing: A phenomenon by which microorganisms communicate and coordinate their behavior by emitting signaling molecules. When the bacteria reach a critical number such that sufficient hormone is produced, quorum sensing occurs and the bacteria act in unison.

radiometric dating: The technique by which the ages of ancient rocks and organic matter are estimated by determining the relative proportions of radioactive isotopes present in a sample.

reactive oxygen species: Chemically active molecules containing oxygen that are produced as a by-product of aerobic metabolism. They include superoxide anions, hydroxyl radicals, and hydrogen peroxide, and are thought to play a role in damaging DNA and contributing to aging.

recessive gene: A gene that expresses the variant of a trait only when the homologous gene (allele) also expresses the same variant of the trait.

reciprocal altruism: Behavior in which one individual performs an act to benefit the welfare of another, with the expectation that the other individual will return the favor.

recombination: The process of exchanging genes between homologous chromosomes, resulting in new genetic combinations. Recombination occurs in meiotic cell division and in bacterial conjugation (DNA transfer).

red giant: A star in the late stages of its life cycle when hydrogen fuel is nearly depleted as it enlarges and turns a reddish color.

regulatory genes: Genes that control—turn on or turn off—the expression of other genes, usually structural genes.

rhizobia: Nitrogen-fixing bacteria usually found in root nodules in leguminous plants.

rhodopsin: A photosensitive pigment coupled to an opsin protein that senses light and generates a signal. Rhodopsin is universal to all light-sensing organisms and is embedded in rods of the vertebrate retina.

ribosomes: small particles in the cytoplasm of cells where protein synthesis occurs through the process of translation.

ribozymes: RNA with enzymatic properties facilitating self-replication.

RNA (ribonucleic acid): A single-stranded macromolecule that copies genetic information from DNA and uses this information in the process of making proteins.

RNA world hypothesis: The hypothesis that the formation of a self-replicating molecule similar to RNA was the initial stage in the origin of life.

rods: photosensitive cells in the vertebrate retina that sense light and generate signals that are transmitted to neurons. Rods can sense low levels of light but cannot distinguish color.

RuBisCO: An enzyme used in the Calvin cycle of photosynthesis that facilitates carbon fixation, a process by which carbon dioxide and water are converted to organic molecules. RuBisCo may be the most abundant protein on Earth.

runaway greenhouse effect: A process in which a positive feedback between increasing surface temperatures and the increasing atmospheric concentration of greenhouse gases results in severe uncontrolled overheating of a planet.

runaway refrigerator effect: A process in which a positive feedback between lower surface temperature and a loss of atmospheric greenhouse gases results in severe uncontrolled cooling of a planet.

sarcomere: The structural unit of a myofibril in striated muscle. Sarcomeres contain actin and myosin protein filaments, which slide over one another to shorten the sarcomere and result in muscle contraction.

sauropods: Giant herbivorous dinosaurs with long necks, long tails, and small heads, known for their enormous size. They are an infraorder of saurischian ("lizard-hipped") dinosaurs and include the well-known giant herbivore genus *Argentinosaurus*.

sauropsids: A clade of amniotes that gave rise to dinosaurs, birds, and all current-day reptiles.

Scala naturae: A hierarchy, also called the Great Chain of Being, handed down from medieval times, in which all living things are arranged in a divine order of perfection.

secondary sexual characteristics: Physical characteristics in males and females that distinguish the sexes and enhance reproductive success but are not directly involved in reproduction. These characteristics evolve through sexual selection.

sensitization: A simple form of learning and memory in which the response to sequential stimuli is enhanced by a separate, intervening, usually noxious, stimulus.

sentient: The capacity to feel, perceive, or experience subjectively. Sentience is the simplest form of consciousness.

sexual selection: A special case of natural selection in which organisms with traits that enhance their ability to mate and reproduce (secondary sexual characteristics) will pass these traits to the next generation.

singularity (technological): The point at which artificial intelligence exceeds human intelligence, triggering runaway technological growth resulting in unfathomable changes to human civilization.

Snowball Earth: An extreme glacial epoch in which ice covers the entire planet.

species: A group of living organisms with similar characteristics that is capable of interbreeding. Species is the taxonomic unit ranking below genus.

sperm: A male gamete that unites with a female egg to form a zygote. The sperm is smaller than the egg by definition.

spores: Reproductive (haploid) cells produced by algae, bryophytes and seedless vascular plants through the process of meiosis that are capable of developing into a mature plant (gametophyte) without fusion with another reproductive cell.

sporophyte: The diploid generation in the life cycle of algae and plants.

stereochemistry: That branch of chemistry dealing with spatial arrangements of atoms in molecules and their effects on the properties and chemistry of molecules.

stomata: Dynamic openings in leaves that allow for exchange of carbon dioxide, oxygen, and water.

stromatolites: Sedimentary mushroom-shaped rocks formed by and inhabited by cyanobacteria or other microorganisms.

structural genes: Genes that code for proteins used for the structure and function of cells.

supernova: A star in the last stage of its life cycle that undergoes a final titanic explosion, appearing as a sudden "new" bright star and sending its fragments across neighboring galaxies to seed new solar systems.

supervised machine learning: Machine learning that requires human supervision to create algorithms that correlate input variables with an output variable.

synapsids: A clade of amniotes that includes extinct pelycosaurs, mammal-like reptiles (therapsids) and all their descendants, which include all mammals.

tectum: A structure in the dorsal midbrain of vertebrates responsible for processing visual and auditory sensory data and for focusing attention. It is believed to be essential for basic sensory consciousness.

telomerase: An enzyme that catalyzes the lengthening of telomeres.

telomeres: Short pieces of noncoding DNA at the end of chromosomes that act as buffers to protect the information carrying part of the chromosomes. Shortening of telomeres is believed to contribute to aging.

tetrapods: A superclass composed of the first vertebrates with four limbs to reach land and all their descendants (amphibians, reptiles, birds, and mammals).

theory of mind (ToM): The ability to attribute mental states to oneself and others and to understand that others have beliefs, desires, intentions, and perspectives that are different from one's own.

therapsids: Extinct mammal-like reptiles of the late Permian and Triassic periods that gave rise to all mammals. They are part of the clade synapsids.

theropods: Carnivorous dinosaurs with short forelimbs that walked on hind legs. They are the ancestors of all birds and include *Tyrannosaurus rex* and *Velociraptor*.

Tiktaalik roseae: A transitional animal between lobe-finned fishes and terrestrial tetrapods, often called a "fishapod."

transcription factors: Regulator proteins produced by regulatory genes that activate or deactivate structural genes.

transcription: The process by which genetic information is transcribed from DNA to messenger RNA to be used in making proteins.

translation: The process by which the genetic code of messenger RNA is decoded to produce a specific sequence of amino acids in protein synthesis.

tribal instincts: Innate traits that promote cooperative behavior toward members within the group or tribe and antagonistic or hostile behavior toward those outside the tribe. These instincts evolved in our hunter-gatherer ancestors and persist in us today.

trilobite: An ancient marine arthropod with a hard shell, multiple segments, and jointed legs that became an icon of the Cambrian period.

universal genetic code: The algorithm that defines how the genetic code in DNA specifies the amino acid sequence in protein synthesis. Each three-nucleotide sequence, called a codon, specifies an amino acid or serves as a signal for starting or stopping protein synthesis.

unsupervised machine learning: Machine learning in which intelligent machines analyze large volumes of input data and identify patterns and categories in the data. There are no correct answers and no supervision or teachers.

vitalism doctrine: The doctrine that living organisms are governed by a "vital principle" distinct from the laws of physics and chemistry.

weak anthropic principle: The concept that the values of all physical and cosmological constants in the universe must allow for the origin and evolution of carbon-based intelligent life, because if they were different, we would not be here to observe them.

weathering: The breaking down of rocks and minerals through contact with atmospheric CO_2 and water. Carbon dioxide and water vapor in the atmosphere form carbonic acid, which rains to Earth and reacts with and erodes rocks and minerals. Carbon dioxide is removed from the atmosphere in the process.

Wood–Ljungdahl pathway: A primitive biochemical pathway, also known as the Acetyl-CoA pathway, used by some bacteria and archaea to produce organic molecules from carbon dioxide using energy from hydrogen gas.

xylem: One type of vascular tissue in vascular plants that is used to transmit water and nutrients from roots throughout the plant.

zygote: A diploid cell resulting from the fusion of a sperm and an egg.

REFERENCES

Chapter 1
Books

Alberts, B., D. Bry, J. Lewis, M. Raff, K. Roberts, and J. D. Watson. 1994. *Molecular Biology of the Cell*. 3rd ed. New York, NY: Garland Publishing, Inc.

Darwin, C. 1859. *On the Origin of Species by Natural Selection*. Republished 2004. New York, NY: Barnes and Noble Books.

Voet, D., J. G. Voet, and C. W. Pratt. 2013. *Fundamentals of Biochemistry: Life at the Molecular Level*. 4th ed. Hoboken, NJ: John Wiley & Sons, Inc.

Other References

Choi, C. Q. 2013. "'Left-Handed' Molecules brought to Earth by Meteorites, Nebula Study Suggests." Retrieved January 2017 from http://www.huffingtonpost.com/2013/05/01/left-handed-molecules-earth-meteorites-nebula-chiral_n_3191994.html.

Kwon, J., M. Tamura, P. W. Lucas, J. Hashimoto, N. Kusakabe, R. Kandori, Y. Nakajima, T. Nagayama, T. Nagata, and J. H. Hough. 2013. "Near Infrared circular polarization images of NGC 6334-V." *Astrophysical Journal Letters* 765 (1). doi:10.1088/2041-8205/765/1/L6.

Mosher, D. 2008. "How Life Became Left-Handed." Retrieved January 2017 from http://www.livescience.com/7480-life-left-handed.html.

Pray, L. A. 2008. "Eukaryotic genome complexity." *Nature Education* 1 (1), 96.

Watson, J. D. and F. H. C. Crick. 1953. "Molecular structure of nucleic acids: A structure for deoxyribose nucleic acid." *Nature* 171, 737–38.

Woese, C. R., O. Kandler, and M. L. Wheelis. 1990. "Towards a natural system of organisms: Proposal for the domains Archaea, Bacteria, and Eucarya." *Proc. Natl. Acad. Sci.* 87, 4576–79.

Zhang, Y., and V. N. Gladyshev 2006. "High content of proteins containing 21st and 22nd amino acids, selenocysteine and pyrrolysine, in a symbiotic deltaproteobacterium of gutless worm Olavius algarvensis." *Nucleic Acids Res* 35 (15), 4952–63.

Chapter 2
Books

Bendall, D. S. 1983. *Evolution from Molecules to Men*. Cambridge, UK: Cambridge University Press.

Cziko, G. 1995. *Without Miracles: Universal Selection Theory and the Second Darwinian Revolution*. Cambridge, MA: MIT Press.

Darwin, C. 1859. *On the Origin of Species by Natural Selection*. Lyndhurst, NJ: Barnes and Noble Books. Republished 2004.

Darwin, C. 1871. *The Descent of Man, and Selection in Relation to Sex*. London: John Murray. Retrieved September 2017 from http://darwin-online.org.uk/content/frameset?pageseq=1&itemID=F937.1&viewtype=text

De Camp, L. S. 1968. *The Great Monkey Trial*. Garden City, NY: Doubleday & Company, Inc.

Other References

Heard, E., and R. Martienssen, 2014. "Transgenerational epigenetic inheritance: Myths and mechanisms." *Cell* 157 (1), 95–109.

Heijmans, B. T., E. W. Tobi, A. D. Stein, H. Putter, G. J. Blauw, E. S. Susser … L. H. Lumey. 2008. "Persistent epigenetic differences associated with prenatal exposure to famine in humans." *Proc Natl Acad Sci USA*, 105 (44), 17046–49.

Kuipers, B. 2013. "Why do we believe in electrons, but not in fairies?" Retrieved September 2017 from http://web.eecs.umich.edu/~kuipers/opinions/electrons-vs-fairies.html.

LaBracio, L. 2016. "What's the difference between a scientific law and theory? (in TED-Ed GIFs)." TED-Ed Blog. Retrieved March 2018 from https://blog.ed.ted.com/2016/06/07/whats-the-difference-between-a-scientific-law-and-theory-in-ted-ed-gifs/.

Lim, U. and M. A. Song, 2012. "Dietary and lifestyle factors of DNA methylation." *Methods Mol Biol* 863, 59–76.

Pew Research Center. 2013. "Public's Views on Human Evolution." Retrieved September 2017 from http://www.pewforum.org/files/2013/12/Evolution-12-30.pdf.

Sen, A., N. Heredia, M.-C. Senut, S. Land, K. Hollocher, X. Lu … and D. M. Ruden. 2015. "Multigenerational epigenetic inheritance in humans: DNA methylation changes associated with maternal exposure to lead can be transmitted to the grandchildren." *Sci. Rep.* 5, 14466. doi:10.1038/srep14466.

Skinner, M. 2016. "Unified theory of evolution." Aeon. Retrieved September 2017 from https://aeon.co/essays/on-epigenetics-we-need-both-darwin-s-and-lamarck-s-theories.

Chapter 3
Books

Barrow, J.D. 2002. *The Constants of Nature: The Numbers that Encode the Deepest Secrets of the Universe.* New York, NY: Random House, Inc.

Hawking, S. 1996. *A Brief History of Time.* New York, NY: Bantam Books.

Hawking, S., and L. Mlodinow. 2010. *The Grand Design.* New York, NY: Bantam Books.

Krauss, L. 2012. *A Universe from Nothing.* New York, NY: Simon & Schuster, Inc. Free Press.

Sagan, C. 1980. *Cosmos.* New York, NY: Random House, Inc.

Tyson, N. D. 2007. *Death by Black Hole and Other Cosmic Quandaries.* New York, NY: W. W. Norton & Company.

Ward, P. D., and D. Brownlee. 2004. *Rare Earth: Why Complex Life Is Uncommon in the Universe.* New York, NY: Copernicus Books.

Other References

Anzellini, S., A. Dewaele, M. Mezouar, P. Loubeyre, and G. Morard. 2013. "Melting of Iron at Earth's Inner Core Boundary Based on Fast X-ray Diffraction." *Science* 340 (6131), 464–66.

Hille, K. 2017. "Hubble Reveals Observable Universe Contains 10 Times More Galaxies Than Previously Thought." NASA. Retrieved June 2017 from https://www.nasa.gov/feature/goddard/2016/hubble-reveals-observable-universe-contains-10-times-more-galaxies-than-previously-thought.

NASA. 2017. "Universe is expanding." Retrieved September 2017 from https://cosmictimes.gsfc.nasa.gov/teachers/guide/1929/guide/universe_expanding.html

US Geological Survey. 2019. "How do we know the age of the Earth? Radiometric dating." Retrieved January 2019 from https://geomaps.wr.usgs.gov/parks/gtime/ageofearth.html.

Chapter 4
Books

Alberts, B., D. Bry, J. Lewis, M. Raff, K. Roberts, and J. D. Watson. 1994. *Molecular Biology of the Cell.* 3rd ed. New York, NY: Garland Publishing, Inc.

Darwin, F. 1911. *The Life and Letters of Charles Darwin.* Volume II. New York: D. Appleton & Co.

de Duve, C. 1995. *Vital Dust: The Origin and Evolution of Life on Earth.* New York, NY: Basic Books.

Lane, N. 2009. *Life Ascending: The Ten Great Inventions of Evolution.* London: W. W. Norton & Company, Inc.

———. 2015. *The Vital Question: Energy, Evolution, and the Origins of Complex Life.* New York, NY: W. W. Norton & Company.

Lurquin, P. E. 2003. *The Origins of Life and the Universe.* New York, NY: Columbia University Press.

Smith, J. M. and E. Szathmary. 1999. *The Origins of Life: From the Birth of Life to the Origins of Language.* New York, NY: Oxford University Press.

Voet, D., J. G. Voet, and C.W. Pratt, 2013. *Fundamentals of Biochemistry: Life at the Molecular Level.* 4th ed. Hoboken, NJ: John Wiley & Sons, Inc.

Other References

Cech, T. R. 2018. "Exploring the New RNA World." The Nobel Prize. Retrieved August 2018 from https://www.nobelprize.org/nobel_prizes/chemistry/laureates/1989/cech-article.html.

Clinton, W. 1996. "Statement Regarding Mars Meteorite Discovery." NASA Jet Propulsion Laboratory. Retrieved September 2017 from http://www2.jpl.nasa.gov/snc/clinton.html

Crick, F. H. C., and L. E. Orgel. 1973. "Directed Panspermia." *Icarus* 19, 341–46.

Dodd, M.S., D. Papineau, T. Grenne, J. F. Slack, M. Rittner, F. Pirajno … C. T. S. Little. 2017. "Evidence for early life in Earth's oldest hydrothermal vent precipitates." *Nature* 543, 60–64.

Exploring Life's Origins. "Fatty Acids." 2018. Retrieved August 2018 from http://exploringorigins.org/fattyacids.html.

Gibard, C., S. Bhowmik, M. Karki, E-K. Kim, and R. Krishnamurthy. 2018. "Phosphorylation, oligomerization and self-assembly in water under potential prebiotic conditions." *Nature Chemistry* 10, 212–17.

Hawking, S. 2008. "Why We Should Go into Space." NASA's 50th Anniversary Lecture Series. Retrieved from http://www.nasa.gov/pdf/223968main_HAWKING.pdf.

Kohler, R. 1971. "The background to Eduard Buchner's discovery of cell-free fermentation." *Journal of the History of Biology* 4 (1), 35–61.

Lincoln, T. A., and G. F. Joyce. 2009. "Self-sustained replication of an RNA enzyme." *Science* 323 (5918), 1229–32.

Line, M. A. 2007. "Panspermia in the context of the timing of the origin of life and microbial phylogeny." *Inter J Astrobiology* 6 (3), 249–54.

Martin, W., and M. J. Russell. 2007. "On the origin of biochemistry at an alkaline hydrothermal vent." *Phil Trans R Soc* 362 (1486), 1887–1926.

Miller, S. L., and H. C. Urey. 1959. "Organic compound synthesis on the primitive earth." *Science* 130 (3370), 245–51.

Schmitt-Kopplin, P., Z. Gabelica, R. Gougeon, A. Fekete, B. Kanawati, M. Harir … N. Hertkorn. 2010. "High molecular diversity of extraterrestrial organic matter in Murchison meteorite revealed 40 years after its fall." *Proceedings of the National Academy of Sciences* 107 (7), 2763–68.

Schopf, J. W. 1993. "Microfossils of the early archean apex chert: New evidence of the antiquity of life." *Science* 260 (5108), 640–46.

Schrope, M. 2009. "The Immortal Molecule." Retrieved September 2017 from https://www.scripps.edu/newsandviews/e_20090112/joyce.html.

Vreeland, R. H., W. D. Rosenzweig, and D. W. Powers. 2000. "Isolation of a 250 million-year-old halotolerant bacterium from a primary salt crystal." *Nature* 407, 897–900.

Chapter 5
Books

de Duve, C. 1995. *Vital Dust: The Origin and Evolution of Life on Earth.* New York, NY: Basic Books.

Lane, N. 2009. *Life Ascending: The Ten Great Inventions of Evolution.* London: W. W. Norton & Company, Inc.

———. 2015. *The Vital Question: Energy, Evolution, and the Origins of Complex Life.* New York, NY: W. W. Norton & Company.

Voet, D., J. G. Voet, and C. W. Pratt, 2013. *Fundamentals of Biochemistry: Life at the Molecular Level.* 4th ed. Hoboken, NJ: John Wiley & Sons, Inc.

Other References

Boetius, A. 2005. "Lost City Life." *Science* 307 (5714), 1420–22.

Lane, N., and W. F. Martin. 2012. "The origin of membrane bioenergetics." *Cell* 151 (7), 1406–16.

Mitchell, P. 1961. "Coupling of phosphorylation to electron and hydrogen transfer by a chemi-osmotic type of mechanism." *Nature* 191, 144–48.

Roskoski Jr., R. 2004. "Book Review: *Wandering in the Gardens of the Mind: Peter Mitchell and the Making of Glynn.*" *Biochemistry and Molecular Biology Education* 32 (1), 62–66.

Russell, M. J., R. M. Daniel, A. J. Hall, and J. A. Sherringham. 1994. "A hydrothermally precipitated catalytic iron sulphide membrane as a first step toward life." *J Mol Evol* 39, 231–43.

Russell, M. J., A. J. Hall, A. G. Cirns-Smith, and P. S. Bratrman. 1988. "Submarine hot springs and the origin of life." *Nature* 336, 117.

Sojo, V., A. Pomiankowski, and N. Lane. 2014. "A Bioenergetic Basis for Membrane Divergence in Archaea and Bacteria." *PLoS Biol* 12 (8): e1001926. doi: 10.1371/journal.pbio.1001926.

Chapter 6
Books

de Duve, C. 1995. *Vital Dust: The Origin and Evolution of Life on Earth.* New York, NY: Basic Books.

Gould, S. J. 1989. *Wonderful Life: The Burgess Shale and the Nature of History.* New York, NY: W. W. Norton & Company.

Lane, N. 2009. *Life Ascending: The Ten Great Inventions of Evolution.* New York, NY: W. W. Norton & Company, Inc.

Voet, D., J. G. Voet, and C. W. Pratt, 2013. *Fundamentals of Biochemistry: Life at the Molecular Level.* 4th ed. Hoboken, NJ: John Wiley & Sons, Inc.

Other References

American Chemical Society International Historic Chemical Landmarks. 2017. "Joseph Priestley and the Discovery of Oxygen." Retrieved December 2017 from https://www.acs.org/content/acs/en/education/whatischemistry/landmarks/josephpriestleyoxygen.html.

Blankenship, R. E. 2010. "Early evolution of photosynthesis." *Plant Physiology* 154 (2), 434–38.

Buick, R. 2008. "When did oxygenic photosynthesis evolve?" *Philos Trans R Soc Lond B Biol Sci.* 363 (1504), 2731–43.

Cardona, T. 2017. "Evolution of Photosynthesis." Wiley Online Library. Retrieved December 2017 from http://onlinelibrary.wiley.com/doi/10.1002/9780470015902.a0002034.pub3/abstract.

Holland, H. D. 2006. "The oxygenation of the atmosphere and oceans." *Phil Trans R Soc B* 361, 903–15.

Nutman, A. P., V. C. Bennett, C. R. L. Friend, M. J. Van Kranendonk, and A. R. Chivas. 2016. "Rapid emergence of life shown by discovery of 3,700-million-year-old microbial structures." *Nature.* doi:10.1038/nature19355.

Schopf, J. W. 1993. "Microfossils of the early archean apex chert: New evidence of the antiquity of life." *Science* 260 (5108), 640–46.

Shih, P. M. 2015. "Photosynthesis and early Earth." *Current Biology* 25 (19), R855-R859.

Sparks, W. B., S. DasSarma, and I. N. Reid, 2006. "Evolutionary Competition Between Primitive Photosynthetic Systems: Existence of an Early Purple Earth?" *Bulletin of the American Astronomical Society* 38, 901.

Chapter 7
Books

Alberts, B., D. Bry, J. Lewis, M. Raff, K. Roberts, and J. D. Watson. 1994. *Molecular Biology of the Cell.* 3rd ed. New York, NY: Garland Publishing, Inc.

Dawkins, R. 2004. *The Ancestor's Tale: A Pilgrimage to the Dawn of Evolution.* New York, NY: Houghton Mifflin Company.

Lane, N. 2005. *Power, Sex, Suicide: Mitochondria and the Meaning of Life.* New York, NY: Oxford University Press.

———. 2009. *Life Ascending: The Ten Great Inventions of Evolution.* London: W. W. Norton & Company, Inc.

———. 2015. *The Vital Question: Energy, Evolution and the Origins of Complex Life.* New York, NY: W. W. Norton & Company.

Smith, J. M., and E. Szathmáry. 1999. *The Origins of Life: From the Birth of Life to the Origins of Language.* Oxford, UK: Oxford University Press.

Other References

Anderson S. G. E., O. Karlberg, B. Canback, and C. G. Kurland. 2003. "On the origin of mitochondria: a genomics perspective." *Philos. Trans. R. Soc. B.* 358 (1429), 165–79.

El Albani, E., S. Bengtson, D. E. Canfield, A. Bekker, R. Macchiarelli, A. Mazurier … A. Meunier. 2010. "Large colonial organisms with coordinated growth in oxygenated environments 2.1 Gyr ago." *Nature* 466, 100–104.

Embley, T. M., M. van der Giezen, D. S. Horner, P. L. Dyal, S. Bell, and P. G. Foster. 2003. "Hydrogenosomes, mitochondria and early eukaryotic evolution." *Life* 55 (7), 387–95.

Emelyanov, V. V. 2001. "REVIEW: Rickettsiaceae, *Rickettsia*-like endosymbionts, and the origin of mitochondria." *Bioscience Reports* 21, 1–17.

Martin, W., and Mentel, M. 2010. "The Origin of Mitochondria." *Nature Education* 3 (9), 58.

Woese, C. R., O. Kandler, and M. L. Wheelis. 1990. "Toward a natural system of organisms: Proposal for the domains Archaea, Bacteria, and Eucarya." *Proc. Natl. Acad. Sci.* 87, 4576–79.

Chapter 8
Books

Conway Morris, S. 1998. *The Crucible of Creation: The Burgess Shale and the Rise of Animals.* Oxford, UK: Oxford University Press.

Gould, S. J. 1989. *Wonderful Life: The Burgess Shale and the Nature of History.* New York, NY: W. W. Norton & Company.

Parker, A. 2003. *In the Blink of an Eye: How Vision Sparked the Big Bang of Evolution.* Cambridge, MA: Perseus Pub.

Ward, P. D., and D. Brownlee. 2004. *Rare Earth: Why Complex Life Is Uncommon in the Universe.* New York, NY: Copernicus Books.

Other References

Angier, N. 2014. "When Trilobites Ruled the World." The *New York Times.* March 3. Retrieved November 2014 from https://www.nytimes.com/2014/03/04/science/when-trilobites-ruled-the-world.html?_r=0.

Berner, R. A. 1999. "Atmospheric oxygen over Phanerozoic time." *Proc. Natl. Acad. Sci. USA* 96 (20), 10955–57.

El Albani, E., S. Bengston, D. E. Canfield, A. Bekker, R. Macchiarelli, A. Mazurier ... A. Meunier. 2010. "Large colonial organisms with coordinated growth in oxygenated environments 2.1 Gyr ago." *Nature* 466, 100-104.

Fortey, R. A. 1998. "Shock Lobsters." Review of *The Crucible of Creation: The Burgess Shale and the Rise of Animals* by Simon Conway Morris. *London Review of Books* 20 (19), 24–25.

———. 2004. "The Lifestyles of the Trilobites." *American Scientist* 92, 46–53.

Gaines, R. R., E. U. Hammarlund, X. Hou, C. Qi, S. E. Gabbott, Y. Zhao ... D. E. Canfield. 2012. "Mechanism for Burgess Shale-type preservation." *Proc Natl Acad Sci USA* 109 (14), 5180–84.

Gannon, M. 2015. "Toothy 'Penis Worm' from Cambrian Period discovered." Livescience. May 6. Retrieved July 2018 from https://www.livescience.com/50748-cambrian-penis-worm-discovered.html.

Han, T. M., and B. Runnegar. 1992. "Megascopic eukaryotic algae from the 2.1-billion-year-old negaunee iron-formation, Michigan." *Science* 257 (5067), 232–35.

Knell, R. J., and R. A. Fortey. 2005. "Trilobite spines and beetle horns: Sexual selection in the Palaeozoic?" *Biol Lett* 1 (2), 196–99.

Peters, S. E., and R. R. Gaines. 2012. "Formation of the 'Great Unconformity' as a trigger for the Cambrian explosion." *Nature* 484 (7394), 363–66.

Sperling, E. A., C. A. Frieder, A. V. Raman, P. R. Girgis, L. A. Levin, and A. H. Knoll. 2013. "Oxygen, ecology, and the Cambrian radiation of animals." *Proc Natl Acad Sci USA* 110 (33), 13446–51.

Speyer, S. E., and C. E. Brett. 1985. "Clustered trilobite assemblages in the middle Devonian Hamilton Group." *Lethaia* 18, 85–103.

Va, M. 2007. "Sex and the Cambrian Explosion." Live Journal. September 24. Retrieved September 2018 from https://endcreationism.livejournal.com/195812.html.

Whittington, H. B. 1971. "Redescription of *Marella splendens* (Trilobitoidea) from the Burgess Shale, Middle Cambrian, British Columbia." *Geological Survey of Canada Bulletin* 209, 1–24.

York, R., and B. Clark. 2011. "Stephen Jay Gould's Critique of Progress." Monthly Review. February 1. Retrieved July 2018 from https://monthlyreview.org/2011/02/01/stephen-jay-goulds-critique-of-progress/.

Chapter 9
Books

Alberts, B., D. Bry, J. Lewis, M. Raff, K. Rogerts, and J. D. Watson, 1994. *Molecular Biology of the Cell.* 3rd ed. New York, NY: Garland Publishing, Inc.

Bonner, J. T. 2000. *First Signals: The Evolution of Multicellular Development.* Princeton, NJ: Princeton University Press.

de Duve, C. 1995. *Vital Dust:* The Origin and Evolution of Life on Earth. New York, NY: Basic Books.

Smith, J. M., and E. Szathmary. 1999. *The Origins of Life: From the Birth of Life to the Origins of Language.* New York, NY: Oxford University Press.

Stearns, S. C., and R. F. Hoekstra. 2005. *Evolution: An Introduction.* 2nd ed. New York, NY: Oxford University Press.

Other References

Akam, M. 1995. "Hox genes and the evolution of diverse body plans." *Philos Trans R Soc Lond B Biol Sci.* 349 (1329), 313–19.

Bassler, B. 2009. "How Bacteria 'Talk.'" TED Talk. Retrieved September 2017 from https://www.ted.com/talks/bonnie_bassler_on_how_bacteria_communicate.

Bianconi, E., A. Piovesan, F. Facchin, A. Beraudi, R. Casadei, F. Frabetti ... S. Canaider. 2013. "An estimation of the number of cells in the human body." *Annals of Human Biology* 40 (6), 463–71.

El Albani, E., S. Bengston, D. E. Canfield, A. Bekker, R. Macchiarelli, A. Mazurier ... A. Meunier. 2010. "Large colonial organisms with coordinated growth in oxygenated environments 2.1 Gyr ago." *Nature* 466, 100–104.

McGinnis, W., and R. Krumiauft. 1992. "Homeobox Genes and Axial Patterning." *Cell* 66, 283–302.

Han, T. M., and B. Runnegar. 1992. "Megascopic eukaryotic algae from the 2.1-billion-year-old negaunee iron-formation, Michigan." *Science* 257 (5067), 232–35.

Herron, M. D., J. M. Borin, J. C. Boswell, Walker, C. A. Knox, M. Boyd … W. C. Ratcliff. 2018. "De novo origin of multicellularity in response to predation." *bioRxiv.* doi:https://doi.org/10.1101/247361.

Miller, S. M. 2010. "Volvox, Chlamydomonas, and the Evolution of Multi-cellularity." *Nature Education* 3 (9), 65.

Pennisi, E. 2018. "The Momentous Transition to Multicellular Life May Not Have been So Hard After All." *Science Magazine.* Retrieved June 2018 from http://www.sciencemag.org/news/2018/06/momentous-transition-multicellular-life-may-not-have-been-so-hard-after-all.

Phillips, T. 2008. "Regulation of transcription and gene expression in eukaryotes." *Nature Education* 1 (1), 199.

Ullmann, A. 2009. "*Escherichia coli* Lactose Operon." Encyclopedia of Life Sciences, John Wiley & Sons. Retrieved September 2017 from http://onlinelibrary.wiley.com/doi/10.1002/9780470015902.a0000849.pub2/full.

Chapter 10
Books

Alberts, B., D. Bry, J. Lewis, M. Raff, K. Rogerts, and J. D. Watson. 1994. *Molecular Biology of the Cell.* 3rd ed. New York, NY: Garland Publishing, Inc.

de Duve, C. 1995. *Vital Dust: The Origin and Evolution of Life on Earth.* New York, NY: Basic Books.

Smith, J. M. and E. Szathmary. 1999. *The Origins of Life: From the Birth of Life to the Origins of Language.* New York, NY: Oxford University Press.

Other References

Crow, J. F. 2005. "Hermann Joseph Muller, Evolutionist." *Nature Reviews Genetics* 6, 941–45.

Droser, M. L., and J. G. Gehling. 2008. "Synchronous aggregate growth in an abundant new Ediacaran tubular organism." *Science* 319 (5870), 1660–62.

Haigh, J. 1978. "The accumulation of deleterious genes in a population—Muller's Ratchet." *Theoretical Population Biology*, 14 (2), 251–267.

Ramesh, M. A., S. B. Malik, and J. M. Logsdon Jr. 2005. "A phylogenomic inventory of meiotic genes; evidence for sex in Giardia and an early eukaryotic origin of meiosis." *Current Biology* 15 (2), 185–91.

Scoville, H. 2017. "What is the Red Queen Hypothesis?" ThoughtCo. Retrieved November 2017 from https://www.thoughtco.com/red-queen-hypothesis-1224710.

Chapter 11
Books

Darwin, C. 1871. *The Descent of Man and Selection in Relation to Sex*. London: John Murray.

———. 1873. *The Expression of the Emotions in Man and Animals*. London: John Murray.

Ghiselin, M. T. 1969. *The Evolution of Hermaphroditism among Animals*. Chicago, IL: The University of Chicago Press.

Long, J. A. 2012. *The Dawn of the Deed: The Prehistoric Origins of Sex*. Chicago, IL: The University of Chicago Press.

Other References

Camperio Ciani, A. S., L. Fontanesi, F. Lemmola, E. Giannella, C. Ferron, and L. Lombardi. 2012. "Factors associated with higher fecundity in female maternal relatives of homosexual men." *J Sex Med* 9, 2878–87.

Fletcher, G. J. O., J. A. Simpson, L. Campbell, and N. C. Overall. 2014. "Pair-Bonding, Romantic Love, and Evolution: The Curious Case of *Homo sapiens*." *Perspectives on Psychological Science* 10 (1), 20–36.

Holland, D., L. Chang, T. M. Ernst, M. Curran, S. D. Buchthal, D. Alicata … A. M. Dale. 2014. "Structural Growth Trajectories and Rates of Change in the First 3 Months of Infant Brain Development." *JAMA Neurol* 71 (10), 1266–74.

Jabr, F. 2010. "The Evolution of Emotion: Charles Darwin's Little-known Psychology Experiment." *Scientific American*. May 24. Retrieved September 2018 from https://blogs.scientificamerican.com/observations/the-evolution-of-emotion-charles-darwins-little-known-psychology-experiment/.

Jannini, E. A., R. Blanchard, A. Camperio-Ciani, and J. Bancroft. 2010. "Male Homosexuality: Nature or Culture." *J Sex Med* 7, 3245–53.

Long, J. A., E. Mark-Kurik, Z. Johanson, M. S. Y. Lee, G. C. Young, Z. Min … K. Trinajstic. 2015. "Copulation in antiarch placoderms and the origin of gnathostome internal fertilization." *Nature* 517, 196–99.

Long, J. A., K. Trinajstic, G. C. Joung, and T. Senden. 2008. "Live birth in the Devonian period." *Nature* 453, 650–52.

Luo, Z., C. Yuan, Q. Meng, and Q. Ji. 2011. "Jurassic eutherian mammal and divergence of marsupials and placentals." *Nature* 476, 442–45.

Maestripieri, D. 2012. "The Evolutionary History of Love. What love is and where it comes from." Psychology Today. Retrieved November 2017 from https://www.psychologytoday.com/blog/games-primates-play/201203/the-evolutionary-history-love

Watts, P. C., K. R. Buley, S. Sanderson, W. Boardman, C. Ciofi, and R. Gibson. 2006. "Parthenogenesis in Komodo dragons." *Nature* 444, 1021–22.

Chapter 12
Books

Dawkins, R. 2004. *The Ancestor's Tale: A Pilgrimage to the Dawn of Evolution.* New York, NY: Hougton Miffflin Company.

de Duve, C. 1995. *Vital Dust: The Origin and Evolution of Life on Earth.* New York, NY: Basic Books.

Other References

Carroll, S. B. 2010. "In a single cell predator, clues to the animal kingdom's birth." The *New York Times.* December 13. Retrieved December 2017 from http://www.nytimes.com/2010/12/14/science/14creatures.html.

Dominguez, R., and K. C Holmes. 2011. "Actin Structure and Function." *Annual Review of Biophysics* 40, 169–86.

Hartman, M. A. and Spudich, J. A. 2012. "The myosin superfamily at a glance." *J Cell Science* 125, 1627–32.

Huxley, H. E., and J. Hanson. 1954. "Changes in the cross-striations of muscle during contraction and stretch and their structural interpretation." *Nature* 173, 973–76.

Huxley, A. F., and R. Niedergerke. 1954. "Structural changes in muscle during contraction: Interference microscopy of living muscle fibres." *Nature* 173, 971–73.

King, N. 2004. "The unicellular ancestry of animal development." *Dev Cell 7,* 313–25.

King, N., J. M. Westbrook, S. L. Young, A. Kuo, M. Abedin, J. Chapman … D. Rokhsar. 2008. "The genome of the choanoflagellate *Monosiga brevicollis* and the origin of metazoans." *Nature* 451 (7180), 783–88.

Krans, J. L. 2010. "The sliding filament theory of muscle contraction." *Nature Education* 3 (9), 66.

Monahan-Earley R., A. M. Dvorak, and W. C. Aird. 2013. "Evolutionary origins of the blood vascular system and endothelium." *J Thromb and Haemost* 11 (Suppl. I), 46–66.

Morris S. C., and J-B Caron. 2014. "A primitive fish from the Cambrian of North America." *Nature* 512, 419–22.

Nielsen, C. 2008. "Six major steps in animal evolution: are we derived sponge larvae?" *Evolution and Development,* 10 (2), 241–57.

Rivero, F. and F. Cyrckova. 2007. "Origins and evolution of the actin cytoskeleton" In G. Jekely, editor. *Eukaryotic membranes and cytoskeleton: origins and evolution.* New York, NY: Springer/Landes Bioscience.

Van den Ent, F., L. A. Amos, and J. Lowe, 2001. Prokaryotic origin of the actin cytoskeleton. *Nature* 413, 39–44.

Wickstead, B., and K. Gull. 2011. "The evolution of the cytoskeleton." *J Cell Biol* 194 (4), 513–25.

Chapter 13
Books

Darwin, C. 1859. *On the Origin of Species by Natural Selection.* Lyndhurst, NJ: Barnes and Noble Books. Republished 2004.

Lane, N. 2009. *Life Ascending: The Ten Great Inventions of Evolution.* London: W. W. Norton & Company, Inc.

Parker, A. 2003. *In the Blink of an Eye: How Vision Sparked the Big Bang of Evolution.* Cambridge, MA: Perseus Publishing.

Other References

De Jong, W. W., J. A. Leunissen, and C. E. Voorter. 1993. "Evolution of the alpha-crystallin/ small heat-shock protein family." *Mol Biol Evol* 10 (1), 103–26.

Friedman, A. L. 1998. "A Review of the Highly Conserved PAX6 Gene in Eye Development Regulation." *Journal of Young Investigators* 1, issue 1. Retrieved March 2018 from http://legacy.jyi.org/volumes/volume1/issue1/articles/friedman.html.

Gehring, W. J. 2005. "New perspectives on eye development and the evolution of eyes and photoreceptors." *Journal of Heredity* 96 (3), 171–84.

Lamb, T. D., S. P. Collin, and E. N. Pugh Jr. 2007. "Evolution of the vertebrate eye: opsins, photoreceptors, retina and eye cup." *Nat Rev Neuroscience* 8 (12), 960–76.

Schichida, Y., and T. Matsuyama. 2009. "Evolution of opsins and phototransduction." *Phil Trans R Soc B* 364, 2881–95.

Shen, L., C. Chen, H. Zheng, and L. Jin. 2013. "The evolutionary relationship between microbial rhodopsins and metazoan rhodopsins." *The Scientific World Journal* 2013, 1–10.

Slingsby, C., G. J. Wistow, and A. R. Clark. 2013. "Evolution of crystallins for a role in the vertebrate eye lens." *Protein Science* 22, 367–80.

Chapter 14
Books

Bownds, M. D. 1999. *The Biology of Mind: Origins and Structures of Mind, Brain, and Consciousness.* Bethesda, MD: Fitzgerald Science Press, Inc.

Linden, D. J. 2007. *The Accidental Mind: How Brain Evolution Has Given Us Love, Memory, Dreams, and God.* Cambridge, MA: Harvard University Press.

Other References

Begley, S. 2007. "The Brain: How the Brain Rewires Itself." *Time.* January 19. Retrieved January 2018 from http://content.time.com/time/magazine/article/0,9171,1580438,00.html.

Byrne, J. H. 1997. "Synaptic Plasticity." Chapter 7 in *Neuroscience Online.* Retrieved January 2017 from http://nba.uth.tmc.edu/neuroscience/s1/chapter07.html.

Ellwanger, K., and M. Nickel. 2006. "Neuroactive substances specifically modulate rhythmic body contractions in the nerveless metazoan *Tethya wilhelma* (Demospongiae, Porifera)." *Front Zool* 27, 3–7.

Finn, J. K., T. Tregenza, and M. D. Norman. 2009. "Defensive tool use in a coconut-carrying octopus." *Current Biology* 19 (23), R1069–70.

Finn, J. 2009. "Coconut carrying octopus." Retrieved December 2017 from https://www.youtube.com/watch?v=1DoWdHOtlrk.

Ghysen, A. 2003. "The origin and evolution of the nervous system." *Int. J. Dev. Biol.* 47, 555–62.

Goldman, B. 2010. "Stunning details of brain connections revealed." *ScienceDaily.* November 17. Retrieved December 31, 2017, from www.sciencedaily.com/releases/2010/11/101117121803.htm.

Hochner, B., T. Shomrat, and G. Fiorito. 2006. "The octopus: a model for a comparative analysis of the evolution of learning and memory mechanisms." *Biol. Bull.* 210 (3), 308–317.

Kimata, T., H. Sasakura, N. Ohnishi, N. Nishio, and I. Mori. 2012. "Thermotaxis of *C. elegans* as a model for temperature perception, neural information processing and neural plasticity." *Worm* 1, issue 1, 31–41.

Kolb, B. and I. Q. Whishaw, 1998. "Brain plasticity and behavior." *Annual Review of Psychology* 49, 43–64.

Kristan, W. B. Jr. 2016. "Early evolution of neurons." *Current Biology* 26 (20), R949–54.

Lamprecht, R., and J. LeDoux. 2004. "Structural plasticity and memory." *Nature Reviews Neuroscience* 5, 45–54.

Leys, S. P., and R. W. Meech. 2006. "Physiology of coordination in sponges." *Canadian Journal of Zoology* 84 (2), 288–306.

Mayford, M., S. A. Siegelbaum, and E. R. Kandel. 2012. "Synapses and Memory Storage." *Cold Spring Harb Perspect Biol* 4 (6), a005751.

Nuwer, R. 2013. "Severed Octopus Arms Have a Mind of Their Own." *Smithsonian Magazine.* August 29. Retrieved September 2018 from https://www.smithsonianmag.com/smart-news/severed-octopus-arms-have-a-mind-of-their-own-2403303/.

Perlmutter, D. 2013. "The Gift of Neuroplasticity." Retrieved January 2018 from https://www.drperlmutter.com/gift-neuroplasticity/

Renard, E., J. Vacelet, E. Gazave, P. Lapebie, C. Borchiellini, and A. V. Ereskovsky. 2009. "Origin of the neuro-sensory system: New and expected insights from sponges." *Integr Zool.* 4 (3), 294–308.

Sakarya, O., K. A. Armstrong, M. Adamska, M. Adamski, I. Wang, B. Tidor … and K. S. Kosik. 2007. "A post-synaptic scaffold at the origin of the animal kingdom." *PLoS ONE* 2 (6), e506.

Satterlie, R. A. 1985. "Control of swimming in the hydrozoan jellyfish *Aequorea aequorea*: direct activation of the subumbrella." *J Neurobiol.* 16 (3), 211–26.

———. 2011. "Do jellyfish have central nervous systems?" *J Experimental Biology* 214, 1215–23.

von Bartheld, C. S., J. Bahney, and S. Herculano-Houzel. 2016. "The search for true numbers of neurons and glial cells in the human brain: A review of 150 years of cell counting." *J Comp Neurol*, 524 (18), 3865–95.

Walker, R. J., H. L. Brooks, and L. Holden-Dye. 1996. "Evolution and overview of classical transmitter molecules and their receptors." *Parasitology* 113, supplement S3–33.

Chapter 15
Books

Beerling, D. 2007. *The Emerald Planet: How Plants Changed Earth's History*. New York, NY: Oxford University Press.

de Duve, C. 1995. Vital Dust: The Origin and Evolution of Life on Earth. New York, NY: Basic Books.

Mason, K. A., J. B. Losos, and S. R. Singer. 2008. *Biology*. 11th ed. New York, NY: McGraw-Hill Education.

Voet, D., J. G. Voet, and C. W. Pratt. 2013. *Fundamentals of Biochemistry: Life at the Molecular Level*. 4th ed. Hoboken, NJ: John Wiley and Sons, Inc.

Willis, K. J., and McElwain, J. C. 2010. *The Evolution of Plants*. New York, NY: Oxford University Press.

Other References

Bennici, A. 2008. "Origin and early evolution of land plants." *Commun Integr Biol* 1(2), 212–18.

Bhattacharya, D., and L. Medlin. 1998. "Algal Phylogeny and the Origin of Land Plants." *Plant Physiology* 116(1), 9–15.

Boyd, E. S., and J. W. Peters. 2013. "New insights into the evolutionary history of biological nitrogen fixation." *Front Microbiol* 4, 201.

Edwards, D., K. L. Davies, and L. Axe. 1992. "A vascular conducting strand in the early land plant *Cooksonia*." *Nature* 357 (6380), 683–85.

Gillespie, W. H., Roghwell, G. W., and Scheckler, S. E. 1981. "The earliest seeds." *Nature* 293, 462–64.

Hochuli, P. A., and S. Feist-Burkhardt. 2013. "Angiosperm-like pollen and Afropollis from the Middle Triassic (Anisian) of the Germanic Basin (Northern Switzerland)." *Frontiers in Plant Science* 4, 344.

Horodyski, R. J., and P. L. Knauth. 1994. "Life on Land in the Precambrian." *Science* 263, 494–98.

Howe, C. J., A. C. Barbrook, R. E. R. Nisbet, P. J. Lockhart, and A. W. D. Larkum. 2008. "The origin of plastids." *Phil Trans R Soc B* 363, 2675–85.

Kneip C., P. Lockhart, C. Voss, and U. G. Maier. 2007. "Nitrogen fixation in eukaryotes—new models for symbiosis." *BMC Evol Biol* 7, 55.

Mishler, F. D. 2000. "Deep phylogenetic relationships among 'plants' and their implications for classification." *Taxon* 49, 661–83.

Remy, W., T. Taylor, H. Hass, and H. Kerp. 1994. "Four hundred-million-year-old vesicular arbuscular mycorrhizae." *Proc Natl Acad Sciences USA* 91 (25), 11841–43.

Rubenstein, C. V., P. Gerrienne, G. S. de la Puente, R. A. Astini, and P. Steemans. 2010. "Early middle Ordovician evidence for land plants in Argentina (eastern Godwana)." *New Physiologist* 188, 365–69.

Steemans, P., A. Le Herisse, J. Melvin, M. A. Miller, F. Paris, J. Verniers, and C. H. Wellman. 2009. "Origin and radiation of the earliest vascular land plants." *Science* 324 (5925), 353.

Taylor, T. N., H. Hass, W. Remy, and H. Kerp. 1995. "The oldest fossil lichens." *Nature* 378, 244.

Wellman, C. H., P. L. Osterloff, and U. Mohiuddin. 2003. "Fragments of the earliest land plants." *Nature* 425 (6955), 282–285.

Chapter 16
Books

Boudreaux, H. B. 1979. *Arthropod Phylogeny with Special Reference to Insects*. New York, NY: Wiley.

Dawkins, R. 2004. *The Ancestor's Tale: A Pilgrimage to the Dawn of Evolution*. New York, NY: Hougton Mifflin Company.

de Duve, C. 1995. *Vital Dust: The Origin and Evolution of Life on Earth*. New York, NY: Basic Books.

Other References

Blom, H. 2005. "Taxonomic revision of the Late Devonian tetrapod Ichthyostega from East Greenland." *Palaeontology* 48(1), 111–34.

Carroll, R. L. 1964. "The earliest reptiles." *Journal of the Linnean Society of London, Zoology* 45, 61–83.

King, H. M., N. H. Shubin, M. I. Coates, and M. E. Hale. 2011. "Behavioral evidence for the evolution of walking and bounding before terrestriality in Sarcopterygian fishes." *Proc Natl Acad Sci USA* 108(52), 21146–51.

Kutschera, U., and J. M. Elliott. 2013. "Do mudskippers and lungfishes elucidate the early evolution of four-limbed vertebrates?" *Evolution: Education and Outreach* 6, 8.

Shubin, N. H., E. B. Daeschler, and F. A. Jenkins Jr. 2006. "The pectoral fin of *Tiktaalik roseae* and the origin of the tetrapod limb." *Nature* 440, 764–71.

Strauss, B. 2017. "Why Are Amphibians in Decline?" ThoughtCo. Retrieved September 2018 from https://www.thoughtco.com/why-amphibians-are-in-decline-129435.

Wilford, J. N. 2006. "Fossil Called Missing Link from Sea to Land Animals." The *New York Times*. April 6. Retrieved January 2018 from http://www.nytimes.com/2006/04/06/science/06fossil.html.

Wilson, H. M., and Anderson, L. I. 2004. "Morphology and taxonomy of Paleozoic millipedes (*Diplopoda*: *Chilognatha*: *Archipolypoda*) from Scotland." *Journal of Paleontology* 78, 169–84.

Chapter 17
Books

Dawkins, R. 2004. *The Ancestor's Tale: A Pilgrimage to the Dawn of Evolution*. New York, NY: Hougton Mifflin Company.

Lane, N. 2009. *Life Ascending: The Ten Great Inventions of Evolution*. New York, NY: W. W. Norton and Company, Inc.

Strauss, B. 2015. *A Field Guide to the Dinosaurs of North America and Prehistoric Megafouna*. Landham, MD: Rowman and Littlefield.

Ward, P. D. 2006. *Out of Thin Air: Dinosaurs, Birds, and Earth's Ancient Atmosphere*. Washington, DC: Joseph Henry Press.

Other References

BBC Earth. "Argentinosaurus 'Argentina's Lizard.'" 2018. BBC Earth: Walking with Dinosaurs. Retrieved January 2018 from http://www.bbcearth.com/walking-with-dinosaurs/modal/argentinosaurus/.

Bennett, A. F., and J. A. Ruben. 1979. "Endothermy and activity in vertebrates." *Science* 206 (4419), 649–54.

Castro, J. 2016, March 18. "Stegosaurus: Bony Plates and Tiny Brain." Livescience. Retrieved September 2018 from http://www.livescience.com/24184-stegosaurus-facts.html.

Clarke, A., and H. O. Portner. 2010. "Temperature, metabolic power and the evolution of endothermy." *Biol Rev Camb Philos Soc* 85(4), 703–27.

D'Emic, M. D. 2015. Comment on "Evidence for mesothermy in dinosaurs." *Science* 348(6238), 982.

Erickson G. M., O. W. M. Rauhut, Z. Zhou, A. H. Turner, B. D., Inouye, D. Hu, and M. A. Norell. 2009. "Was Dinosaurian Physiology Inherited by Birds? Reconciling Slow Growth in *Archaeopteryx*." *PLoS ONE* 4 (10), e7390.

Geggel, L. 2015. "Were Dinosaurs Warm-Blooded? New Study Fuels Debate." *Live Science*. June 10. Retrieved January 2018 from https://www.livescience.com/51162-dinosaurs-warm-blooded-growth-rates.html.

Grady, J. M., B. J. Enquist, E. Dettweiler-Robinson, N. A. Wright, and F. A. Smith. 2014. "Dinosaur physiology. Evidence for mesothermy in dinosaurs." *Science* 344(6189), 1268–72.

Luo, Z-X., A. W. Crompton, and A-L. Sun. 2001. "A new mammaliaform from the early Jurassic and evolution of mammalian characteristics." *Science* 292, 1535–40.

Nedergaard, J., D. Ricquier, and L. P. Kozak, 2005. "Uncoupling proteins: current status and therapeutic prospects." *EMBO Rep* 6 (10), 917–21.

Ritchison, G. 2018. "Avian Respiration." Avian Biology. Retrieved July 2018 from http://people.eku.edu/ritchisong/birdrespiration.html.

Rogers, P. 2014. "Why Were Dinosaurs So Successful?" Forbes. Retrieved January 2018 from https://www.forbes.com/sites/paulrodgers/2014/09/08/the-secret-of-dinosaur-success/#1aa33baf7cc0.

Sookias, R. B., R. J. Butler, and R. B. J. Benson. 2012. "Rise of dinosaurs reveals major body-size transitions are driven by passive processes of trait evolution." *Proc R. Soc. B* 279, 2180–87.

Strauss, B. 2018, June 8. "What Is the Scientific Definition of a Dinosaur?" ThoughtCo. Retrieved September 2018 from https://www.thoughtco.com/definition-of-a-dinosaur-1091930.

Stricherz, V. 2003. "Ultra-low oxygen could have triggered mass extinctions, spurred bird breathing system." *UW News.* October 31. Retrieved September 2018 from http://www.washington.edu/news/2003/10/31/ultra-low-oxygen-could-have-triggered-mass-extinctions-spurred-bird-breathing-system/.

Watanabe, M. E. 2005. "Generating heat: New twists in the evolution of endothermy." *BioScience* 55 (6), 470–75.

Chapter 18
Books

Benton, M. J. 2003. *When Life Nearly Died: The Greatest Mass Extinction of All Time.* London, UK: Thames and Hudson Ltd.

Raup, D. M. 1991. *Extinction: Bad Genes or Bad Luck?* New York, NY: W. W. Norton and Company.

Rothschild, L. J., and A. M. Lister, editors. 2003. *Evolution on Planet Earth: The Impact of the Physical Environment.* San Diego, CA: Academic Press.

Ward, P. 2006. *Out of Thin Air: Dinosaurs, Birds, and Earth's Ancient Atmosphere.* Washington, DC: Joseph Henry Press.

Ward, P., and Brownlee, D. 2000. *Rare Earth: Why Complex Life is Uncommon in the Universe.* New York, NY: Copernicus Books.

Wilson, E. O. 2002. *The Future of Life.* New York, NY: Vintage Books.

Other References

Alvarez, L. W., W. Alvarez, F. Asaro, and H. V. Michel. 1980. "Extraterrestrial cause for the Cretaceous–Tertiary extinction." *Science* 208 (4448), 1095–108.

Beerling, D. J., M. Harfoot, B. Lomax, and J. A. Pyle. 2007. "The stability of the stratospheric ozone layer during the end-Permian eruption of the Siberian Traps." *Philosophical Transactions of the Royal Society A* 365, 1843–66.

Berner, R. A. 2006. "GEOCARBSULF: A combined model for Phanerozoic atmospheric O2 and CO2." *Geochim et Cosmochimica Acta* 70, 5653–64.

Berner, R. A., J. M. VandenBrooks, and P. D. Ward. 2007. "Oxygen and Evolution." *Science* 316 (5824), 557–58.

Bond, D. P. G., and S. E. Grasby. 2017. "On the causes of mass extinctions." Palaeogeography, Palaeoclimatology, Palaeoecology, 478, 3–29.

Brannen, P. 2017. "When Life on Earth Was Nearly Extinguished." *The New York Times*. July 29. Retrieved January 2018 from https://www.nytimes.com/2017/07/29/opinion/sunday/when-life-on-earth-was-nearly-extinguished.html.

Darwin, C. 1859. AZQuotes.com. Retrieved January 2018 from http://www.azquotes.com/quote/1095185.

Gould, S. J., and N. Eldredge. 1977. "Punctuated Equilibria: The Tempo and Mode of Evolution Reconsidered." *Paleobiology* 3 (2), 115–51.

Kaplan, M. 2009. "Why big eruptions don't always fuel mass extinctions." *Nature*. doi:10.1038/news.2009.380.

Kiehl, J. T., and C. A. Shields. 2005. "Climate simulation of the latest Permian: Implications for mass extinction." *Geology* 33 (9), 757–60.

Kump, L. R., A. Pavlov, and M. A. Arthur 2005. "Massive release of hydrogen sulfide to the surface ocean and atmosphere during intervals of oceanic anoxia." *Geology* 33 (5), 37–400.

Mora, C., D. P. Tittensor, S. Adl, A. G. B. Simpson, and B. Worm. 2011. "How Many Species Are There on Earth and in the Ocean?" *PLoS Biol* 9 (8): e1001127.

Puiu, T. 2012. "During the greatest mass extinction in Earth's history the world's oceans reached 40°C – lethally hot." Retrieved January 2018 from https://www.zmescience.com/research/studies/great-pre-permian-mass-extinction-temperature-too-hot-941432/.

Renne, P. R., and A. R. Basu. 1991. « Rapid eruption of the Siberian Traps flood basalts at the Permo-Triassic boundary." *Science* 253 (5016), 176–79.

Renne, P. R., C. J. Sprain, M. A. Richards, S. Self, L. Vanderkluysen, and K. Pande. 2015. "State shift in Deccan volcanism at the Cretaceous-Paleogene boundary, possibly induced by impact." *Science* 350 (6256), 76–78.

Richards, M. A., W. Alvarez, S. Self, L. Karlstrom, P. R. Renne, M. Manga, M … S. A. Gibson. 2015. "Triggering of the largest Deccan eruptions by the Chicxulub impact." *Geological Society of America Bulletin* 127 (11–12), 1507–20.

Rothman, D. H., G. P. Fournier, K. L. French, E. J. Alm, E. A. Boyle, C Cao, and R. E. Summons. 2014. "Methanogenic burst in the end-Permian carbon cycle." *Proceedings of the National Academy of Sciences* 111 (15), 5462–67.

Sun, Y., M. M. Joacimski, P. B. Wignall, C. Yan, Y. Chen, H. Jiang … X. Lai. 2012. "Lethally Hot Temperatures During the Early Triassic Greenhouse." *Science* 338 (6105), 336–70.

Chapter 19
Books

Diamond, J. 2006. *The Third Chimpanzee: The Evolution and Future of the Human Animal.* New York, NY: Harper Perennial.

Other References

Amos, W., and J. I. Hoffman. 2010. Evidence that two main bottleneck events shaped modern human genetic diversity. *Proc Biol Sci* 277(1678), 131–37.

Ambrose, S. H. 1998. "Late Pleistocene human population bottlenecks, volcanic winter, and differentiation of modern humans." *Journal of Human Evolution* 34, 623–51.

———. 2003. "Did the super-eruption of Toba cause a human population bottleneck? Reply to Gathorne-Hardy and Harcourt-Smith." *J Human Evolution* 45, 231–37.

Bowden R., T. S. MaacFie, S. Myers, G. Hellenthal, E. Nerrienet, R. E. Bontrop … N. I. Mundy. 2012. Genomic tools for evolution and conservation in the chimpanzee: *Pan troglodytes ellioti* is a genetically distinct population. *PLoS Genet* 8, e1002504.

Cann, R. L., M. Stoneking, and A. C. Wilson. 1987. "Mitochondrial DNA and human evolution." *Nature* 325 (6099), 31–36.

Gibbons, A. 2009. "How We Lost Our Diversity." *Science Magazine.* October 8. Retrieved February 2018 from http://www.sciencemag.org/news/2009/10/how-we-lost-our-diversity.

Green, R. E., J, Krause, A. W. Briggs, T. Maricic, U. Stenzel, M. Kircher … S. Pääbo. 2010. "A Draft Sequence of the Neanderthal Genome." *Science* 328 (5979), 710–22.

Hershkovitz, I., G. W. Weber, R. Quam, M. Duval, R. Grun, L. Kinsley … M. Weinstein-Evron. 2018. "The earliest modern humans outside Africa." *Science*, 359 (6374), 456–59.

Hublin, J-J., A. Ben-Ncer, S. E. Bailey, S. E. Freidline, S. Neubauer, M. M. Skinner … P. Gunz. 2017. "New fossils from Jebel Irhoud, Morocco and the pan-African origin of *Homo sapiens*." *Nature* 546, 289–92.

Johanson, D. 2001. "Origins of Modern Humans: Multiregional or Out of Africa?" Actiobioscience. Retrieved June 2017 from http://www.actionbioscience.org/evolution/johanson.html.

Jorde, L. B., and S. P. Wooding. 2004. "Genetic variation, classification and 'race.'" *Nature Genetics* 36, S28–33.

Kimbel, W. H. "Lucy's Story." Institute of Human Origins. Retrieved June 2017 from https://iho.asu.edu/about/lucys-story.

Klein, R. G. 2003. "Whither the Neanderthals?" *Science* 299 (5612), 1525–27.

Lane, C. S., B. T. Chorn, and T. C. Johnson. 2013. "Ash from the Toba super eruption in Lake Malawi shows no volcanic winter in East Africa at 75 ka." *Proc Natl Acad Sci* 110 (20), 8025–29.

Leakey, L. S. B., P. V. Tobias, and J. R. Napier. 1964. "A new species of genus *Homo* from Olduvai Gorge." *Nature* 202, 7–9.

Lemonick, M. D. 1987. Everyone's Genealogical Mother. *Time*. January 26. Retrieved March 2018 from http://content.time.com/time/magazine/article/0,9171,963320,00.html.

Levy, S., G. Sutton, P. C. Ng, L., Feuk, A. L. Halpern, B. P. Walenz … J. C. Venter. 2007. "The Diploid Genome Sequence of an Individual Human." PLoS Biology, 5 (10), 2113–44.

Li, H., and R. Durbin. 2011. "Inference of human population history from individual whole-genome sequences." *Nature* 475, 493–97.

Manzi, G. 2011. "Before the emergence of *Homo sapiens*: Overview on the Early-to-middle Pleistocene fossil record (with a proposal about *Homo heidelbergensis* at the subspecific level)." *Int J Evol Biol* 2011, 582678.

McDougall, I., F. H. Brown, and I. G. Fleagle, 2005. "Stratigraphic placement and age of modern humans from Kibish, Ethiopia." *Nature* 433 (7027) 733–36.

NIH. 2017. "Understanding Human Genetic Variation." NIH Curriculum Supplement Series. Retrieved September 2018 from https://www.ncbi.nlm.nih.gov/books/NBK20363/.

NIH Research News. 2010. "Neanderthal Genome Sequenced." *NIH Research News.* May 24. Retrieved February 2018 from https://www.nih.gov/news-events/nih-research-matters/neanderthal-genome-sequenced.

Ni, X., D. L. Gebo, M. Dagosto, J. Meng, P. Tafforeau, J. J. Flynn, and K. C. Beard. 2013. "The oldest known primate skeleton and early haplorhine evolution." *Nature* 498 (7452), 60–64.

Pinhasi, R., T. F. G. Higham, L. V. Golovanova, and V. B. Doronichev. 2011. "Revised age of late Neanderthal occupation and the end of the Middle Paleolithic in the northern Caucasus." *Proc Natl Acad Sci USA*, 108 (21), 8611–16.

Pontzer, H. 2012. "Overview of Hominin Evolution." *Nature Education Knowledge* 3 (10), 8.

Pruitt, S. 2018. "Stunning New Fossil Suggests Humans Left Africa Far Earlier Than We Thought." History.com. Retrieved March 2018 from https://www.history.com/news/oldest-human-fossil-found-outside-africa.

Reich, D. 2018. "How Genetics is Changing Our Understanding of Race." *New York Times*. March 23. Retrieved March 2018 from https://www.nytimes.com/2018/03/23/opinion/sunday/genetics-race.html.

Stevens, N. J., E. R. Seiffert, P. M. O'Connor, E. M. Roberts, M. D. Schmitz, C. Krause … J. Temu. 2013. "Palaeontological evidence for an Oligocene divergence between Old World monkeys and apes." *Nature* 497, 611–14.

Van Arsdale, A. P. 2013. "*Homo erectus* - A bigger, smarter, faster hominin lineage." *Nature Education Knowledge* 4 (1), 2.

Villmoare, B., W. H. Kimbel, C. Deyoum, C. J. Campisano, E. N. DiMaggio, J. Rowan … K. E. Reed. 2015. "Early Homo at 2.8 Ma from Ledi-Geraru, Afar, Ethiopia." *Science*, 347 (6228), 1352–55.

Wolpoff, M. H., J. N. Spuhler, F. H. Smith, J. Radovcic, G. Pope, D. W. Frayer … G. Clark. 1988. "Modern Human Origins." *Science* 241 (4867), 772–74.

Chapter 20
Books

Corning, P. 2018. *Synergistic Selection: How Cooperation Has shaped evolution and the Rise of Humankind.* London, UK: World Scientific Publishing Co. Pte. Ltd.

Cziko, G. 1995. *Without Miracles: Universal Selection Theory and the Second Darwinian Revolution.* Boston, MA: The MIT Press.

Dawkins, R. 1989. *The Selfish Gene.* New York, NY: Oxford University Press.

Diamond, J. 1992. *The Third Chimpanzee: The Evolution and Future of the Human Animal.* New York, NY: Harper Perennial.

Harari, Y. N. 2015. *Sapiens: A Brief History of Humankind.* New York, NY: HarperCollins Publishers.

Joyce, R. 2007. *The Evolution of Morality.* Cambridge, MA: The MIT Press.

Pinker, S. 2007. *The Language Instinct: How the Mind Creates Language.* New York, NY: Harper Perennial Modern Classics.

Smith, J. M., and E. Szathmary. 1999. *The Origins of Life: From the Birth of Life to the Origins of Language.* New York, NY: Oxford University Press.

Wilson, E. O. 2012. *The Social Conquest of Earth.* New York, NY: W. W. Norton and Company.

Other References

Constable, G. W. A., T. Rogers, A. J. McKane, and C. E Tarnita. 2016. "Strength in numbers: Demographic noise can reverse the direction of selection." *Proc Natl Acad Sci USA* 113 (32), E4745–54.

Enard, W., M. Przeworski, S. E. Fisher, C. S. L. Lai, V. Wiebe, T. Kitano … S. Pääbo. 2002. "Molecular evolution of *FOXP2*, a gene involved in speech and language." *Nature* 483, 869–72.

Green, R. E., J. Krause, A. W. Briggs, T. Maricic, U. Stenzel, M. Kircher … S. Pääbo. 2010. "A Draft Sequence of the Neanderthal Genome." *Science* 328 (5979), 710–22.

NIH Research News. 2010. "Neanderthal Genome Sequenced." *NIH Research News*. May 24. Retrieved February 2018 from https://www.nih.gov/news-events/nih-research-matters/neanderthal-genome-sequenced.

Nowak, M. A., and K. Sigmund. 2005. "Evolution of indirect reciprocity." *Nature* 437, 1291–98.

Viegas, J. 2017. "Comparison of Primate Brains Reveals Why Humans Are Unique." Retrieved July 2018 from https://www.seeker.com/health/mind/comparison-of-primate-brains-reveals-why-humans-are-unique.

Wloch-Salamon, D. M. 2013. "Sociobiology of the Budding Yeast." *J Biosci* 38, 1–12.

Chapter 21
Books

Boehm, C. 2001. *Hierarchy in the Forest: The Evolution of Egalitarian Behavior*, Cambridge, MA: Harvard University Press.

Dawkins, R. 1989. *The Selfish Gene*. New York, NY: Oxford University Press.

de Waal, F. 1996. *Good Natured*. Cambridge, MA: Harvard University Press.

Diamond, J. 1992. *The Third Chimpanzee: The Evolution and Future of the Human Animal*. New York, NY: Harper Perennial.

Ingold, T., D. Riches, and J. Woodburn, editors. 1988. *Hunters and Gatherers 2: Property, Power and Ideology*. Oxford, UK: Berg.

Richerson, P. J., and R. Boyd, 2005. *Not by Genes Alone: How Culture Transformed Human Evolution*. Chicago, IL: The University of Chicago Press.

Other References

Boyd, R., and Richerson, P. J. 2009. "Culture and the evolution of human cooperation." *Philos Trans R Soc Lond B Biol Sci* 364 (1533), 3281–88.

Gavrilets, S. 2012. "On the evolutionary origins of the egalitarian syndrome." *Proc Natl Acad Sci USA* 109 (35), 14069–74.

Gray, P. 2011. "How hunter-gatherers maintained their egalitarian ways." Libcom.org. Retrieved December 2016 from http://libcom.org/history/how-hunter-gatherers-maintained-their-egalitarian-ways-peter-gray.

Hawks, J., E. T. Wang, G. M. Cochran, H. C. Harpending, and R. K. Moyzis. 2007. "Recent acceleration of human adaptive evolution." *Proc Natl Acad Sci USA* 104 (52), 20753–58.

UNESCO. "Neolithic Site of Çatalhöyük." 2012. UNESCO. Retrieved July 2018 from https://whc.unesco.org/en/list/1405.

Rubin, P. H. 2000. "Hierarchy." *Human Nature* 11, 259.

Chapter 22
Books

Bloom, P. 2013. *Just Babies: The Origins of Good and Evil.* New York, NY: Crown.

de Waal, F. 2006. *Primates and Philosophers.* Princeton, NJ: Princeton University Press.

———. 2009. *The Age of Empathy: Nature's Lessons for a Kinder Society.* New York, NY: Three Rivers Press.

Harris, S. 2010. *The Moral Landscape: How Science Can Determine Human Values.* New York, NY: Free Press.

Joyce, R. 2007. *The Evolution of Morality.* Cambridge, MA: The MIT Press.

Pinker, S. 2003. *The Blank Slate: The Modern Denial of Human Nature.* New York, NY: Penguin Books.

Smith, D. L. 2007. *The Most Dangerous Animal: Human Nature and the Origins of War.* New York, NY: St. Martin's Press.

Wilson, E. O. 1975. *Sociobiology: The New Synthesis.* Cambridge, MA: Harvard University Press.

———. 2012. *The Social Conquest of Earth.* New York, NY: W. W. Norton & Co.

Other References

Bloom, P. 2010. "The Moral Life of Babies." *The New York Times Magazine.* May 5. Retrieved February 2018 from http://www.nytimes.com/2010/05/09/magazine/09babies-t.html.

de Waal, F. 2011. "Moral Behavior in Animals." TED Talk. Retrieved February 2018 from http://www.ted.com/talks/frans_de_waal_do_animals_have_morals.

Dostoevsky, F. 2012. In "If God Does Not Exist Everything is Permitted." The-Philosophy.com. Retrieved February 2018 from https://www.the-philosophy.com/god-exist-permitted-dostoevsky.

Hamlin, J. K. 2013. "Moral Judgment and Action in Preverbal Infants and Toddlers: Evidence for an Innate Moral Core." *Cur Dir in Psychol Science* 22, 186–93.

Hamlin, J. K., K. Wynn, and P. Bloom. 2007. "Social evaluation by preverbal infants." *Nature* 540, 557–59.

Prinz, J. 2011. "Morality is a Culturally Conditioned Response." *Philosophy Now* 86, 6–9. Retrieved February 2018 from https://philosophynow.org/issues/82/Morality_is_a_Culturally_Conditioned_Response.

Sage, J. 2017. "The Evolutionary Basis of Self-Deception." Retrieved October 2017 from https://www.uwsp.edu/philosophy/FacultyStaffDocs/jSage/Sage%20Evolutionary%20Basis%20of%20Self-Deception.pdf.

Stewart-Williams, S. 2010, May 2. "Did Morality Evolve? Human nature informs morality, but morality sometimes counteracts human nature." *Psychology Today*. Retrieved September 2018 from https://www.psychologytoday.com/us/blog/the-nature-nurture-nietzsche-blog/201005/did-morality-evolve.

United to End Genocide. "Atrocities Against Native Americans." 2018. Retrieved June 2018 from http://endgenocide.org/learn/past-genocides/native-americans/.

von Hippel, W., and R. Trivers. 2011. "The evolution and psychology of self-deception." *Behavioral and Brain Sciences* 34 (1), 1–16.

Chapter 23
Books

Bownds, M. D. 1999. *The Biology of Mind: Origins and Structures of Mind, Brain, and Consciousness*. Bethesda, MD: Fitzgerald Science Press.

Harari, Y. N. 2016. *Homo Deus: A Brief History of Tomorrow*. London, UK: Harvill Secker.

Harris, S. 2012. *Free Will*. New York, NY: Free Press.

Humphrey, N. 2011. *Soul Dust: The Magic of Consciousness*. Princeton, NJ: Princeton University Press.

Koch, C. 2012. *Consciousness: Confessions of a Romantic Reductionist*. Cambridge, MA: The MIT Press.

———. 2004. *The Quest for Consciousness: A Neurobiological Approach*. Englewood, CO: Roberts and Company Publishers.

Ornstein, R. 1991. *The Evolution of Consciousness: The Origins of the Way We Think*. New York, NY: Simon & Schuster.

Wright, R. 2001. *Nonzero: The Logic of Human Destiny*. New York: Vintage Books.

Other References

Allen, C., and M. Trestman. 2016. "Animal Consciousness." In E. N. Zalta, ed. *The Stanford Encyclopedia of Philosophy*. Winter 2016 edition. Retrieved April 2017 from https://plato.stanford.edu/archives/win2016/entries/consciousness-animal/.

Askenasy, J., and J. Lehmann. 2013. "Consciousness, brain, neuroplasticity." *Front Psychol* 4, 412.

Barash, D. P. 2012. "Mind Readers: Human awareness of our own minds and others' is unlike that of any other animal. But why did consciousness evolve?" Retrieved June 2017 from https://aeon.co/essays/the-self-conscious-animal-how-human-minds-evolved.

Barr, S., P. R. Laming, J. T. A. Dick, and R. W. Elwood. 2008. "Nociception or pain in a decapod crustacean?" *Animal Behavior* 75 (3), 745–51.

Block, N. 1995. "On a Confusion About a Function of Consciousness." *Behavioral and Brain Sciences* 18, 227–47.

Boesch, C., and H. Boesch. 1984. "Mental map in wild chimpanzees: An analysis of hammer transports for nut cracking." *Primates* 2, 160–70.

Feinberg, T. E., and J. Mallat. 2013. "The evolutionary and genetic origins of consciousness in the Cambrian Period over 500 million years ago." *Front Psychol* 4, 667.

Graziano, M. 2016, June 6. "A New theory explains how consciousness evolved: A neuroscientist on how we came to be aware of ourselves." *Atlantic*. Retrieved February 2108 from https://www.theatlantic.com/science/archive/2016/06/how-consciousness-evolved/485558/.

Kipling, R. 1943. "If." Retrieved July 2018 from https://www.poetryfoundation.org/poems/46473/if---.

Libet, B., C. A. Gleason, E. W. Wright, and, D. K. Pearl. 1983. "Time of Conscious Intention to Act in Relation to Onset of Cerebral Activity (Readiness-Potential) - The Unconscious Initiation of a Freely Voluntary Act." *Brain* 106, 623–42.

Logothetis, N. K., D. A. Leopold, and D. L. Sheinberg. 1996. What is rivaling during binocular rivalry? *Nature*, 380, 621-624.

Lutkenhoff, E. S., J. Chiang, L. Tshibanda, E. Kamau, M. Kirsch, J. D. Pickard … M. M. Monti. 2015. "Thalamic and extrathalamic mechanisms of consciousness after severe brain injury." *Ann Neurol* 78 (1), 68–76.

Magee, B., and R. W. Elwood. 2013. "Shock avoidance by discrimination learning in the shore crab (Carcinus maenas) is consistent with a key criterion for pain." *J Experimental Biology* 216, 353–58.

Prior, H., A. Schwarz, and O. Gunturkun. 2008. "Mirror-induced behavior in the magpie (*Pica pica*): Evidence of self-recognition." *PLoS Biol* 6 (8), e202.

Soon, C. S., M. Brass, H-J. Heinze, and J-D. Haynes. 2008. "Unconscious determinants of free decisions in the human brain." *Nature Neuroscience* 11, 543–45.

Van Gulick, R. 2017. "Consciousness." In E. N. Zalta, ed. *The Stanford Encyclopedia of Philosophy*. Summer 2017 Edition. Retrieved April 2017 from https://plato.stanford.edu/entries/consciousness/.

Chapter 24
Books

Arking, R. 2006. *The Biology of Aging: Observations and Principles*. New York, NY: Oxford University Press.

Clark, W. R. 2002. *A Means to an End: The Biological Basis of Aging and Death.* New York, NY: Oxford University Press.

Dawkins, R. 1989. *The Selfish Gene.* New York, NY: Oxford University Press.

Mitteldorf, J., and D. Sagan. 2016. *Cracking the Aging Code: The New Science of Growing Old – and What It Means for Staying Young.* New York, NY: Flatiron Books.

Other References

Adams, J. U. 2008. "Genetic Control of Aging and Life Span." *Nature Education* 1 (1), 130.

Anderson, R. M., D. Shanmuganayagam, and R. Weindruch. 2009. "Caloric Restriction and Aging: Studies in Mice and Monkeys." *Toxicologic Pathology* 37, 47–51.

Arias, E. 2015. "United States Life Tables, 2011." *National Vital Statistics Reports* 64 (11), 1-60. Retrieved March 2018 from http://www.cdc.gov/nchs/data/nvsr/nvsr64/nvsr64_11.pdf.

Baur, J. A., K. J. Pearson, N. L. Price, H. A. Jamieson, C. Lerin, A. Kalra ... D. A. Sinclair. 2006. "Resveratrol improves health and survival of mice on a high-calorie diet." *Nature* 444, 337–42.

Ben-Porath, I., and R. A. Weinberg. 2005. "The signals and pathways activating cellular senescence." *Inter. J. Biochemistry and Cell Biology* 37, 961–76.

Boehm, A. M., K. Khalturin, F. Anton-Erxleben, et al. 2012. "FoxO is a critical regulator of stem cell maintenance in immortal Hydra." *Proc. Natl. Acad. Sci. US* 109 (48), 19697–702.

Borra, M. T., B. C. Smith, and J. M. Denu. 2005. "Mechanism of Human SIRT1 Activation by Resveratrol." *J Biol Chem*, 280 (17), 17187–95.

Browner, W. S., A. J. Kahn, E. Ziv, A. P. Reiner, J. Oshima, R. M. Cawthon ... Cummings, S. R. 2004. "The Genetics of Human Longevity." *Am J Med* 117 (11), 851–60.

Buttner, S., T. Eisenberg, E. Herker, D. Carmona-Gutierrez, G. Kroemer, and F. Madeo. 2006. "Why yeast cells can undergo apoptosis: death in times of peace, love, and war." *J Cell Biol* 175 (4), 521–25.

Canto, C., and J. Auwerx. 2009. "Caloric restriction, SIRT1 and longevity." *Trends Endocrinol Metab* 20 (7), 325–31.

Childs, B. G., D. J. Baker, J. L. Kirkland, J. Campisi, and J. M. van Deursen. 2014. "Senescence and apoptosis: Dueling or complementary cell fates?" *EMBO reports* 15 (11), 1139–53.

Fabian, D., and T. Flatt. 2011. "The Evolution of Aging." *Nature Education Knowledge* 3 (10), 9.

Favaloro, B., N. Allocati, V. Graziano, C. Di Ilio, and V. De Laurenzi. 2012. "Role of apoptosis in disease." *Aging* 4 (5), 330–49.

Flachsbart, F., A. Caliebe, R. Kleindorp, H. Blanche, H. von Eller-Eberstein, S. Nikolaus ... A. Nebel. 2009. "Association of FOXO3A variation with human longevity confirmed in German centenarians." Proc. Natl. Acad. Sci. 106 (8), 2700-2705.

Gavrilov, L. A., and N. S. Gavrilova. 2002. "Evolutionary theories of aging and longevity." *Scientific World Journal* 2, 339–56.

Guarente, L., and C. Kenyon. 2000. "Genetic pathways that regulate ageing in model organisms." *Nature* 408, 255–62.

Hayflick, L., and P. S. Moorhead. 1961. "The serial cultivation of human diploid cell strains." *Exp Cell Res* 25, 585–621.

Herker, E., H. Jungwirth, K. A. Lehmann, C. Maldener, K. U. Frohlich, S. Wissing ... F. Madeo. 2004. "Chronological aging leads to apoptosis in yeast." *J Cell Biol* 164 (4), 501–7.

Howitz, K. T., K. J. Bitterman, H. Y. Cohen, D. W. Lamming, S. Lavu, J. G. Wood ... D. A. Sinclair. 2003. "Small molecule activators of sirtuins extend Saccharomyces cerevisiae lifespan." *Nature* 425 (694), 191–96.

Lagouge, M., and N-G. Larsson. 2013. "The role of mitochondrial DNA mutations and free radicals in disease and ageing." *J Intern Med* 273 (6), 529–43.

Leadem, R. 2016, October 5. Steve Jobs Commencement Speech, Stanford University, June 2005. Transcript. *Entrepreneur*. Retrieved December 2017 from https://www.entrepreneur.com/article/283250#.

Leonhardt, D. 2006. "Life Expectancy Data." The *New York Times*. September 27. Retrieved October 2018 from https://www.nytimes.com/2006/09/27/business/27leonhardt_sidebar.html.

Longo, V. D., J. Mitteldorf, and V. P. Skulachev. 2005. "Programmed and altruistic ageing." *Nature Reviews Genetics* 6, 866–72.

Mitchell, S. J., A. Martin-Montalvo, E. M. Mercken, H. H. Palacios, T. M. Ward, G. Abulwerdi ... R. de Cabo. 2014. "The SRIT1 activator SRT1720 extends lifespan and improves health of mice fed a standard diet." *Cell Rep* 6 (5), 836–43.

Mitteldorf, J. 2018. "Telomerase: Update and Downgrade." Retrieved March 2018 from https://joshmitteldorf.scienceblog.com/2018/03/19/telomerase-update-and-downgrade/.

Morris, B. J., D. C. Wilcox, T. A. Donlon, and B. J. Wilcox. 2015. "FOXO3: A major gene for human longevity – a mini-review." *Gerontology* 61, 515–25.

Paleothea. "Eos and Tithonus." 2007. Retrieved December 2017 from https://www.paleothea.com/Myths/Eos.html.

Rodier, F., and J. Campisi, 2011. "Four faces of cellular senescence." *J of Cell Biology* 192 (4), 547–56.

Spalding, K. L., R. D. Bhardwaj, B. A. Buchholz, and D. H. Frisen. 2005. "Retrospective birth dating of cells in humans." *Cell* 122 (1), 133–43.

Tissenbaum, H. A., and L. Guarente. 2001. "Increased dosage of a sir-2 gene extends lifespan in Caenorhabditis elegans." *Nature* 410 (6825), 227–30.

Tome-Carneiro, J., M. Larrosa, A. Gonzalez-Sarrias, F. A. Tomas-Barberan, M. T. Garcia-Conesa, and J. C. Espin, 2013. "Resveratrol and clinical trials: the crossroad from in vitro studies to human evidence." *Curr Pharm Des* 19 (34), 6064–93.

Wade, N. 2006. "Yes, Red Wine Holds Answer. Check Dosage." The *New York Times*. November 2. Retrieved December 2017 from http://www.nytimes.com/2006/11/02/science/02drug.html.

Wilcox, B. J., T. A. Donlon, O. He, R. Chen, J. S. Grove, K. Yano … J. D. Curb. 2008. "FOXO3A genotype is strongly associated with human longevity." *Proc Natl Acad Sci USA* 105 (37), 13987–92.

Chapter 25
Books

Conway Morris, S. 1998. *The Crucible of Creation: The Burgess Shale and the Rise of Animals.* Oxford, UK: Oxford University Press.

Gould, S. J. 1989. *Wonderful Life: The Burgess Shale and the Nature of History.* New York, NY: W. W. Norton and Company.

Sagan, C. 1980. *Cosmos.* New York, NY: Ballantine Books.

Ward, P. D., and D. Brownlee, 2000. *Rare Earth: Why Complex Life Is Rare in the Universe.* New York, NY: Copernicus Books.

Other References

Astronomy Notes. "Earth-Venus-Mars: A Comparison of Earth and Its Two Closest Neighbors." 2016. Retrieved October 2017 from http://www.astronomynotes.com/solarsys/s10.htm.

Bostrom, N. 2012. "Existential Risk Prevention as Global Priority." *Global Policy* 4 (1), 15–31.

Cain, F. 2015. "Venus Greenhouse Effect." Retrieved October 2017 from https://www.universetoday.com/22577/venus-greenhouse-effect/.

Catling, D. C. 2009. "Atmospheric Evolution of Mars." In V. Gornitz, editor. *Encyclopedia of Paleoclimatology and Ancient Environments.* Dordrecht, Netherlands: Springer, 66–75.

Chou, F., and M. Johnson. 2015. "NASA's Kepler Mission Discovers Bigger, Older Cousin to Earth." Retrieved September 2017 from https://www.nasa.gov/press-release/nasa-kepler-mission-discovers-bigger-older-cousin-to-earth.

Cofield, C. 2015. "Aliens Could Destroy Humanity, But Let's Search Anyway." Space.com. Retrieved February 2018 from https://www.space.com/29999-stephen-hawking-intelligent-alien-life-danger.html.

ESA. "New Launch Date for James Webb Space Telescope." 2018. Retrieved August 2018 from https://www.esa.int/Our_Activities/Space_Science/ New_launch_date_for_James_Webb_Space_Telescope.

Gillon, M., A. H. M. J. Triaud, B. O. Demory, E. Jehin, E. Agol, K. M. Deck … E. J. Kotze. 2017. "Seven temperate terrestrial planets around the nearby ultracool dwarf star TRAPPIST-1." *Nature* 542, 456–60.

Gray, R. H. 2015. "The Fermi Paradox is Neither Fermi's nor a Paradox." *Astrobiology* 15 (3), 195–99.

Hansen, C. J., D. E. Shemansky, L. W. Esposito, A. I. F. Stewart, B. R. Lewis, J. E. Colwell, … B. A. Magee. 2011. The composition and structure of the Enceladus plume, *Geophysical Research Letters*, 38(11), L11202, doi:10.1029/2011GL047415.

Hanson, R. 1998. "The Great Filter – Are We Almost Past It?" Retrieved December 2016 from http://mason.gmu.edu/~rhanson/greatfilter.html.

Howell, E. 2017. "How Many Stars Are in the Universe?" Space.com. Retrieved August 2018 from https://www.space.com/26078-how-many-stars-are-there.html

Jenkins, J.M., J. D. Twicken, N. M. Batalha, D. A. Caldwell, W. D. Cochran, M. Endl … W. J. Borucki. 2015. "Discovery and validation of Kepler-452b; A 1.6-R super Earth exoplanet in the habitable zone of a G2 star." *The Astronomical Journal* 150 (2), 1–20.

Khan, A. 2013. "Milky Way may host billions of Earth-size planets." *Los Angeles Times*. November 4. Retrieved October 2017 from http://www.latimes.com/science/la-sci-earth-like-planets-20131105-story.html.

Losos, J. 2017. "Would Aliens Look Like Us?" NPR.org. Cosmos & Culture. Retrieved October 2017 from http://www.npr.org/sections/13.7/2017/10/04/555044820/would-aliens-look-like-us.

Petigura, E. A., A. W. Howard, and G. W. Marcy. 2013. "Prevalence of Earth-size planets orbiting Sun-like stars." *Proc Natl Acad Sci USA*, 110 (48), 19273–78.

Roth, L., J. Saur, K. D. Retherford, D. F. Strobel, P. D. Feldman, M. A. McGrath, and F. Nimmo. 2014. "Transient Water Vapor at Europa's South Pole." *Science* 343 (6167), 171–74.

Sagan, C. 1995. "The Abundance of Life-Bearing Planets." The *Bioastronomy News* 7 (4). Retrieved December 2016 from http://www2.hawaii.edu/~pine/sagan.html.

Seti Institute. 2017. "Project Phoenix." Retrieved December 2016 from http://www.seti.org/seti-institute/project/details/project-phoenix.

Strobel, N. 2013. "Nick Strobel's Astronomy Notes." Retrieved December 2016 from http://www.astronomynotes.com/solarsys/s9.htm.

Tarter, J. 2017. "The Allen Telescope Array." Seti Institute. Retrieved October 2017 from https://www.seti.org/ata

Chapter 26
Books

Archer, D. 2009. *The Long Thaw: How Humans are Changing the Next 100,000 Years of Earth's Climate*. Princeton, NJ: Princeton University Press.

Fagan, B. 2004. *The Long Summer: How Climate Changed Civilization*. Cambridge, MA: Basic Books.

Kaku, M. 2012. *Physics of the Future: How Science Will Shape Human Destiny and Our Daily Lives by the Year 2100*. New York, NY: Anchor Books.

Nordhaus, W. 2013. *The Climate Casino: Risk, Uncertainty, and Economics for a Warming World*. New Haven, CT: Yale University Press.

Ward, P. D., and Brownlee, D. 2000. *Rare Earth: Why Complex Life is Uncommon in the Universe*. New York, NY: Copernicus Books.

Other References

Cook, J., D. Nuccitelli, S. A. Green, M. Richardson, B. Winkler, R. Painting … A. Skuce. 2013. "Quantifying the consensus on anthropogenic global warming in the scientific literature." *Environmental Research Letters* 8, 024024.

EPICA Community Members. 2004. "Eight glacial cycles from an Antarctic ice core." *Nature* 429, 623–28.

Fleming, J. R. 1999. "Joseph Fourier, the 'greenhouse effect' and the quest for a universal theory of terrestrial temperatures." *Endeavour* 23 2, 72–75.

Global Greenhouse Gas Reference Network. 2018. "Trends in Atmospheric Carbon Dioxide." Retrieved February 2018 from https://www.esrl.noaa.gov/gmd/ccgg/trends/.

Graham, S. 2000. "Milutin Milankovitch (1879-1958)." NASA. Retrieved December 2016 from http://earthobservatory.nasa.gov/Features/Milankovitch/.

Gu, G., G. R. Dickens, G. Bhatnagar, F. S. Colwell, G. J. Hirasaki, and W. G. Chapman. 2011. "Abundant Early Paleocene marine gas hydrates despite warm deep-ocean temperatures." *Nature Geoscience* 4, 848–51.

Harrington, G. J., J. Eberle, B. A. Le-Page, M. Dawson, and J. H. Hutchison. 2011. "Arctic plant diversity in the early Eocene greenhouse." *Proc. R. Soc. B.* doi:10.1098/rspb.2011.1704.

Hays, J. D., J. Imbrie, and N. J Shackleton. 1976. "Variations in the Earth's Orbit: Pacemaker of the Ice Ages." *Science* 194 (4270), 1121–32.

Huber, M., and R. Caballero. 2011. "The early Eocene equable climate problem revisited." *Clim Past* 7, 603–33.

IPCC (Intergovernmental Panel on Climate Change). "Climate Change 2014: Synthesis Report." Retrieved July 2017 from http://ipcc.ch/report/ar5/syr/.

Kerrick, D. M. 2001. "Present and Past Non-anthropogenic CO_2 degassing from the solid Earth." *Reviews of Geophysics* 39 (4) 565–85.

NASA. 2018. "Scientific consensus: Earth's climate is warming." Retrieved October 2018 from http://climate.nasa.gov/scientific-consensus/.

———. 2018. "The Relentless Rise of Carbon Dioxide." Retrieved October 2018 from http://climate.nasa.gov/climate_resources/24/.

NASA Goddard Institute for Space Studies. 2017. "Surface Temperature Analysis." Retrieved July 2017 from http://data.giss.nasa.gov/gistemp/.

———. 2018. "Global Land-Ocean Temperature Index." Retrieved October 2018 from https://www.giss.nasa.gov/research/news/20100121/418335main_land-ocean-full.jpg.

NOAA. 2018. "The Data: The Story Told from CO_2 Samples." Retrieved October 2018 from http://www.esrl.noaa.gov/gmd/outreach/isotopes/mixing.html.

Pearson, P. N., and M. R. Palmer. 2000. "Atmospheric carbon dioxide concentrations over the past 60 million years." *Nature* 406, 695–99.

Raftery, A. E., A. Zimmer, D. M. W. Frierson, R. Startz, and P. Liu. 2017. "Less than 2°C warming by 2100 unlikely." *Nature Climate Change*. doi:10.1038/nclimate3352.

Skeptical Science. 2018. "CO_2 Lags Temperature – What Does It Mean?" Retrieved October 2018 from https://www.skepticalscience.com/co2-lags-temperature-intermediate.htm.

Sloan, L. C., C. G. Walker, T. C. Moore Jr., D. K. Rea, and J. C. Zachos. 1992. "Possible methane-induced polar warming in the early Eocene." *Nature* 357, 1129–31.

Tang, H., and Y. Chen. 2013. "Global glaciations and atmospheric change at ca. 2.3 Ga." *Geoscience Frontiers* 4 (5), 583–96.

University of Michigan Global Change Program. "Analysis of Vostok Ice Core Data." 2017. Retrieved July 2017 from https://globalchange.umich.edu/globalchange1/current/labs/Lab10_Vostok/Vostok.htm.

US EPA. 2017. "Climate Change Science. Future of Climate Change." Retrieved July 2017 from https://19january2017snapshot.epa.gov/climate-change-science/future-climate-change_.html#Sea%20level.

———. 2017. "Getting to the core: The link between temperature and carbon dioxide." Retrieved December 2017 from https://www3.epa.gov/climatechange//kids/documents/temp-and-co2.pdf.

Chapter 27
Books

Eldredge, N. 2000. *Life in the Balance: Humanity and the Biodiversity Crisis*. Princeton, NJ: Princeton University Press.

Gould, S. J. 1985. *The Flamingo's Smile: Reflections in Natural History.* New York, NY: W. W. Norton and Company.

Kolbert, E. 2014. *The Sixth Extinction: An Unnatural History.* New York, NY: Henry Holt and Company, LLC.

Leakey, R., and Lewin, R. 1995. *The Sixth Extinction: Patterns of Life and the Future of Humankind.* New York, NY: Anchor Books.

Wilson, E. O. 2002. *The Future of Life.* New York, NY: Vintage Books.

Other References

Barnosky, A. D. 2014. "10 Ways You Can Help Stop the Sixth Mass Extinction." Huffington Post. Retrieved August 2017 from http://www.huffingtonpost.com/anthony-d-barnosky/10-ways-you-can-help-stop_b_5968774.html.

Barnosky, A. D., N. Matzke, S. Tomiya, G. O. U. Wogan, B. Swartz, G. B. Quental ... E. A. Ferrer. 2011. "Has the Earth's sixth mass extinction already arrived? *Nature* 471, 51–57.

Ceballos, G., P. R. Ehrlich, A. D. Barnosky, A. Garcia, R. M. Pringle, and T. M. Palmer. 2015. "Accelerated modern human-induced species losses: Entering the sixth mass extinction." *Sci Adv* 1, e1400253.

Ceballos, G., P. R. Ehrlich, and R. Dirzo, 2017. "Biological annihilation via the ongoing sixth mass extinction signaled by vertebrate population losses and declines." *Proc Natl Acad Sci USA.* doi:10.1073/pnas.1704949114.

Cuda, H. S., and E. Glazner. 2015. "The Turtle that Became the Anti-plastic Straw Poster Child." Plastic Pollution Coalition. Retrieved August 2018 from http://www.plasticpollutioncoalition.org/pft/2015/10/27/the-turtle-that-became-the-anti-plastic-straw-poster-child.

Fengler, W. 2014. "The Rapid Slowdown of Population Growth." The World Bank. Retrieved August 2017 from http://blogs.worldbank.org/futuredevelopment/rapid-slowdown-population-growth#comments

Foreman, D. 2004. "The Pleistocene-Holocene Event: The Sixth Great Extinction." Rewilding Earth. Retrieved July 2017 from www.rewilding.org/thesixthgreatextinction.htm.

Gall, S. C., and R. C. Thompson. 2015. "The impact of debris on marine life." *Marine Pollution Bulletin* 92 (1–2), 170–79.

Good, K. 2018. "700 Marine Species Might Go Extinct Because of Plastic Pollution. Here are 5 Ways You Can Help." One Green Planet. Retrieved September 2018 from http://www.onegreenplanet.org/environment/marine-species-extinction-and-plastic-pollution/.

IUCN. 2016. "Poaching behind worst African elephant losses in 25 years – IUCN report." Retrieved August 2017 from https://www.iucn.org/news/species/201609/poaching-behind-worst-african-elephant-losses-25-years---iucn-report.

IUCN Red List of Threatened Species. 2001. Retrieved July 2017 from http://www.iucnredlist.org/technical-documents/categories-and-criteria/2001-categories-criteria.

Joyce, C. 2015. "8 Million Tons of Plastic Clutter our Seas." NPR.org. Retrieved July 2018 from https://www.npr.org/2015/02/12/385752248/8-million-tons-of-plastic-clutter-our-seas.

Kahn, J. 2018. "Should some Species Be Allowed to Die?" The *New York Times Magazine*. March 13. Retrieved March 2018 from https://www.nytimes.com/2018/03/13/magazine/should-some-species-be-allowed-to-die-out.html.

Karlekar, H. 2018. "The Sixth Extinction?" *The Pioneer*. Retrieved October 2018 from https://www.dailypioneer.com/2018/columnists/the-sixth-extinction-.html.

Lindsey, R. 2007. "Tropical Deforestation." Retrieved August 2017 from https://earthobservatory.nasa.gov/Features/Deforestation/.

Mora, C., D. P. Tittensor, S. Adl, A. G. B. Simpson, and B. Worm. 2011. "How Many Species Are There on Earth and in the Ocean?" *PLoS Biol* 9 (8), e1001127.

Nature Conservancy. 2017. "Rainforests: Facts about rainforests." Retrieved August 2017 from https://www.nature.org/ourinitiatives/urgentissues/land-conservation/forests/rainforests/rainforests-facts.xml.Share to Facebook.

Pimm, S. L., C. N. Jenkins, R. Abell, T. M. Brooks, J. L. Gittleman, L. N. Joppa … J. O. Sexton 2014. "The biodiversity of species and their rates of extinction, distribution, and protection." *Science* 344 (6187). doi:10.1126/science.1246752.

Pyron, R. A. 2017. "We don't need to save endangered species. Extinction is part of evolution." The *Washington Post*. November 22. Retrieved November 2017 from https://www.washingtonpost.com/outlook/we-dont-need-to-save-endangered-species-extinction-is-part-of-evolution/2017/11/21/57fc5658-cdb4-11e7-a1a3-0d1e45a6de3d_story.html?utm_term=.39a3f87c85ad.

Rewilding Institute. 2017. "The Pleistocene-Holocene Event: The Sixth Great Extinction." Retrieved July 2017 from http://rewilding.org/extinction.pdf.

Spacequotations.com. 2018. "Eyes Turned Skyward Looking Back at Earth." Retrieved February 2018 from http://www.spacequotations.com/earth.html.

Sutter, J. D. 2016. "How to stop the sixth mass extinction." CNN. Retrieved August 2017 from http://www.cnn.com/2016/12/12/world/sutter-vanishing-help/index.html.

Think Global Green. "Deforestation. 2017. Retrieved August 2017 from https://www.thinkglobalgreen.org/deforestation.html.

United Nations Department of Economic and Social Affairs. 2017."World Population Prospects: The 2017 Revision." Retrieved August 2017 from https://esa.un.org/unpd/wpp/Publications/Files/WPP2017_KeyFindings.pdf.

Wilcove, D. S., D. Rothstein, J. Dubow, A. Phillips, and E. Losos. 1998. "Quantifying Threats to Imperiled Species in the United States." *BioScience* 48, 607–16.

Wise, J. 2013. "About That Overpopulation Problem." Slate. Retrieved August 2017 from http://www.slate.com/articles/technology/future_tense/2013/01/world_population_may_actually_start_declining_not_exploding.html.

World Bank. 2018. "Forest Area (% of Land Area)." The World Bank. Retrieved March 2018 from https://data.worldbank.org/indicator/AG.LND.FRST.ZS.

Chapter 28
Books

Barrat, J. 2013. *Our Final Invention: Artificial Intelligence and the End of the Human Era*. New York, NY: St Martin's Press.

Bostrom, N. 2014. *Superintelligence: Paths, Dangers, Strategies*. Oxford, UK: Oxford University Press.

Dormehl, L. 2017. *Thinking Machines: The Quest for Artificial Intelligence and Where It's Taking Us Next*. New York, NY: Penguin Random House LLC.

Enriquez, J., and Gullans, S. 2015. *Evolving Ourselves: How Unnatural Selection and Nonrandom Mutation Are Changing Life on Earth*. New York, NY: Penguin Group.

Kaku, M. 2012. *Physics of the Future: How Science Will Shape Human Destiny and Our Daily Lives by the Year 2100*. New York, NY: Anchor Books.

Kurzweil, R. 2006. *The Singularity is Near: When Humans Transcend Biology*. New York, NY: Penguin Group (USA).

Yonck, R. 2017. *Heart of the Machine: Our Future in a World of Artificial Emotional Intelligence*. New York, NY: Arcade Publishing.

Other References

Allen, P. 2011. "The Singularity Isn't Near." MIT Technology Review. Retrieved July 2017 from https://www.technologyreview.com/s/425733/paul-allen-the-singularity-isnt-near/.

Aung, L. 2017. "Deep Blue: The History and Engineering behind Computer Chess." Illumin. Retrieved March 2017 from https://illumin.usc.edu/printer/188/deep-blue-the-history-and-engineering-behind-computer-chess/.

Campbell, K. H. S., J. McWhir, W. A. Ritchie, and I. Wilmut. 1996. "Sheep cloned by nuclear transfer from a cultured cell line." *Nature* 380, 64–66.

Frey, C. B., and Osborne, M. A. 2013. "The Future of Employment: How susceptible are Jobs to Computerization?" Retrieved June 2017 from http://www.oxfordmartin.ox.ac.uk/downloads/academic/The_Future_of_Employment.pdf.

Geitgey, A. 2016. "Machine Learning is Fun Part 6: How to do Speech Recognition with Deep Learning." Medium. Retrieved January 2018 from https://medium.com/@ageitgey/machine-learning-is-fun-part-6-how-to-do-speech-recognition-with-deep-learning-28293c162f7a.

Gershgorn, D. 2017. "Researchers are using Darwin's theories to evolve AI so only the strongest algorithms survive. Retrieved March 2017 from https://qz.com/933695/researchers-are-using-darwins-theories-to-evolve-ai-so-only-the-strongest-algorithms-survive/.

Gibson, D. G., J. I. Glass, C. Lartigue, V. N. Noskov, R. Y. Chuang, M. A. Algire ... C. Venter. 2010. "Creation of a Bacterial Cell Controlled by a Chemically Synthesized Genome." *Science* 329 (5987) 52–56.

Guarente, L. and C. Kenyon. 2000. "Genetic pathways that regulate ageing in model organisms." *Nature* 408, 255–62.

Hintze, A. 2016. "Understanding the four types of AI, from reactive robots to self-aware beings." The Conversation. Retrieved May 2017 from http://theconversation.com/understanding-the-four-types-of-ai-from-reactive-robots-to-self-aware-beings-67616.

History.com editors. 2009. "Deep Blue beats Kasparov." History.com. Retrieved April 2017 from http://www.history.com/this-day-in-history/deep-blue-beats-kasparov-at-chess.

Lai, L., J. X. Kang, R. Li, J. Wang, W. T. Witt, H. Y. Yong ... Dai, Y. 2006. "Generation of cloned transgenic pigs rich in omega-3 fatty acids." *Nature Biotechnology* 24, 435–36.

Ledford, H. 2013. "Transgenic salmon nears approval." *Nature.* 497 (7447), 17–18.

Lee, K-F. 2017. "The Real Threat of Artificial Intelligence." The *New York Times.* June 24. Retrieved June 2017 from https://www.nytimes.com/2017/06/24/opinion/sunday/artificial-intelligence-economic-inequality.html?_r=0.

Modis, T. 2006. "The Singularity Myth." Retrieved July 2017 from http://www.growth-dynamics.com/articles/Kurzweil.htm.

Pollack, A. 2015. "Genetically Engineered Salmon Approved for Consumption." The *New York Times.* November 19. Retrieved May 2017 from https://www.nytimes.com/2015/11/20/business/genetically-engineered-salmon-approved-for-consumption.html?_r=0.

Reardon, S. 2016. "Welcome to the CRISPR Zoo: Birds and Bees are Just the Beginning for a Burgeoning Technology." *Nature* 531 (7593), 160–63.

Tang, Y-P., E. Shimizu, G. R. Dube, C. Rampon, G. A Kerchner, M. Zhuo ... J. Z. Tsien. 1999. "Genetic enhancement of learning and memory in mice." *Nature* 401, 63–69.

Tissenbaum, H. A., and L. Guarente. 2001. "Increased dosage of a sir-2 gene extends lifespan in *Caenorhabditis elegans.*" *Nature* 410 (6825), 227–30.

Ward, L. 2010. "Craig Venter Boots Up First Synthetic Cell." *Popular Mechanics.* Retrieved January 2017 from http://www.popularmechanics.com/science/health/a5761/synthetic-cell-breakthrough/.

Yang, H., H. Wang, and R. Jaenisch. 2014. "Generating genetically modified mice using CRISPR/Cas-mediated genome engineering." *Nature Protocols* 9, 1966–68.

Chapter 29
Books

Barrat, J. 2013. *Our Final Invention: Artificial Intelligence and the End of the Human Era*. New York, NY: St Martin's Press.

Bostrom, N. 2014. *Superintelligence: Paths, Dangers, Strategies*. Oxford, UK: Oxford University Press.

Christian, D., C. S. Brown, and C. Benjamin. 2014. *Big History: Between Nothing and Everything*. New York, NY: McGraw-Hill Education.

Kurzweil, R. 2006. *The Singularity Is Near: When Humans Transcend Biology*. London, England: Penguin Books.

Pinker, S. 2011. *The Better Angels of Our Nature: Why Violence Has Declined*. New York, NY: Penguin Books.

Other References

Adams, T. 2016. "Artificial Intelligence: 'We're like children playing with a bomb.'" The *Guardian*.Retrieved June 2017 from https://www.theguardian.com/technology/2016/jun/12/nick-bostrom-artificial-intelligence-machine.

Aken, J. V., and E. Hammond. 2003. "Genetic engineering and biological weapons." *EMBO Rep* 4s(supplement 1) S57–70.

Betuel, E. 2018. "40 Genes Linked to aggression and Violence Are Also Crucial for Survival." *Inverse*. July 9. Retrieved July 2018 from https://www.inverse.com/article/46830-genes-linked-to-violence-and-aggression.

Bostrom, N. 2002. "Existential Risks: Analyzing Human Extinction Scenarios and Related Hazards." *J Evolution and Technology* 9 (1).

———. 2015. "What happens when our computers get smarter than we are?" TED Talk. Retrieved from https://www.ted.com/talks/nick_bostrom_what_happens_when_our_computers_get_smarter_than_we_are#t-975288.

Foley, M. 2013. "Genetically Engineered Bioweapons: A New Breed of Weapons for Modern Warfare." Dartmouth Undergraduate Journal of Science. Retrieved May 2017 from http://dujs.dartmouth.edu/2013/03/genetically-engineered-bioweapons-a-new-breed-of-weapons-for-modern-warfare/#.WSNCt6OZM0o.

Gill, H. B. Jr. 2004. "Colonial Germ Warfare." Colonial Williamsburg. Retrieved May 2017 from http://www.history.org/foundation/journal/spring04/warfare.cfm.

Joy, B. 2000. "Why the future doesn't Need Us." *Wired*. April 1. Retrieved February 2018 from https://www.wired.com/2000/04/joy-2/.

Kazan, C. 2009. "Planet's Experts on Space Colonization – Our Future or Fantasy?" *The Daily Galaxy* (blog). Retrieved January 2018 from http://www.dailygalaxy.com/my_weblog/2009/04/space-colonizat.html.

Marcus, G. 2017. "Artificial Intelligence Is Stuck. Here's How to Move It Forward." The *New York Times*. July 29. Retrieved December 2017 from https://www.nytimes.com/2017/07/29/opinion/sunday/artificial-intelligence-is-stuck-heres-how-to-move-it-forward.html?_r=1.

Moskowitz, C. 2012. "Speed of Universe's Expansion Measured Better Than Ever." Space.com. Retrieved July 2017 from https://www.space.com/17884-universe-expansion-speed-hubble-constant.html.

Roser, M., and E. Ortiz-Ospina.2017. "Future World Population Growth." Our World in Data. Retrieved December 2017 from https://ourworldindata.org/world-population-growth/.

Sainato, M. 2015. "Stephen Hawking, Elon Musk, and Bill Gates Warn About Artificial Intelligence." Observer. Retrieved June 2017 from http://observer.com/2015/08/stephen-hawking-elon-musk-and-bill-gates-warn-about-artificial-intelligence/.

Sandberg, A. 2014. "The five biggest threats to human existence." The Conversation. Retrieved May 2017 from http://theconversation.com/the-five-biggest-threats-to-human-existence-27053.

Schlosser, E. 2018. "Dangers of the New Nuclear-Arms Race." The *New Yorker*. May 24. Retrieved August 2018 from https://www.newyorker.com/news/news-desk/the-growing-dangers-of-the-new-nuclear-arms-race.

Treder, M., and C. Phoenix 2006. "Nanotechnology and Future WMD." Center for Responsible Nanotechnology. Retrieved May 2017 from http://crnano.org/Paper-FutureWMD.htm.

Wheelis, M. 2002. "Biological Warfare at the 1346 Siege of Caffa." *Emerging Infectious Diseases* 8 (9), 971–75.

Williams, M. 2016. "Will Earth survive when the sun becomes a red giant?" Phys.org. Retrieved July 2017 from https://phys.org/news/2016-05-earth-survive-sun-red-giant.html.

World Future Fund. 2017. Retrieved June 2017 from http://www.worldfuturefund.org/wffmaster/Reading/war.crimes/World.war.2/Jap%20Bio-Warfare.htm.

Chapter 30
Books

Bostrom, N. 2014. *Superintelligence: Paths, Dangers, Strategies*. Oxford, UK: Oxford University Press.

Enriquez, J., and S. Gullans. 2015. *Evolving Ourselves: How Unnatural Selection and Nonrandom Mutation Are Changing Life on Earth*. New York, NY: Penguin Group.

Kurzweil, R. 2006. *The Singularity Is Near: When Humans Transcend Biology*. London, England: Penguin Books.

Simborg, D. 2017. *What Comes After Homo Sapiens? When and How Our Species Will Evolve into Another Species*. Mill Valley, CA: DWS Publishing.

Other References

Bostrom, N. 2005. "Ethical Principles in the Creation of Artificial Minds." Nickbostrom.com. Retrieved July 2017 from http://www.nickbostrom.com/ethics/aiethics.html.

Hawks, J., E. T. Wang, G. M. Cochran, H. C. Harpending, and R. K. Moyzis, 2007. "Recent acceleration of human adaptive evolution." *Proc Natl Acad Sci USA*, 104 (52), 20753–58.

Levy, S. 2017. "Why you will one day have a chip in your brain." *Wired*. Retrieved July 2017 from https://www.wired.com/story/why-you-will-one-day-have-a-chip-in-your-brain/.

Mattmiller, B. 2007. "Genome study places modern humans in the evolutionary fast lane." University of Wisconsin-Madison News. Retrieved January 2017 from https://news.wisc.edu/genome-study-places-modern-humans-in-the-evolutionary-fast-lane/.

Rajendra. 2017. "Stephen Hawking Fears A. I. May Replace Humans, and He's Not Alone." House of Bots. Retrieved October 2018 from http://houseofbots.com/news-detail/1401-1-stephen-hawking-fears-ai-may-replace-humans-and-he-is-not-alone.

Rao, R. P. N., and J. Wu. 2017. "How Close Are We Really to Connecting Human Minds to Artificial Intelligence?" Smithsonian.com. Retrieved May 2017 from http://www.smithsonianmag.com/innovation/melding-mind-and-machine-how-close-are-we-180962857/. S

Epilogue
Books

Adams, D. 1980. *The Hitchhiker's guide to the Galaxy*. New York, NY: Harmony Books.

Camus, A. 1995. *The Myth of Sisyphus and Other Essays*. New York: Vintage International.

Corning, P. 2018. *Synergistic Selection: How Cooperation Has shaped Evolution and the Rise of Humankind*. London, UK: World Scientific Publishing Co. Pte. Ltd.

Enriquez, J., and S. Gullans. 2015. *Evolving Ourselves: How Unnatural Selection and Nonrandom Mutation Are Changing Life on Earth*. New York, NY: Penguin Group.

Fromm, E. 1956. *The Art of Loving*. New York: Harper Perennial Modern Classics.

Haidt, J. 2006. *The Happiness Hypothesis: Finding Modern Truth in Ancient Wisdom*. New York: Basic Books.

Sagan, C. 1994. *Pale Blue Dot: A Vision of the Human Future in Space*. New York, NY: Ballantine Books.

References

Druyan, A. 2003. "Ann Druyan Talks About Science, Religion, Wonder, Awe … and Carl Sagan." *Skeptical Inquirer*, November/December. Retrieved May 2018 from https://www.csicop.org/si/show/ann_druyan_talks_about_science_religion.

IMDb. "The Meaning of Life: Quotes." Retrieved October 2017 from http://www.imdb.com/title/tt0085959/quotes.

Landau, E. 2015. "'Pale Blue Dot' Images Turn 25." NASA Jet Propulsion Laboratory. Retrieved June 2017 from https://voyager.jpl.nasa.gov/news/pale_blue_25.html.

White, R. B. 1959. "Motivation reconsidered: The concept of competence." *Psychological Review* 66, 297–333.

INDEX

angiosperms, 145*f*, 148, 149, 152, 154, 346

animals. *See also specific animals*
    evolutionary tree of, 112*f*
    genetic modification of, 308
    geological timeline for land animals, 156*f*

anions, 255

aniridia, 125

anisogamous sex, 102, 346

Anomalocaris, 74–75, **74**, 346

Anthropocene extinction, 301

anthropocentricity, 301

anticodon, 346

antioxidants, 255

apes, 107, 192, 193*f*, 194, 195, 243–244, 324

Apollo mission, 21

Apollo spacecraft, 300

apoptosis, 256, 346

Apple's Siri, 322

AquaBounty Technologies, 308

archaea, 7, 46, 47, 50, 65–66, 346

Archaean Eon, 23–24

Archaeopteryx, 171–172

Archicebus, 192

archosaurs, 163, 165–166, 166*f*, 167, 176, 183, 346

Argentinosaurus genus, 167, **167**, 168, 169

Armstrong, Eugene, 229

The Art of Loving (Fromm), 340

arthropods, 74, 78, 79, 116, 125, 156–157, 346

artificial intelligence (AI)
    and cyborgs, 332–334
    darker side of, 323
    Deep Blue and Garry Kasparov example of, 310–311, **311**
    defined, 346
    development of, 245, 273
    and emotions, 313
    Hinton as "Godfather" of, 312
    impact of on society, 314–315
    as often abbreviated by GNR, 305
    as opportunity as well as risk, 317
    photos of robot, **314**, **323**
    potential benefits of, 342–343, 344
    potential future scenario, 322–324

    as potential threat, 324
    as potential way advanced civilizations could destroy themselves, 272

artificial neural network, 312, 346

ascariasis, 115

Ascaris genus, 115

asexual reproduction, 91–92, 98, 99, 346

asexuality, 94

assistance, international, 327

asteroid collision, 165, 172, 176, 177–180, 192, 282, 320, 325

asteroid impact, as existential threat, 320

asteroids, 41, 46, 179, 320

ATP (adenosine triphosphate)
    defined, 346
    production/creation of, 6, 43–45, 50, 55, 56, 59, 60, 175

ATP synthase, 44, 48, 50, 346

Aurora australis, **27**

Australian magpie, **244**

Australopithecus afarensis, 195

automation, 315

axon terminal, 131, 137

B

baboons, 192

bacteria. *See also specific bacteria*
    analysis of by Koonin, 47
    bacterial slaves, 67–68
    bacterial "talk," 83–84
    cell membranes, 50
    defined, 346
    as example of machinery-fixing magic of life, 252
    as first to venture out of seas, 142
    free-living bacteria, 67
    as one of three domains of living organisms, 7, 65–66
    reproduction of, 94
    slime bacteria, 83–84
    solar-powered, 53–61
    as using fermentation, 46

bacterial conjugation, 346

L

M

vertebrate nervous systems, 133–135

vervet monkeys, 209–210

violence, 200, 224, 230, 231–232, 326

"virgin births," 106

Vital Dust (de Duve), 31, 37, 60, 148–149

"vital engagement," 341

vitalism doctrine, 32, 360

The Vital Question (Lane), 48, 51, 69

VO$_2$ max (aerobic capacity), 174–175, 345

voice recognition, 312

volcanic eruptions/activity, 24, 181, 184, 185, 204, 205, 284

volcanic winter, 204, 205

Volvox, 84, 85, 89

Vonnegut, Kurt, 275, 291

Voyager I spacecraft, 337, 338

W

Walcott, Charles, 76

Ward, Peter, 22, 24, 165, 170, 182, 269–270, 284, 329

warfare, 221, 231–232. *See also* biological warfare; germ warfare

warm-bloodedness, 162, 163, 169, 170, 173–176

Watson, James, 2, 3

weak anthropic principle, 28, 360

The Wealth of Nations (Smith), 222

weapons of mass destruction, as existential threat, 321

weathering, 22, 185, 271, 281, 284, 360

Weismann, August, 252

Whittington, Harry Blackmore, 76

"Why Do We Believe in Electrons, but Not in Fairies?" (Kuipers), 14–15

"Why the future doesn't need us" (Joy), 322

wildlife trafficking, illegal, 304

Wilmut, Ian, 307

Wilson, E. O., 212, 217, 225, 228, 301, 302

Wired magazine, 322

Woese, Carl, 7, 34, 65

women, subordination of, 234

Wonderful Life: The Burgess Shale and the Nature of History (Gould), 61, 73, 76, 79

Wood-Ljungdahl pathway, 40, 360

woolly mammoth, 198, 276, 297

World Trade Organization, 327

World Wildlife Fund, 297, 304

Wright, Robert, 240–241

X

xenophobic behavior, 229, 230, 339–340, 343

xenophobic traits, 223

xenophobic trends, 343

xylem, 145, 360

Y

"Yes, Red Wine Holds the Answer," 259

Yoho National Park (British Columbia, Canada), 75

Yonck, Richard, 313

Z

zero population growth (ZPG), 302, 303, 324

zombies, 244–245

zygote, 86, 92, 93, 95, 102, 103, 129, 151, 152, 308, 360

ACKNOWLEDGMENTS

First, and most importantly, I'd like to thank my dear, patient wife, Dora Brodie. She retired from a busy gastroenterology practice recently and took up Russian, which she learned as a child in Czechoslovakia under the Communist regime. Her Russian studies have been a good antidote to my busy reading, researching, and writing—keeping us studying together. She has been supportive and patient throughout this long journey and has carried far more than her share of our mutual duties. I'd also like to give great thanks to my editor, Megan McGrath, who has stuck with me over these past five to six years on our journey together. Megan is a graduate student in biology and has provided not only great editorial support but also critical evaluation of content. I'd also like to thank the people at iUniverse who have helped guide me through the tedious process of transforming my written word into a mature publication.

Finally, I'd like to thank a number of people who have provided critical evaluation and feedback along the way. My son Dr. Mark Brodie, a past psychology major at Dartmouth College, provided helpful insights and contributions to the chapter on the evolution of morality. My son Ben Brodie, a computer scientist, provided great perspective on the future impact of artificial intelligence. My colleague Dr. Jim Adelman provided helpful suggestions throughout the entire process—especially suggestions regarding how I am able to reach my target audience. I also appreciated the insightful reviews, feedback, and suggestions from Robert E. Cannon, professor emeritus in the Biology Department at the University of North Carolina–Greensboro; Dr. Karl Fields, professor of medicine at the University of North Carolina and head of Sports Medicine at Cone Health in Greensboro, North Carolina; Matt Oechsli, past CEO of the Oechsli Institute in Greensboro, North Carolina; my good friends Jane and Lloyd Peterson; and my sister and brother-in-law, Joyce and Mike Patton.

CREDITS FOR IMAGES
AND FIGURES

Image 1.1: Chimpanzees and *E. coli*. Photograph of chimpanzees by author (Brodie). Scanning electron microscopy of E. coli by NIAID. CC BY 2.0. Retrieved and reproduced with permission from https://www.flickr.com/photos/niaid/16598492368.

Image 1.2: DNA by Qimono. Creative Commons CCO. Retrieved and reproduced with permission from https://pixabay.com/en/dna-string-biology-3d-1811955/.

Image 2.1: Charles Darwin. Downloaded from Pixabay free images. COO Public Domain. Retrieved and reproduced with permission from https://pixabay.com/en/charles-robert-darwin-scientists-62911/.

Figure 3.1: Data from Hubble Space Telescope Key project by John Huchra. Retrieved and reproduced with permission from https://cosmictimes.gsfc.nasa.gov/teachers/guide/1929/guide/universe_expanding.html.

Figure 3.2: Moon's tidal forces. Image produced by author (Brodie).

Figure 3.3: Goldilocks zone. Keplar 69 and the Solar System, NASA Ames/JPL CalTech, adapted. Retrieved and reproduced with permission from https://www.nasa.gov/mission_pages/kepler/multimedia/images/kepler-69-diagram.html.

Image 3.1: Aurora Australis. Spacecraft Pictures Aurora, by NASA. Retrieved and reproduced with permission from https://earthobservatory.nasa.gov/IOTD/view.php?id=6226.

Image 4.1: Stromatolites, Hamelin Pool, Shark Bay, Australia, by Paul Morris. CC BY-SA 2.0. Retrieved and reproduced with permission from https://www.flickr.com/photos/aa3sd/3028449763.

Figure 4.1: The Interaction of a ribosome with mRNA. Process of initiation of translation. By Designua. Purchased from Shutterstock (ID 285444794).

Figure 4.2: Lipid vesicle by Henrik Skov Midtiby. CC BY 2.5. Retrieved and reproduced with permission from http://www.texample.net/tikz/examples/lipid-vesicle/.

Image 5.1: Carbonate spires in the Lost City vent field, Mid Atlantic Ridge (IFE, URI-IAO, UW, Lost City Science Party; NOAA/OAR/OER; The Lost City 2005 Expedition). CC BY 2.0. Retrieved and reproduced with permission from http://www.photolib.noaa.gov/htmls/expl2230.htm.

Figure 6.1: Photosynthesis. Figure produced by author (Brodie).

Image 6.1: Magnetite banded iron formation (Soudan Iron-Formation, Neoarchean, 2.722 Ga; Rt. 169 road cut between Soudan & Robinson, Minnesota, USA) by James St. John. CC BY 2.0. Retrieved and reproduced with permission from https://www.flickr.com/photos/jsjgeology/18853053918.

Figure 7.1: Eukaryotic and prokaryotic cells: Eukaryotic cell by biology science, Flickr Photo Sharing, Creative Commons (CC BY-SA 2.0), Retrieved September 2017 from https://www.flickr.com/photos/ajc1/11820433176. Prokaryote purchased from Shutterstock (ID 739786744) 12/26/18. Figure adapted by author.

Figure 7.2: Phagocytosis and endosymbiosis: Figure prepared by author (Brodie).

Image 8.1: Trilobite by Catmando. Purchased from Shutterstock (ID 584461975).

Image 8.2: *Anomalocaris.* Painting courtesy UNE Photos and Flickr. CC BY 2.0. Retrieved and reproduced with permission from https://www.flickr.com/photos/unephotos/6786859303.

Image 8.3: *Opabinia.* Purchased from CanStockPhoto (csp22122198).

Image 8.4: *Pikaia.* Purchased from CanStockPhoto (csp32736497).

Image 9.1: Scanning electron microscopic image *Chlamydomonas reinhardtii* by Louisa Howard, Dartmouth College. Courtesy National Science Foundation. Public domain. Retrieved and reproduced with permission from https://nsf.gov/news/mmg/mmg_disp.jsp?med_id=79712&from=.

Figure 9.1: How *E. coli* digest food. Figure prepared by author (Brodie).

Figure 10.1: *Chlamydomonas* life cycle. Figure prepared by author (Brodie). The scanning electron microscopic image of *Chlamydomonas reinhardtii* is courtesy of Dartmouth Microscope Facility (public domain). Retrieved December 2017 from https://nsf.gov/news/mmg/mmg_disp.jsp?med_id=79712&from=.

Image 11.1: Anemone fish protecting its spawn by Prilfish. CC BY 2.0. Retrieved and reproduced with permission from https://www.flickr.com/photos/silkebaron/4735960070.

Figure 11.1: Amniotic egg by Designua. Purchased from Shutterstock (ID 233712685).

Image 11.2: Short-beaked Echidna by Laurens. CC BY-ND 2.0. Retrieved and reproduced with permission from https://www.flickr.com/photos/47456200@N04/4421411452.

Image 11.3: Komodo dragon-*Varanus komodoensis* by S. Rohrlach. CC BY 2.0. Retrieved and reproduced with permission from https://www.flickr.com/photos/95098864@N08/32576452894.

Image 11.4: Lilly. Photo by author (Brodie).

Image 11.5: Peacock and Tail. Photos by author (Brodie).

Figure 12.1: Animal evolutionary tree. Image designed and created by author (Brodie).

Image 12.1 Purple stovepipe sponges. Photo by Laura Juul, used with permission.

Image 12.2: *Hydra* with ingested copepod by E.R. Degginger. Purchased from Science Source.

Figure 12.2: Muscle tissue showing actin and myosin by Blamb. Purchased from Shutterstock (ID 215720644).

Image 13.1: Human eye. Single eye close-up by DenisNata. Purchased from Shutterstock (ID 314152310).

Figure 13.1: Human eye anatomy. Purchased from CanStockPhoto (csp18033789).

Image 13.2: Chambered nautilus. Purchased from CanStockPhoto (csp7244328).

Image 13.3: Bald eagle. Photograph taken by author (Brodie).

Image 13.4: Small yellow-white squid by Lozzy Squire. Purchased from Shutterstock (ID 692184838).

Figure 14.1: Neuron synapse. Purchased from CamStockPhoto (csp12495188).

Image 14.1: Jellyfish. Photograph by author (Brodie).

Figure 14.2: Vertebrate brain. Figure 3-2, "Vertebrate Brain," in *Biology of Mind* by M. Deric Bownds. Reproduced with permission.

Figure 15.1: Geologic time scale. Purchased from Shutterstock (ID 670187668) and modified by author (Brodie).

Figure 15.2: Phylogenetic tree land plants. Image prepared by author (Brodie).

Image 15.1: Moss. Photo by author (Brodie).

Image 15.2: Fern. Photo by author (Brodie).

Image 15.3: Pine cones. Photo by author (Brodie).

Figure 15.3: Green algae's sex life. Image prepared by author (Brodie).

Image 15.4: Root nodules. Purchased from Shutterstock (Image ID 729057532).

Figure 16.1: Timeline: animals' invasion of the continents. Geologic time scale. Purchased from Shutterstock (ID 670187668) and modified by author (Brodie).

Image 16.1: *Tiktaalik roseae,* pencil drawing, digital coloring, by Nobu Tamura. CC BY-SA 3.0. Retrieved and reproduced with permission from http://spinops.blogspot.com/2015/03/tiktaalik-roseae.html.

Image 16.2: Lungfish. Purchased from Canstockphoto (csp22217453).

Figure 16.2: Chordate phylogenetic tree. Figure designed and prepared by author (Brodie).

Image 16.3: *Mastodonsaurus.* Purchased from Depositphotos (2020537).

Figure 16.3: Pangaea. Position of Continents of Carboniferous and Permian Periods by Mila Kananovych. Purchased from Shutterstock (ID 367271882 and ID 365744177).

Image 16.4: *Dimetrodon* by Herschel Hoffmeyer. Purchased from Shutterstock (ID 462675379).

Image 16.5: *Cynognathus crateronotus* by Nobu Tamura. CC BY-SA 3.0. Retrieved and reproduced with permission from http://spinops.blogspot.com/2012/05/cynognathus-crateronotus.html.

Image 17.1: *Tyrannosaurus rex and Triceratops.* Scene of the giant dinosaur by metha1819. Purchased from Shutterstock (ID 654522793).

Figure 17.1: Amniote phylogenetic tree. Figure designed and prepared by author (Brodie).

Image 17.2: *Argentinosaurus* by David Roland. Purchased from Shutterstock (ID 386999533).

Image 17.3: *Stegosaurus* by kamui29. Purchased from Shutterstock (ID 435291058).

Figure 17.2: Dinosaur size by Zachi Enevor. CC BY-SA 2.0. Retrieved and reproduced with permission from https://www.flickr.com/photos/zachievenor/14378304528.

Figure 17.3: Bird respiration. Figure designed and prepared by author (Brodie).

Image 18.1 Dinosaurs' extinction. Purchased from Shutterstock (ID 471399755) 11/5/18.

Figure 18.1: Geological timeline mass extinctions. Geologic Time Scale. Purchased from Shutterstock (ID 670187668) and modified by author (Brodie).

Image 19.1: Lemur of Madagascar. Public domain. Retrieved and reproduced with permission from http://www.publicdomainpictures.net/view-image.php?image=151418&picture=lemur-of-madagascar.

Figure 19.1: Primate Evolutionary Tree. Figure designed and created by author (Brodie).

Figure 19.2: The hominin lineage. Figure designed and prepared by author (Brodie).

Image 19.2: Lucy. *Australopithecus afaensis* by Tim Evanson. Modified to B&W. CC BY-SA 2.0. Retrieved and reproduced with permission from https://www.flickr.com/photos/timevanson/7283201084.

Image 19.3: *Homo erectus,* Sterkfontein Caves Exhibition. CC BY 2.0. Retrieved and reproduced with permission from https://www.flickr.com/photos/flowcomm/4258109849/.

Image 19.4: Neanderthal and woman. Recreation face of Neanderthal purchased from Canstockphoto (csp702876). Young attractive woman in blowing silk purchased from Canstockphoto (csp26133917).

Figure 19.3: Out-of-Africa map. World map. Creative Commons CC0. Retrieved and reproduced with permission from https://pixabay.com/en/world-map-continent-country-117174/. Modified by author to show migration of *Homo sapiens*.

Image 20.1: Chimpanzee. Photo taken by author (Brodie).

Figure 20.1: Larynx and pharynx. Nose and throat. Purchased from CanStockPhoto (csp14738388).

Image 21.1: Lower Manhattan by Jack Delano. Public domain. CC BY 2.0. Retrieved February 2018 from https://www.flickr.com/photos/trialsanderrors/3092854029.

Image 22.1: Human Infant. C00 Public Domain. Retrieved and reproduced with permission from http://www.publicdomainpictures.net/view-image.php?image=98607&picture=baby-boy-face.

Image 22.2: Capuchin monkey by joelfotos. Creative Commons C00. Retrieved and reproduced with permission from https://pixabay.com/en/capuchin-monkey-looking-habitat-1187570/.

Image 23.1: Functional MRI. NIDA(NIH). US government work. Retrieved February 2018 from https://www.flickr.com/photos/nida-nih/7882374582.

Figure 23.1: Where does consciousness reside? Figure prepared by author (Brodie).

Image 23.2: Australian magpie. Free download from Pixabay. C00 Public domain. Retrieved February 2018 from https://pixabay.com/p-432765/?no_redirect.

Image 24.1: *Hydra vulgaris* dark field by Lebendkulturen.de. Purchased from Shutterstock (ID 92793046).

Image 25.1: Keplar-452b. NASA/Ames/JPL-Caltech/T. Pyle. Public domain. Created by NASA. Retrieved and reproduced with permission from https://www.nasa.gov/sites/default/files/thumbnails/image/452b_artistconcept_beautyshot.jpg.

Image 25.2: Enrico Fermi at the blackboard. US government work. Retrieved and reproduced with permission from https://www.flickr.com/photos/departmentofenergy/10542564734.

Figure 26.1: Greenhouse effect. Designed and prepared by author (Brodie).

Figure 26.2: Temperature and CO_2 cycles. Figure prepared by author (Brodie) from Vostok Antarctica ice core data. ("Getting to the Core"). Data Retrieved with permission from https://www3.epa.gov/climatechange//kids/documents/temp-and-co2.pdf.

Image 26.3: Geological timeline glacial epochs. Geologic time scale. Purchased from Shutterstock (ID 670187668) and modified by author (Brodie).

Figure 26.4: Global temperature trends. Global land-ocean temperature index. NASA's Goddard Institute for Space Studies. Public domain. Retrieved and reproduced with permission from https://www.giss.nasa.gov/research/news/20100121/418335main_land-ocean-full.jpg.

Image 26.1: Gray Glacier, Torres del Paine National Park, Chile. Retrieved and reproduced with permission from https://www.nasa.gov/mission_pages/station/multimedia/gray_glacier.html.

Image 27.1: African elephant by Roelroelofs. COO. Retrieved and reproduced with permission from https://pixabay.com/en/amboseli-national-park-kenya-2063592/.

Image 27.2: Ivory. A pile of ivory by Joe Mercier. Purchased from Shutterstock (ID 17942068).

Image 27.3: Threatened polar bear (Ursis maritimus) by Josette Prinsen. CC-BY-2.0. Retrieved and reproduced with permission from https://www.flickr.com/photos/usfwsendsp/5038885231/.

Image 27.4: 1968 Earthrise (Apollo 8). NASA. Retrieved and reproduced with permission from https://explorer1.jpl.nasa.gov/galleries/earth-from-space/#gallery-9.

Figure 27.1: World population growth. From World Population Prospects: The 2017 Revision by Department of Economic and Social Affairs, Population Division, United Nations, Ó 2017 United Nations. Retrieved and reproduced with permission from the United Nations from https://esa.un.org/unpd/wpp/Publications/Files/WPP2017_KeyFindings.pdf.

Image 28.1: Genetic engineering by igorstevanovic. Purchased from Shutterstock (ID 577548931).

Image 28.2: Chessboard, by Felix Mittermeier. Creative Commons CCO. Retrieved and reproduced with permission from https://pixabay.com/en/chess-black-white-chess-pieces-king-2730034/.

Image 28.3: Artificial intelligence. Robot learning and solving problems by Phonlamai Photo. Purchased from Shutterstock (ID 680929729).

Image 29.1: Hydrogen bomb. "Ivy Mike" atmospheric nuclear test November 1952. CC BY 2.0. Public domain. Retrieved and reproduced with permission from https://www.flickr.com/photos/ctbto/6476282811.

Image 29.2: Artificial intelligence / robots. Purchased from Shutterstock (ID 70155106).

Image 30.1: CRISPR/Cas9 system for editing. Purchased from CanStockPhoto (csp33713236).

Image 30.2: Brain–computer interface. Artificial Intelligence. Creative Commons CCO. Retrieved and reproduced with permission from https://pixabay.com/en/header-banner-head-display-dummy-915122/.

Image E1: Earth as "Pale Blue Dot" by NASA Jet Propulsion Laboratory (9/12/96). NASA/JPL-Caltech. Retrieved and reproduced with permission from https://photojournal.jpl.nasa.gov/catalog/PIA00452.

Image E2: Sisyphus. Man pushing up heavy stone by zhekakopylov. Purchased from Shutterstock (ID 290077539).

ABOUT THE AUTHOR

Dr. Bruce Brodie is a retired clinical professor of medicine at the University of North Carolina Teaching Service in Greensboro, North Carolina, a retired interventional cardiologist, and cofounder and past chairman of the LeBauer-Brodie Center for Cardiovascular Research and Education at Cone Health in Greensboro, North Carolina. He has been a leader in research in the treatment of heart attack patients with balloons, stents, drugs, and other devices and has authored or coauthored more than two hundred peer-reviewed manuscripts in leading cardiology and medical journals. He has devoted the last six years to researching, studying, and exploring the origins and evolution of life on our planet in a search for answers to the universal question, Why are we here?

Brodie

Printed in the United States
by Baker & Taylor Publisher Services